TEXT-BOOK

OF

ZOOLOGY

FOR SCHOOLS AND COLLEGES.

BY

H. ALLEYNE NICHOLSON,

M. D., D. Sc., M. A., Ph. D., F. R. S. E., F. G. S., Etc.,

PROFESSOR OF NATURAL HISTORY AND BOTANY IN UNIVERSITY COLLEGE, TORONTO; FORMERLY LECTURER ON NATURAL HISTORY IN THE MEDICAL SCHOOL OF EDINBURGH; AUTHOR OF "MANUAL OF ZOOLOGY FOR THE USE OF STUDENTS," "TEXT-BOOK OF GEOLOGY," "INTRODUCTORY TEXT-BOOK OF ZOOLOGY," "GEOLOGY OF CUMBERLAND AND WESTMORELAND," ETC., ETC.

NEW YORK:
D. APPLETON AND COMPANY,
549 & 551 BROADWAY.
1875.

PREFACE.

In bringing out a Text-book of Natural History, intended mainly for the use of schools, there are a few remarks which it may be as well to make by way of preface, if only to explain the principles upon which the work has been written.

In the first place, more space has been devoted, comparatively speaking, to the Invertebrate Animals than has usually been the case in introductory works, upon the belief that any practical Zoological work likely to be undertaken by young students will certainly be in connection with these rather than with the Vertebrate Animals.

Secondly, the Author has devoted considerable space to a discussion of the principles of Zoological classification, believing that it is of paramount importance that the student should have a clear idea of the principles upon which the Animal Kingdom has been systematically divided. At the same time, the introductory portion of the work is more especially intended for the teacher; and there is much in it that the learner may perhaps hardly understand till he has arrived at some clear idea of Natural History as a whole.

Thirdly, while the Author trusts that the style of the work will be found clear and intelligible, he does not believe in the existence of any royal road to learning in Natural History, any more than in any other department of human knowledge. If Natural History is ever to be taught in schools, with any satisfaction to the teacher, or any profit to the learner, it must be taught as systematically and as unflinchingly as Mathematics and Greek have been taught for many generations. The Author is one of those who believe that the time is now approaching, if it be not already here, when the Natural Sciences will take their true place in school education, as second to no other branch of knowledge, either as regards their intrinsic value and interest, or regarded merely as a means of developing the mental powers. Acting upon this belief, the Author has, therefore, treated his subject in a purely scientific spirit; and, while avoiding as much as possible the use of technicalities, he has not endeavored to lend his subject any false glitter or embellishment; firmly believing that there is even a certain mental training involved in the recognition that a strictly scientific description is not without its own charms and beauties. While the use of technical terms has been as far as possible restricted, it is believed that an explanation of every unavoidable term will be found in the glossary, or is appended in the text.

Lastly, the illustrations, with few exceptions, have been drawn on the wood by the Author, and he has thought it wise to wholly eschew the use of pictorial illustrations, as unnecessary in a scientific work, however elementary it may be.

TABLE OF CONTENTS.

INTRODUCTION.

CHAPTER I.

CHAPTER II.

CHAPTER III.

CHAPTER IV.

CHAPTER V.

CHAPTER VI.

CHAPTER VII.

CHAPTER XIX.

CHAPTER XX.

CHAPTER XXI.

CHAPTER XXII.

CHAPTER XXIII.

CHAPTER XXIV.

CHAPTER XXV.

CHAPTER XXVI.

CHAPTER XXVII.

CHAPTER XXVIII.

CHAPTER XXIX.

CHAPTER XXX.

CHAPTER XXXI.

CHAPTER XXXII.

LIST OF ILLUSTRATIONS.

ZOOLOGY.

INTRODUCTION.

1. Definition of Biology and Zoology.

All natural objects may be roughly divided into three groups constituting the so-called Mineral, Animal, and Vegetable kingdoms. The objects comprised in the mineral kingdom are all devoid of life, and they exhibit the following characters: *a.* Their chemical composition is simple. They consist of either a single element, as is the case, for instance, with native gold; or, if combined, they are almost always in nature in the form of simple compounds, composed of no more than two or three elements—as, for example, common salt, limestone, plaster of Paris, and many others. *b.* Mineral bodies are, when unmixed, composed of similar particles, which have no definite relations to one another, or, in other words, they are *homogeneous*. *c.* The form of mineral bodies is either altogether indefinite, when they are said to be "amorphous;" or, if they have a definite shape, they are crystalline, in which case they are usually bounded by plane surfaces and straight lines. *d.* When mineral bodies increase in size, as crystals may do, the increase is produced simply by the addition of particles from the outside (technically called "accretion"). *e.* Mineral bodies exhibit no phenomena which are not purely physical and chemical, and they show no tendency to periodic changes of any kind.

All the bodies which exhibit these characteristics properly belong to the mineral kingdom, and fall to be treated of by the sciences of Geology, Mineralogy, Chemistry, and Physics. It should be borne in mind, however, that, in the case of what are called "fossils" or "petrifactions," we have mineral bodies which owe their existence and characters to living beings which existed at former periods in the history of the earth. For this reason, fossils, though composed of mineral matter, can hardly be said properly to belong to the mineral kingdom.

On the other hand, the objects which belong to the animal and vegetable kingdoms differ from those which are comprised in the mineral kingdom in the following points: *a.* They are composed of few chemical elements, of which carbon, hydrogen, oxygen, and nitrogen, are the most important; and these elements are combined to form complex organic compounds, which always contain a large proportion of water, are very unstable, and are prone to spontaneous decomposition. *b.* They are composed of diverse or *heterogeneous* parts, which have usually more or less definite relations to one another. These heterogeneous but related parts are termed "organs," and the objects possessing them are said to be "organized." Some of the lowest forms of animals have bodies composed of so uniform a substance that they cannot be said to be organized, as they exhibit no definite organs. This exception, however, does not affect the general value of this distinction. *c.* They are always more or less definite in shape, presenting concave and convex surfaces, and being bounded by curved lines. *d.* When they increase in size, or "grow," they do so, not by the addition of particles from the outside, but by the reception of foreign matter into their interior and its assimilation there (technically called "intussusception"). *e.* They invariably pass through certain periodic changes in a definite and discoverable order; these changes constituting *life.* They are subject to the same physical and chemical laws as those which govern dead matter, but the living body is the seat of something in virtue of which it can override the physical laws which enslave mere dead matter. The living body, so long as it *is* a living body, is the seat of *energy*, and can overcome the primary law of the *inertia* of matter. It has certain relations with the outer world other than the merely passive ones of dead matter. However humble it may be, and even if it be permanently rooted to one place, some part or other of every living body possesses the power of spontaneous and independent motion—a power possessed by nothing that is dead. In the higher animals the relations of the living body to dead nature become still further complicated, and their mastery over the physical forces becomes more and more pronounced, till in man, whose complex organization is wielded by an undying intelligence, we have a being in whose hands the dead matter of the universe is obedient as plastic wax. *f.* If our observation be continued for a sufficient length of time, we always discover that every living body has the power, by more or less complex process, of reproducing its like. That is to say, it has the power, directly or indirectly, of giving origin to minute germs, which can be developed under proper conditions into the likeness of the parent. *g.* Lastly, all living beings alike appear to be primitively composed of a substance which is more or less closely allied to *albumen* or white-of-egg, and which has been termed "protoplasm." Vital phenomena can apparently be manifested by no other form of matter, and protoplasm bears to life the same relation that a conductor does to the electric current. It is the necessary vehicle and medium through which life is brought into relation with the outer world. It does not, however, follow from this, as has been assumed, that protoplasmic matter holds any other or higher relation to life, or that vital phenomena are in any way an inherent property of the matter by which they are manifested.

All the objects, then, which fulfil these conditions, are said to be alive; and they all belong either to the animal or to the vegetable kingdom.* The study of living objects of all kinds, irrespective of which kingdom they belong to, is conveniently called by the general name of *Biology* (Gr. *bios*, life; and *logos*, discourse). As all living objects, however, may be referred to one or other of these two kingdoms, so Biology may be divided into the two sciences of *Botany*, which treats of plants, and *Zoology* (Gr. *zoön*, animal; *logos*, discourse), which treats of animals. The term Natural History, again, is generally understood nowadays as being equivalent to Zoology alone, though originally it was applied to the study of all natural objects indiscriminately.

2. Differences between Animals and Plants.

It now becomes necessary to inquire into the differences which subsist between animals and plants, and which enable us to separate the kindred sciences of Zoology and Botany. It might have been thought that nothing could be easier than to

* As will be mentioned immediately, it has been proposed to form an intermediate kingdom between the animal and vegetable kingdoms for the reception of organisms which cannot certainly be stated to be either plants or animals. There does not appear, however, to be any necessity for this in the mean while.

determine the animal or vegetable nature of any given organism; and such, indeed, was the almost universal belief of older observers. In point of fact, however, no hard-and-fast line can be drawn, in the present state of our knowledge, between the animal and vegetable kingdoms, and it is often a matter of extreme difficulty, or even wholly impossible, to decide positively whether we are dealing with an animal or a plant. So deeply has this difficulty been felt of late, that a most able zoologist—Dr. Ernst Haeckel—has proposed to form an intermediate kingdom, which he calls the *Regnum Protisticum*, and in which he proposes to place all organisms of a doubtful character. Even such a cautious observer as Professor Rolleston, while questioning the propriety of this step, is forced to come to the conclusion that "there are organisms which at one period of their life exhibit an aggregate of phenomena such as to justify us in speaking of them as animals, while at another they appear to be as distinctly vegetable." In the case of the higher members of the two kingdoms there is no difficulty in arriving at a decision. The higher animals are readily separated from the higher plants by the possession of a distinct nervous system, of locomotive power which can be voluntarily exercised, and of an internal cavity fitted for the reception and digestion of solid food. The higher plants, on the other hand, possess no nervous system or organs of sense, are incapable of voluntary changes of place, and are not provided with any definite internal cavity, their food being wholly fluid or gaseous.

The lower animals (*Protozoa*) cannot, however, be separated in many cases from the lower plants (*Protophyta*) by these distinctions, since many of the former have no digestive cavity, and are destitute of a nervous system, and many of the latter possess the power of active locomotion. In determining, therefore, the nature of these ambiguous organisms, the following are the chief points to be attended to:

Firstly, as to mere *form* or external configuration, no certain rules can be laid down for separating animals and plants. Many of the lower plants, either in their earlier stages of existence or when grown up, are exactly similar in form to some of the lower animals. This is the case, for example, in some of the *Algæ*, which closely resemble some of the Infusorian animalcules. Many undoubted animals, again, are rooted to solid objects in their adult state, and are so plant-like in appearance as to be always popularly regarded as vegetables. This is the case with many of the so-called hydroid zoophytes, such as the sea-firs, and also with the much more highly organized sea-mats (*Flustra*), all of which are usually regarded as sea-weeds by seaside visitors. This is also, but less strikingly, the case with the corals and sea-anemones, of which the latter are often spoken of as "sea-flowers."

Secondly, no decided distinction can be drawn between animals and plants as to their minute *internal structure*. Both alike consist essentially of minute solid particles (molecules or granules), of cells, or of fibres.

Thirdly, as regards *chemical composition*, there are some decided, though not universal, differences between plants and animals. As a general rule, it may be stated that plants exhibit a decided predominance of what are known to chemists as "ternary compounds"—that is to say, compounds which, like sugar, starch, and cellulose, are composed of the three elements, carbon, hydrogen, and oxygen. They are, comparatively speaking, poorly supplied with "quaternary" compounds, which contain the fourth element, nitrogen, in addition to the three first mentioned (e. g., gluten and legumin). Animals, on the other hand, are rich in quaternary nitrogenized compounds, such as albumen or fibrine. Still in both kingdoms we find nitrogenized and non-nitrogenized compounds, and it is only in the proportion which these bear to one another in the organism that animals differ in any way from plants. The most characteristic of all vegetable compounds is the one known as *cellulose*, very nearly allied in its chemical composition to ordinary starch. As a general rule it may be stated that the presence of an external envelope of cellulose in any organism raises a strong presumption as to its vegetable nature. Still cellulose is not exclusively confined to plants, as was at one time believed. It is now well known that the outer covering of the so-called sea-squirts or Ascidian Mollusks contains a large quantity of cellulose (as much as 60 per cent. in some cases); and recent researches seem to prove that this substance is present also in some of the lower forms of animal life (coccospheres). Another highly characteristic vegetable product is *chlorophyll*, the green coloring-matter of plants. Any organism which exhibits chlorophyll in any quantity as a proper element of its tissues is most probably vegetable. In this case also, however, the presence of chlorophyll cannot be regarded as a certain test, since it occurs regularly in some undoubted animals (e. g., *Stentor* among the *Infusoria*, and the *Hydra viridis*, or green fresh-water polype, among the *Cœlenterata*).

Fourthly, as regards *locomotive power*, or the ability to effect changes of place at will, the results of observation are singularly at variance with our preconceived notions. Before the invention of the microscope, no instances of independent voluntary movements were known in plants, if we except the voluntary opening and closure of flowers and their turning toward the sun, the drooping of the leaves of sensitive plants under irritation, and some other phenomena of a like nature. Now, however, we know of many plants which are endowed, either when young or throughout life, with the power of effecting voluntary movements apparently as spontaneous and independent as those exhibited by the lower animals. In some cases the movements are brought about by means of little vibrating hairs or cilia with which a part or the whole of the surface is furnished. In other cases the movements seem to be certainly not produced by cilia, but their exact cause is obscure (e. g., in the *Diatomaceæ* and *Desmidiæ*, two of the lower orders of plants, all of which are microscopic in size). When it is added that many animals are permanently fixed and rooted to solid objects in their fully-grown condition, it will be seen that no absolute distinction can be drawn between animals and plants merely on the ground of the presence or absence of independent locomotive power.

Fifthly, we have shortly to consider one of the most reliable of all the tests by which an animal may be separated from a plant—namely, the *nature of the food*, and the products which are formed out of the food within the body.

The differences between animals and plants in this respect may be roughly stated as follows:

1. Plants live upon purely dead or inorganic substances, such as water, carbonic acid, and ammonia—and they have the power of making out of these true organic substances, such as starch, cellulose, sugar, etc. Plants, therefore, take as food very simple bodies, and manufacture them into much more complex substances, so that plants are the great producers in nature.

2. Plants in the process of digestion break up carbonic acid into the two elements of which it is composed—namely, carbon and oxygen, keeping the carbon and setting free the oxygen. As carbonic acid occurs always in the air in small quantities, the result of this is that plants remove carbonic acid from the atmosphere and give out oxygen.

3. Animals, on the other hand, have no power of living on dead or inorganic matters, such as water, carbonic acid, and ammonia. They have no power of converting these into the complex organic substances of which their bodies are composed. On the contrary, animals require to be supplied with ready-made organic compounds if their existence is to be maintained. These they can only get in the first place from plants, and therefore animals are all dependent upon plants for food either directly or indirectly. Animals, therefore, differ from plants in requiring as food complex organic bodies which they ultimately reduce to very much simpler inorganic bodies. While plants, then, are the great manufacturers in nature, animals are the great consumers. Another distinction arising from the nature of their food is, that while plants decompose carbonic acid, keeping the carbon and setting free the oxygen, animals absorb oxygen and give out carbonic acid, so that their reaction upon the atmosphere is the reverse of that of plants.

As regards these general distinctions between plants and animals, there are three points which should be remembered:

1. That, even if universally true, these distinctions can often not be applied in practice to the ambiguous microscopic organisms about which alone any doubt can be entertained.

2. These general laws are certainly not of universal application in the case of plants. Some fungi are known which in the matter of food are animals—that is to say, they cannot live upon inorganic materials alone, but require ready-made organic products for their support.

3. Recent researches have rendered it not unlikely that some of the lower animals have the power of acting as plants, and of manufacturing organic compounds out of inorganic materials.

3. Classification.

By the term classification is understood the arrangement of a number of dissimilar objects of any kind into larger or smaller groups according as they exhibit more or less likeness to one another. The number of different animals is so enormous that it was long ago perceived that some classification of them, or method of arranging them into groups, was absolutely indispensable. Without some such arrangement it would have been utterly impossible to have ever acquired a clear notion of the animal kingdom as a whole. In the older arrangements, animals were grouped in accordance with some particular character, which might or might not be a really essential one; and the result was that these classifications were "artificial," and not "natural," as they are when *all* the characters are taken into consideration. To take a familiar example of this: when we speak of "quadrupeds," we really do so in consequence of our having, consciously or unconsciously, formed something like a rough classification of the animal kingdom. We have a dim idea that all animals with four legs belong together somehow, and form a single group. Our classification, however, is founded upon a single character only—the possession, namely, of four legs; and it is, therefore, a purely artificial arrangement. It will, however, be practically good or bad, just as this single character expresses a genuine and fundamental distinction, or is of a merely trivial and superficial nature. The instance here chosen will serve to illustrate either case. If we insist upon the fact that all the four legs must be externally visible, unmistakable legs, never fewer in number than four, then our classification is a very bad one, in fact entirely "artificial." In this case our group of "quadrupeds" will comprise only the ordinary four-legged mammals, such as oxen, sheep, horses, and such-like—together with the very dissimilar groups of the four-legged reptiles and amphibians, such as tortoises, lizards, crocodiles, frogs, and newts. Now, these different animals have certainly much in common, but we are not justified in placing them together simply upon the ground that they have four conspicuous legs, unless we are willing to put in a vast number of other animals as well. We must, in fact, put in a great number of animals which are not quadrupeds in the sense that they have four legs, but which agree with those that *have* four legs in the other fundamental and essential points of their structure. In this way we may arrive at a very genuine and natural classification by making some concessions. We must allow, for instance, that two of the legs or limbs, ceasing to be fit for walking, may be converted into organs of flight, or *wings*. This will let in the birds. We must allow, again, that all the limbs may be converted into *fins*. This admits most of the fishes. We must further grant that two of the legs may be altogether absent while the remaining two are converted into *swimming-paddles*. This will bring in the whales and dolphins. Lastly —and this is the greatest admission of all—we must allow the total absence of *all* the limbs, provided the animal only show those other essential characters which are invariably found to go along with the possession of four legs in the regular quadrupeds. This will bring in the snakes and some of the fishes. So that, paradoxical as it may seem, it is in one sense scientifically correct to speak of a snake as a quadruped, though in reality it has no legs at all. In other words, there is no reason why a snake should not some day be found with four legs, and in point of fact some snakes show rudiments of these appendages. Making these allowances, and some more of a similar nature, we may ultimately succeed in converting our division of Quadrupeds into a strictly scientific group, comprising the Mammals, the Birds, the Reptiles, the Amphibia, and the Fishes. In fact, our group of Quadrupeds now agrees exactly with the great and natural division of the *Vertebrata* or vertebrate animals. It is true that *all* vertebrate animals have not got four limbs, or not obviously so, but they never have *more* than four under any circumstances; and a closer examination soon shows us that they agree with one another in many other characters which are of much greater importance than the characters of the limbs alone.

We have arrived, then, at the grand principle of all good classification—namely, that we should group together those objects only which are united by essential and fundamental points of similarity, and that in so doing we should ignore all mere superficial resemblances. The question now arises, What are these essential and fundamental points in the case of animals?

If for the moment we look at animals simply as so many machines, we shall not find much difficulty in answering this question. Let us suppose ourselves placed in a gigantic workshop full of an immense number of complicated and curiously-constructed machines of different sorts, and asked to put them in order—to put those of one kind in one place, and those of another kind in a different place. How should we proceed to act? Supposing, in the first place, that all the machines were at a stand-still, all that could be done would be to examine carefully the external form and internal structure of each, and to do our best to pick out some peculiarity which would distinguish some from all the others. In this way, if our mechanical knowledge were sufficiently extensive, we should no doubt ultimately succeed in classing all our machines into something like a rough natural arrangement. We should, for instance, have those made on the principle of the lever in one place, those on the principle of the inclined plane in another, and those on the principle of the pulley in a third.

Still our classification would most certainly be imperfect, and in some cases altogether incorrect. In some instances the parts of the machine would be so complex as to be utterly incomprehensible, and in many cases our ignorance of what each was intended to effect would be an insuperable bar to our arriving at any arrangement. Suppose now, however, that all the machines were suddenly set in motion, so that we could see not only the manner in which they were constructed and the materials of which they were composed, but could also see what they could *do*—could see, in fact, for what *work* each is intended. The task of arrangement now becomes immensely easier. Our previous classification, founded simply upon the structure of the machines, is now supplemented and rectified by our knowledge of what each is able to effect. One machine is found performing one set of actions, another a different set; and in this way not only is our classification rendered much easier, but we now get an insight into the meaning and nature of many points of structure which were formerly obscure.

To make this illustration fully meet the case of the naturalist who deals with living beings only, we have simply to suppose that the machines to be examined are reasonably perfect in their parts and fit for work, and that our imaginary workshop is supplied with a reasonable amount of light, not very brilliant, perhaps, and striking upon some objects more sharply than on others, but still upon the whole moderately steady and uniform. Far worse, however, is the case of the naturalist who has to deal with the remains of extinct generations of animals and plants, whose work lies among those relics of a by-gone world which are known as "fossils" or "petrifactions"—objects in many cases more wonderful and more perplexing and more beautiful than the most ornate and elaborate productions of human skill. In his case the workshop is a vast and gloomy vault or charnel-house, with no internal source of light, and but fitfully illuminated by uncertain gleams from the world without. And what is worse than this, his machines are mutilated and defaced, in many cases wanting their most important parts, in all cases destitute of life and motion, and usually very unlike any thing visible at the present day. Nevertheless it is almost incredible with what certainty and precision a mere fragment of a fossil, a single tooth or bone, can be referred by a skilled worker in this field of science to its proper place in the animal kingdom —with what exactitude the missing parts can be restored—

and what splendid generalizations can be drawn from what at first sight would appear to be the most fragmentary evidence.

This imaginary illustration exactly expresses the points which are to be regarded as essential and fundamental in classifying and arranging animals. We have to look, namely, *firstly*, to the plan upon which each animal is constructed; *secondly*, to the manner in which it discharges its vital functions. These are the two points of view from which every organism may be regarded—in their nature quite distinct, and indeed sometimes apparently opposite. From the one point of view we have to look solely to the laws, form, and arrangement of the *structures* of the organism. This constitutes what is technically called "Morphology," or the science of form (from the Greek words, *morphê*, form; and *logos*, a discourse). From the second point of view, we are concerned simply with the *functions* discharged by the different parts of the organism, and this constitutes what is known as "Physiology." It is most important to remember that there are no other points in which it is possible for one animal to differ from another. If two animals are different, they *must* differ in one or other or in both of these points. Either they differ *morphologically*, in being constructed upon altogether different plans; or they differ *physiologically*, in performing a different amount of vital work in a different manner, and with different instruments; or they differ both morphologically and physiologically. Philosophical classification, therefore, insomuch as it depends entirely upon a due appreciation of what are the real differences between different animals, is nothing more than an attempt to express formally the facts and laws of Morphology and Physiology.

Examining next into the nature and extent of the morphological or structural differences between different animals, we find that these are much less and much fewer than might have been thought. By one not previously acquainted with the subject, it might readily be supposed that every kind of animal was constructed upon a type or plan peculiar to itself and not shared by any other. We should certainly suppose, for example, that animals so different as a lobster and a butterfly were built upon different types or plans of structure. When we come, however, to examine the question, we find that this is not the case. The lobster and the butterfly are constructed upon the same structural plan or morphological type. What is still more remarkable, we find that *all* known animals, in spite of their immense differences in external appearance, are really constructed upon no more than some half-dozen primary plans of structure or morphological types. These types are all different from one another, but there is no animal yet known to us, living or extinct, which cannot be referred to one or other of these six plans. These plans, then, give us the primary basis for a classification of the animal kingdom—all the animals formed upon one plan being grouped together so as to form a single division. The animal kingdom, therefore, is primarily divided into six great sections corresponding to the six morphological types, and these sections are known to naturalists under the name of the "sub-kingdoms." Each of these sub-kingdoms has its special name, and it is the object of the present work to describe the leading characters and more important examples of each.

We have to understand, then, that all the animals belonging to each sub-kingdom agree with one another in their morphological type, or, in other words, in the plan upon which they are constructed; and the question now arises how they can be separated from each other. If they agree morphologically, there is only one other way in which they *can* differ, and that is *physiologically*, in the manner in which they discharge their vital functions. Consequently, all animals which agree with one another in their plan of structure, and which are therefore placed in the same sub-kingdom, are separated from one another solely by their physiological perfection. In other words, as machines, they are constructed of the same fundamental parts, but they do their work in a different way and with different instruments.

Returning to our old illustration, suppose we had separated from the mass of machines before us all those which were intended to mark the lapse of time, and had in this way assembled a large collection of hour-glasses, watches, timepieces, and clocks, and suppose that we wanted to arrange these more minutely, we should soon discover that each of these different time-keepers was formed upon a principle peculiar to itself. The hour-glasses, as the most simple, would form one division; the timepieces and clocks, possessing pendulums, would form another; and the watches would form a third. These, as being constructed upon different plans, would constitute three distinct groups, which we should call classes or sub-kingdoms according to the value we might see fit to place upon the differences between them. But we must further suppose that we wished to divide one of these groups—say the watches—into still smaller groups. If they were all standing, we should probably find this a matter of very great difficulty. The moment, however, that they commenced to go—or, in other words, to perform their own peculiar function—we should soon see that some would be different to the others. Some, for instance, would strike the hours, and these would have to be laid aside in a group by themselves. And we should further discover that, in accordance with the difference in the *function*, there would be an equivalent difference in the *structure*, of these two groups. The striking watches would be formed upon the same fundamental type as those which did not strike; but, in addition to the broad and general details of structure in which all were the same, the striking watches would have a special apparatus or structure fitted for striking the hours. The non-striking watches would be destitute of this apparatus, so that the physiological or functional difference between the two groups would thus entail a corresponding difference in structure.

It is just the same with animals. If we take a lobster, a butterfly, a scorpion, and a spider, we find that, dissimilar as they are in external appearance, they are all constructed upon the same fundamental plan. They agree in morphological type, and they belong to the same sub-kingdom. They lead different lives, however—they are placed under different conditions—and they discharge different functions in the general economy of Nature. They differ, therefore, physiologically; and, as every physiological difference implies a corresponding structural difference, they differ structurally as well. But they differ structurally only *because* they differ physiologically, and in all the really essential details of their structure they are the same. The lobster is aquatic in its habits, and has therefore gills, or organs adapted for breathing air dissolved in water. The butterfly is aërial, and has respiratory organs adapted for breathing air directly, and not through the medium of water. They differ, then, physiologically, and therefore, necessarily, in the corresponding structure. Both, however, have distinct organs set apart and dedicated to the function of respiration. This is an essential and fundamental point in their structure, and in this they both agree with one another, and differ from a large number of animals in which there are *no* distinct breathing-organs. It is only by the combined effect of a number of these physiological differences, taken collectively, that the lobster and the butterfly come ultimately to be so strikingly distinct from one another

It is now possible to comprehend fully the principles upon

which a naturalist proceeds in framing a classification of the animal kingdom. His great primary divisions are founded upon differences in the smaller and fundamental details of structure. His smaller divisions are based upon the less important physiological differences with their corresponding structural distinctions. Of course, in carrying out this programme of a truly philosophical and natural classification, the naturalist works to a great extent in the dark, and is liable to many sources of error. It is by no means always easy to determine what points of structure *are* essential and fundamental, and what are only caused by physiological differences. Such, too, is the constitution of the human mind, that different observers place different values upon the same structures; points which some look upon as of essential value are regarded by others as of a merely superficial nature. Nevertheless there can be no doubt that the progress of Natural History as a science has been strictly conterminous with the development of these great principles of classification.

In the present work an outline is given of the morphological differences between all the larger groups of the animal kingdom, but it may be as well here to say a few words upon the subject of Physiology. As already remarked, Physiology treats of all the functions exercised by living bodies, or discharged by the various definite parts or organs of which most animals are composed. All these various functions come under three great heads: 1. *Functions of Nutrition*, comprising all those functions by means of which an animal is able to live, grow, and maintain its existence as an individual. 2. *Functions of Reproduction*, comprising all the functions by which fresh individuals are produced and the perpetuation of the *species* insured. 3. A series of functions which are known by the somewhat misleading name of the *Functions of Relation* or of *Correlation*. Under this term are included all those functions by means of which external objects are brought into *relation* with the organism, and by which it, in turn, reacts upon the outer world. The functions of nutrition and reproduction are often spoken of collectively as the functions of "organic" or "vegetative" life, as being common to animals and plants alike. The functions of relation, again, are often called the functions of "animal" life, as being most highly developed in animals. These functions, however, though more highly characteristic of animals, are not peculiar to them, but are manifested to a greater or less extent by various plants.

As regards animals, all alike, whatever their structure may

be, perform the three great physiological functions—that is to say, they all nourish themselves, reproduce their like, directly or indirectly, and have certain relations with the external world. When we come, however, to compare animals together physiologically, it is soon seen that the functions of relation stand in quite a different position to that occupied by the functions of nutrition and reproduction. As far as these last are concerned, there can be no difference in the *amount* or *perfection* of the function discharged by the organism. The simplest and most degraded of animals—say a sponge—nourishes itself as perfectly, as far as the result to itself is concerned, as does the highest of animals. Nutrition can do no more than maintain the body of any animal in a healthy and vigorous condition. This is the highest possible perfection of the function, and it is attained as fully and perfectly by the sponge as it is by man himself. The same holds good of reproduction. While the functions of nutrition and reproduction are thus, as regards their essence and results, the same in all animals, it must be remembered that there are enormous differences in the *manner* in which the functions are discharged. The result attained is in all cases the same, but it may be arrived at in the most different ways and with the most different apparatus. As regards the functions of relation, on the other hand, we have every possible grade of perfection exhibited as we ascend from the lowest members of the animal kingdom to the highest. So numerous, in fact, are the changes in these functions, and so great the additions which are made in the higher organisms, that it may be doubted if there exists any common element by which a comparison can be drawn on this head between the higher and lower animals. It may reasonably be doubted whether in this respect a horse or a dog has any thing in common with a sponge.

Instead of giving here a general sketch of each of the great physiological functions as a whole, it may be as well to accompany the morphological account of each primary division of animals with a short account of the manner in which the vital functions are carried out in the same. In this way a clearer view will be obtained of the gradual rise in physiological perfection in passing from the bottom to the summit of the animal series.

Homology and Analogy.—In connection with the morphological and physiological differences between animals, a short explanation may be given of the meaning of the terms Homology and Analogy, which are in constant use in zoological works. When organs in different animals agree with one another in their plan of *structure*, they are said to be "homologous," no matter what may be the functions which they perform. For example, the arm of a man, the fore-leg of a horse, the wing of a bird, and the swimming-paddle of a dolphin or whale, are all composed essentially of the same structural elements, and they are therefore said to be homologous, though they are fitted for altogether different functions.

On the other hand, when organs in different animals perform the same *functions*, they are said to be "analogous," whatever their fundamental structure may be. Thus the wing of a bat, the wing of a bird, and the wing of an insect, all serve for flight, and they are therefore "analogous" organs. They are all, however, constructed upon different plans, and they are, therefore, not "homologous." At the same time, however, it is to be remembered that there are plenty of cases in which organs in different animals are not only constructed upon the same plan, but also perform the same function, so that they are *both* homologous and analogous.

General Divisions of the Animal Kingdom.

As already stated, the entire animal kingdom may be divided into some half-dozen primary plans of structure or morphological types, to one or other of which every known animal is referable. These primary types are known to naturalists as the *sub-kingdoms*, under the following names: *Protozoa*, *Cœlenterata*, *Annuloida*, *Annulosa*, *Mollusca*, and *Vertebrata*. The characters and minor subdivisions of these sub-kingdoms form the subject of the remainder of this work. In the mean while, it is sufficient to state that the first five of these are often grouped together under the collective name of the *Invertebrata*, or "invertebrate animals." The *Invertebrata*, comprising the *Protozoa*, *Cœlenterata*, *Annuloida*, *Annulosa*, and *Mollusca*, are collectively distinguished by the following points among others: The body, if divided transversely, or cut in two, shows only a single tube containing all the vital organs (Fig. 1, A). These organs, in the higher *Invertebrata*, consist of an alimentary or digestive cavity, a circulatory or "hæmal" system, and a nervous or "neural" system. The side of the body on which the "hæmal" or blood-vascular system is placed is called the "hæmal aspect;" while the side of the body on which the main masses of the nervous system are situated is called the "neural aspect." When there is any skeleton, this is *external* (forming an "exo-skeleton"), and it is really nothing more than a hardening of the skin. The limbs, when present, are turned toward the neural aspect of the body.

In the *Vertebrata*, on the other hand, the body, if transversely divided, exhibits *two* tubes. In one (Fig. 1, B) is placed the main mass of the nervous system (the brain and spinal

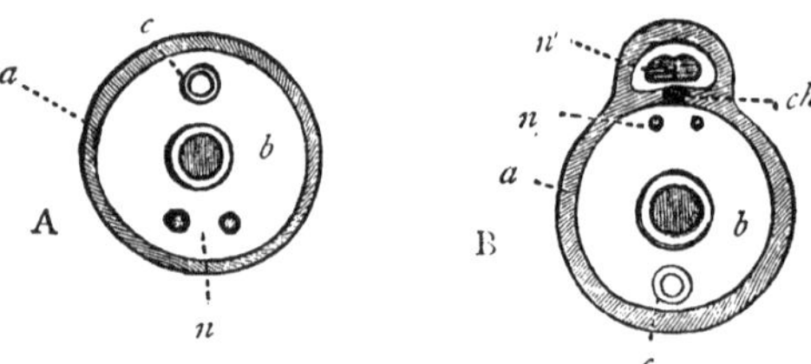

FIG. 1.—Diagrams representing transverse sections of one of the higher Invertebrata, A—and one of the Vertebrata, B. *a* Wall of the body; *b* Alimentary canal; *c* Hæmal or blood-vascular system; *n* Nervous system; *n'* Cerebro-spinal axis, or brain and spinal cord of the Vertebrata, enclosed in a separate tube; *ch* Noto-chord or chorda dorsalis. (Slightly altered from Huxley.)

cord). In the other tube are the alimentary canal, the hæmal or blood-vascular system, and certain other portions of the nervous system, which are known as the "sympathetic" system of nerves, and which correspond to, or are homologous with, the entire nervous system of the Invertebrata. Further, in the *Vertebrata* there is always an *internal* skeleton (or endo-skeleton), the central stem of which is usually constituted by a true backbone or "vertebral column." When this is not present, there is always a structure which will be afterward described as the "noto-chord" or "chorda dorsalis." Lastly, the limbs of the *Vertebrata*, when present, are never more than four in number, and they are always turned away from the neural aspect of the body—away, that is, from the side on which the main masses of the nervous system are placed.

Subjoined is a short tabular view of the main existing divisions of the Animal Kingdom, the characters and smaller divisions of which will be considered hereafter at length:

INVERTEBRATE ANIMALS.

SUB-KINGDOM I.—PROTOZOA.

Animal simple or forming colonies, usually very minute; the body composed of the structureless, jelly-like, albuminous substance called "sarcode;" not divided into regular segments; having no nervous system; no regular circulatory system; usually no mouth; no definite body-cavity, or at most but a short gullet.

CLASS A. GREGARINIDÆ —Minute Protozoa which inhabit the interior of insects and other animals, and which have not the power of throwing out prolongations of their substance (pseudópodia). No mouth.

CLASS B. RHIZOPODA (Root-footed Protozoa).—Protozoa which are simple or compound, and have the power of throwing out and retracting prolongations of the body-substance (the so-called "pseudopodia"). No mouth, in most, if not in all.

Order 1. *Monera.—Ex.* Protogenes.
Order 2. *Amœbea.—Ex.* Proteus Animalcule (Amœba).
Order 3. *Foraminifera.—Ex.* Lagena, Nodosaria, Globigerina.
Order 4. *Radiolaria.—Ex.* Thalassicolla, Polycystina.
Order 5. *Spongida.—Ex.* Fresh-water Sponge (Spongilla), Venus's Flower-Basket (Euplectella).

CLASS C. INFUSORIA (Infusorian Animalcules).—Protozoa with a mouth and short gullet; destitute of the power of emitting pseudopodia; furnished with vibratile cilia or contractile filaments; the body usually composed of three distinct layers.

Order 1. *Ciliata.—Ex.* Bell-animalcule (Vorticella), Paramœcium.
Order 2. *Flagellata.—Ex.* Peranema.
Order 3. *Suctoria.—Ex.* Podophyra.

SUB-KINGDOM II.—CŒLENTERATA.

Animals whose alimentary canal communicates freely with the general cavity of the body; body composed essentially of two layers or membranes, an outer layer or "ectoderm," and an inner layer or "endoderm." No circulatory system or heart, and in most no nervous system. Skin furnished with minute stinging organs or "thread-cells." Distinct reproductive organs in all.

CLASS A. HYDROZOA.—Walls of the digestive sac not separated from those of the general body-cavity, the two coinciding with one another. Reproductive organs external.

Sub-class I. HYDROIDA (Hydroid Zoophytes).
Order 1. *Hydroida.—Ex.* Fresh-water Polype (Hydra).
Order 2. *Corynida.—Ex.* Pipe-coralline (Tubularia).
Order 3. *Sertularida.—Ex.* Sea-firs (Sertularia).

Sub-class II. SIPHONOPHORA (Oceanic Hydrozoa).
Order 4. *Calycophoridæ.—Ex.* Diphyes.
Order 5. *Physophoridæ.—Ex.* Portuguese Man-of-War (Physalia).

Sub-class III. DISCOPHORA (Jelly-fish).
Order 6. *Medusidæ.—Ex.* Trachynema.

Sub-class IV. LUCERNARIDA (Sea-blubbers).
Order 7. *Lucernariadæ.—Ex.* Lucernaria.
Order 8. *Pelagidæ.—Ex.* Pelagia.
Order 9. *Rhizostomidæ.—Ex.* Rhizostoma.

Sub-class V. GRAPTOLITIDÆ (extinct).

CLASS B. ACTINOZOA.—Stomach opening below into the body-cavity, which is divided into a number of compartments by a series of vertical partitions or "mesenteries." Reproductive organs internal.

Order 1. *Zoantharia.*—Tentacles simply rounded, in multiples of five or six.—*Ex.* Sea-Anemones (Actinidæ), Star-corals (Astræidæ), Brain-corals (Meandrina), Madrepores (Madreporidæ).

Order 2. *Alcyonaria.*—Tentacles fringed, in multiples of four.—*Ex.* Dead-man's-toes (Alcyonium), Organ-pipe Coral (Tubipora), Sea-rods (Virgularia), Sea-pens (Pennatula), Red Coral (Corallium).

Order 3. *Rugosa* (extinct).

Order 4. *Ctenophora.*—Animal oceanic, swimming by means of bands of cilia or "ctenophores."—*Ex.* Pleurobrachia, Venus's Girdle (Cestum).

SUB-KINGDOM III.—ANNULOIDA.

Animals in which the alimentary canal is completely shut off from the general cavity of the body, and in which there is a distinct nervous system. A true blood-circulatory system may or may not be present. In all there is a peculiar system of canals, which usually communicate with the exterior, and which constitute what is called the "water-vascular system." The body of the adult is never composed of a succession of definite rings, or provided with successive pairs of appendages disposed symmetrically on the two sides of the body.

The Annuloida are divided into two great classes:

A. ECHINODERMATA.—Integument composed of numerous calcareous plates jointed together, or leathery and having grains, spines, or tubercles of calcareous matter developed in it. Water-vascular system (ambulacral system) mostly employed in locomotion, and generally communicating with the exterior. Adult generally more or less starlike or "radiate" in shape; young mostly showing more or less complete "bilateral symmetry," that is, showing similar parts on the two sides of the body. Nervous system radiate.

Order 1. *Crinoidea.*—(Sea-lilies).—*Ex.* Feather-star (Comatula). Medusa-head Crinoid (Pentacrinus), Stone-lily (Encrinus.)

Order 2. *Blastoidea* (extinct).

Order 3. *Cystoidea* (extinct).

Order 4. *Ophiuroidea* (Brittle-stars).—*Ex.* Sand-stars (Ophiura), Brittle-stars (Ophiocoma).

Order 5. *Asteroidea* (Star-fishes).—*Ex.* Cross-fish (Uraster), Sun-star (Solaster), Cushion-star (Goniaster).

Order 6. *Echinoidea* (Sea-urchins).—*Ex.* Sea-eggs (Echinus), Heart-urchins (Spatangus).

Order 7. *Holothuroidea* (Sea-cucumbers).—*Ex.* Trepangs (Holothuria).

B. SCOLECIDA.—Body usually flattened, or cylindrical and worm-like; integument soft, without lime. Water-vascular system not assisting in locomotion. Nervous system consisting of one or two ganglia or little masses, and not disposed in a radiate manner.

Order 1. *Tæniada.*—*Ex.* Tape-worm (Tænia).

Order 2. *Trematoda* (Suctorial worms).—*Ex.* Liver-fluke (Distoma).

Order 3. *Turbellaria.—Ex.* Planarians (Planaria), Ribbon-worms (Nemertes).
Order 4. *Acanthocephala* (Thorn-headed worms).—*Ex.* Echinorhynchus.
Order 5. *Gordiacea* (Hair-worms).—*Ex.* Gordius.
Order 6. *Nematoda* (Thread-worms).—*Ex.* Round-worm (Ascaris), Guinea-worm (Filaria), Vinegar-eel (Anguillula).
Order 7. *Rotifera* (Wheel-animalcules).—*Ex.* Builder-animalcule (Melicerta), Flexible Creeper (Notommata).

SUB-KINGDOM IV.—ANNULOSA.

Animal composed of numerous definite segments or "somites," arranged longitudinally, one behind the other. Nervous system always present, consisting of a double chain of nervous masses, or ganglia, which are placed along the lower surface of the body, and form a collar around the gullet. Limbs (when present) turned toward that side of the body on which the main masses of the nervous system are situated.

DIVISION A. ANARTHROPODA.—Locomotive appendages, when present, not distinctly jointed or articulated to the body.

CLASS I. GEPHYREA.—*Ex.* Spoon-worms (Sipunculus).

CLASS II. ANNELIDA (Ringed-worms).

Order 1. *Hirudinea.—Ex.* Leeches (Sanguisuga, Hirudo).
Order 2. *Oligochæta.—Ex.* Earth-worms (Lumbricus), Water-worms (Nais).
Order 3. *Tubicola.—Ex.* Tube-worms (Serpula).
Order 4. *Errantia.—Ex.* Sand-worms and Sea-centipedes (Nereis), Lob-worm (Arenicola), Sea-mouse (Aphrodite).

CLASS III. CHÆTOGNATHA (Arrow-worms).—*Ex.* Sagitta.

DIVISION B. ARTHROPODA.—Locomotive appendages jointed or articulated to the body.

CLASS I. CRUSTACEA.—Respiration aquatic, mostly by gills. Two pairs of antennæ. Limbs more than four pairs in number, carried upon the thorax, and generally the abdomen also.

Order 1. *Rhizocephala.—Ex.* Peltogaster.
Order 2. *Ichthyophthira.—Ex.* Lernæa.
Order 3. *Cirripedia.—Ex.* Barnacles (Lepas), Acorn-shells (Balanus).
Order 4. *Ostracoda.—Ex.* Water-fleas (Cypris).
Order 5. *Copepoda. Ex.* Cyclops
Order 6. *Cladocera.—Ex.* Branched-horned Water-fleas (Daphnia).
Order 7. *Phyllopoda.—Ex.* Brine-shrimp (Artemia).
Order 8. *Trilobita* (extinct).
Order 9. *Merostomata.—Ex.* King-crabs (Limulus).
Order 10. *Lœmodipoda.—Ex.* Whale-louse (Cyamus).
Order 11. *Isopoda.—Ex.* Wood-lice (Oniscus), Slaters (Ligia).
Order 12. *Amphipoda.—Ex.* Sandhopper (Talitrus), Fresh-water Shrimp (Gammarus).
Order 13. *Stomapoda.—Ex.* Locust-shrimp (Squilla).
Order 14. *Decapoda.—Ex.* Lobster (Homarus), Cray-fish (Astacus), Shrimps (Crangon); Hermit-crabs (Pagurus); Crabs (Cancer, Carcinus), Land-crabs (Gecarcinus).

CLASS II. ARACHNIDA.—Respiration aërial, by pulmonary chambers or air-tubes (tracheæ) in the higher forms. Antennæ converted into jaws. Head and thorax amalgamated. Four pairs of legs. Abdomen without limbs.

Order 1. *Podosomata* (Sea-spiders).—*Ex.* Pycnogonum.
Order 2. *Monomerosomata.*—*Ex.* Mites (Acarus), Water-mites (Hydrachna), Ticks (Ixodes).
Order 3. *Adelarthrosomata.* — *Ex.* Harvest-spiders (Phalangidæ), Book-scorpions (Chelifer).
Order 4. *Pedipalpi.*—*Ex.* Scorpions (Scorpio).
Order 5. *Araneida.*—*Ex.* House-spiders (Tegenaria), Field-spiders (Epeira).

Class III. Myriapoda.—Respiration aërial, by tracheæ (air-tubes) or by the skin. Head distinct; remainder of body composed of nearly similar segments; legs more than eight pairs in number, and borne partly upon the abdomen. One pair of antennæ.

Order 1. *Chilopoda.*—*Ex.* Centipedes (Scolopendra).
Order 2. *Chilognatha.*—*Ex.* Millipedes (Iulus).
Order 3. *Pauropoda.*—*Ex.* Pauropus.

Class IV. Insecta.—Respiration aërial, by tracheæ. Head, thorax, and abdomen distinct. One pair of antennæ. Three pairs of legs, and generally two pairs of wings on the thorax. No locomotive limbs on the abdomen.

Order 1. *Anoplura.*—*Ex.* Lice (Pediculus).
Order 2. *Mallophaga* (Bird-lice).
Order 3. *Thysanura* (Springtails.)
Order 4. *Hemiptera.*—*Ex.* Plant-lice (Aphides), Field-bug (Pentatoma), Cochineal Insects (Coccus).
Order 5. *Orthoptera.* — *Ex.* Locusts (Acrydium), Grass-hoppers (Gryllus), Crickets (Achetina), Cockroach (Blatta).
Order 6. *Neuroptera.*—*Ex.* White Ants (Termes), Dragon-flies (Libellulidæ), May-flies (Ephemeridæ).
Order 7. *Aphaniptera.*—*Ex.* Fleas (Pulex).
Order 8. *Diptera.*—*Ex.* Gnats (Culex), Crane-flies (Tipula), House-flies and Flesh-flies (Musca).
Order 9. *Lepidoptera* (Butterflies and Moths).
Order 10. *Hymenoptera.*—*Ex.* Bees (Apidæ), Humble-bees (Bombidæ), Wasps (Vespidæ), Ants (Formicidæ), Saw-flies (Tenthredinidæ).
Order 11. *Strepsiptera.*—*Ex.* Stylops.
Order 12. *Coleoptera* (Beetles).

SUB-KINGDOM V.—MOLLUSCA.

Animal soft-bodied, generally with a hard covering or shell. Nervous system consisting of a single ganglion or of scattered pairs of ganglia. A distinct heart and breathing-organ, or neither.

The Mollusca may be divided into the two following primary divisions; containing the following classes:

A. Molluscoida.—Nervous system consisting of a single ganglion or of a principal pair of ganglia. No heart, or an imperfect one.

Class I. Polyzoa.—Animal always forming compound growths or colonies. No heart. The mouth of each zoöid surrounded by a circle or crescent of ciliated tentacles. — *Ex.* Sea-mats (Flustra), Lace-coral (Fenestella).

CLASS II. TUNICATA.—Animal simple or compound, enclosed in a leathery or gristly case. An imperfect heart.—*Ex.* Sea-squirts (Ascidia).

CLASS III. BRACHIOPODA.—Animal always simple; the body enclosed in a bivalve shell. Mouth furnished with two long fringed processes or "arms."—*Ex.* Lamp-shells (Terebratula).

B. MOLLUSCA PROPER.—Nervous system consisting of three principal pairs of ganglia. Heart well developed, consisting of at least two chambers.

CLASS IV. LAMELLIBRANCHIATA (Bivalve Shell-fish):—No distinct head; no teeth. Body enclosed in a shell which is "bivalve," or composed of two distinct pieces. One or two leaf-like gills on each side of the body.—*Ex.* Oyster (Ostrea), Scallop (Pecten), Mussel (Mytilus).

CLASS V. GASTEROPODA.—A distinct head and toothed tongue. Shell absent in some, but mostly present, and consisting of a single piece ("univalve"). Locomotion effected by creeping about on the flattened under surface of the body ("foot"), or by swimming by means of a fin-like modification of the same.—*Ex.* Whelks (Buccinum), Limpets (Patella), Sea-lemons (Doris), Land-snails (Helix), Slugs (Limax).

CLASS VI. PTEROPODA.—Animal oceanic, swimming by means of two wing-like appendages, one on each side of the head. Size minute.—*Ex.* Cleodora.

CLASS VII. CEPHALOPODA.—Animal with eight or more arms, placed in a circle round the mouth. Mouth armed with jaws, and a toothed tongue. Two or four plume-like gills. In front of the body, a muscular tube ("funnel") through which is expelled the water which has been used in respiration. An external shell in some, an internal skeleton in others.—*Ex.* Calamaries (Loligo), Cuttle-fishes or Poulpes (Octopus), Paper-Nautilus (Argonauta), Pearly Nautilus (Nautilus).

VERTEBRATE ANIMALS.

SUB-KINGDOM VI.—VERTEBRATA.

Body composed of a number of definite segments arranged longitudinally or one behind the other. The main masses of the nervous system are placed on the dorsal aspect of the body, and are completely shut off from the general body-cavity. The limbs (when present) are turned away from that side of the body on which the main nervous masses are situated, and are never more than four in number. In most cases, a backbone, or "vertebral column," is present in the fully-grown animal.

CLASS I. PISCES (Fishes).—Breathing-organs in the form of gills; heart usually of two chambers, rarely of three; blood cold; limbs, when present, converted into fins.

Order 1. *Pharyngobranchii.*—*Ex.* Lancelet (Amphioxus).

Order 2. *Marsipobranchii.*—*Ex.* Lamprey (*Petromyzon*), Hag-fish (*Myxine*).

Order 3. *Teleostei* (Bony Fishes).—*Ex.* Eels (Muræenidæ), Herrings (Clupeidæ), Salmon and Trout (Salmonidæ), Cod and Haddock (Gadidæ), Flat-fishes (Pleuronectidæ), Perch (Percidæ), Mackerel (Scomberidæ).

Order 4. *Ganoidei.*—*Ex.* Bony Pike (Lepidosteus), Paddle-fish (Spatularia), Sturgeon (Sturio).

Order 5. *Elasmobranchii.*—*Ex.* Sharks (Carcharidæ), Dog-fishes (Scylliadæ), Saw-fishes (Pristis), Rays and Skates (Raiidæ).

Order 6. *Dipnoi.*—*Ex.* Mud-fish (Lepidosiren).

CLASS II. AMPHIBIA (Amphibians).—Breathing-organs in the young in the form of gills alone, afterward lungs, either alone or associated with gills. Skull jointed to the backbone by two articulating surfaces ("condyles"). Limbs never converted into fins. Heart in the young of two chambers only, in the adult of three chambers. Blood cold.

Order 1. *Labyrinthodontia* (extinct).

Order 2. *Ophiomorpha.*—*Ex.* Cæcilia.

Order 3. *Urodela* (Tailed Amphibians).—*Ex.* Water-newts (Triton), Salamanders (Salamandra), Axolotl (Siredon), Mud-eel (Siren).

Order 4. *Anoura* (Tailless Amphibians).—*Ex.* Frogs (Rana), Tree-frogs (Hyla), Toads (Bufo), Surinam Toads (Pipa).

CLASS III. REPTILIA (Reptiles).—Respiratory organs in the form of lungs, never in the form of gills. Heart three-chambered, rarely four-chambered, the pulmonary and systemic circulations always connected together directly, either in the heart itself or in its immediate neighborhood. Blood cold. Skull jointed to the backbone by a single articulating surface or "condyle." Each half of the lower jaw composed of several pieces. Appendages of the skin in the form of scales or plates.

Order 1. *Chelonia.*—*Ex.* Turtles (Cheloniid)æ, Soft Tortoises (Trionycidæ), Terrapins (Emydidæ), Land Tortoises (Testudinidæ).

Order 2. *Ophidia.*—*Ex.* Vipers (Viperidæ), Rattlesnakes (Crotalidæ), Sea-snakes (Hydrophidæ), Boas and Pythons (Boidæ).

Order 3. *Lacertilia.*—*Ex.* Lizards (Lacerta), Iguanas (Iguanidæ), Monitors (Varanidæ), Chameleons (Chameleontidæ).

Order 4. *Crocodilia.*—*Ex.* Crocodiles, Alligators, Gavials.

Order 5. *Ichthyopterygia* (extinct).—*Ex.* Ichthyosaurus.

Order 6. *Sauropterygia* (extinct).—*Ex.* Plesiosaurus.

Order 7. *Pterosauria* (extinct).—*Ex.* Pterodactylus.

Order 8. *Anomodontia* (extinct).—*Ex.* Dicynodon.

Order 9. *Deinosauria* (extinct).—*Ex.* Iguanodon.

CLASS IV. AVES (Birds).—Respiratory organs in the form of lungs, never in the form of gills. Lungs connected with air-receptacles placed in different parts of the body. Heart four-chambered. Blood warm. Skull connected with the backbone by a single articulating surface or "condyle." Each half of the lower jaw composed of several pieces. Appendages of the skin in the form of feathers. Cavities of the chest and abdomen not separated by a complete partition (diaphragm). Fore-limbs converted into wings. Animal oviparous.

Order 1. *Natatores* (Swimmers).—*Ex.* Penguins (Spheniscidæ), Gulls (Laridæ), Ducks (Anatidæ), Geese (Anserinæ), Flamingos (Phænicopteridæ).

Order 2. *Grallatores* (Waders).—*Ex.* Rails (Rallidæ), Water-hens (Gallinulæ), Cranes (Gruidæ), Herons (Ardeidæ), Storks (Ciconinæ), Snipes and Woodcock (Scolopacidæ), Plovers, Oyster-catchers, and Turnstones (Charadriidæ).

Order 3. *Cursores* (Runners).—*Ex.* Ostrich (Struthio), American Ostrich (Rhea), Emeu (Dromaius), Cassowary (Casuarius), Apteryx.

Order 4. *Rasores* (Scratchers).—*Ex.* Grouse, Ptarmigan, Partridges, Pheasants, Turkey, Guinea-fowl, Domestic Fowl, Peafowl (Gallinacei); Doves, Pigeons, Ground-pigeons (Columbacei).

Order 5. *Scansores* (Climbers).—*Ex.* Cuckoos (Cuculidæ), Woodpeckers (Picidæ), Parrots, Cockatoos, Parrakeets (Psittacidæ), Toucans (Rhamphastidæ), Trogons (Trogonidæ).

Order 6. *Insessores* (Perchers). —*Ex.* Crows, Magpies, and Jays (Corvidæ), Starlings (Sturnidæ), Finches, Grosbeaks, Larks (Fringillidæ), Thrushes, Blackbirds, Orioles (Merulidæ), Creepers and Wrens (Certhidæ), Humming-birds (Trochilidæ), Swallows and Martens (Hirundinidæ), Swifts (Cypselidæ), King-fishers (Alcedinidæ).

Order 7. *Raptores* (Birds of Prey).—*Ex.* Owls (Strigidæ), Falcons and Hawks (Falconidæ), Eagles (Aquilina), Vultures (Vulturidæ).

Order 8. *Saururæ* (extinct).—*Ex.* Archæopteryx.

CLASS V. MAMMALIA (Mammals or Quadrupeds).—Respiratory organs in the form of lungs, which are never connected with air-sacs placed in different parts of the body. Heart four-chambered. Blood warm. Skull united to the backbone by two articulating surfaces or "condyles." Each half of the lower jaw composed of a single piece. Appendages of the skin in the form of hairs. Young nourished by means of a special fluid—the milk—secreted by special glands—the mammary glands. Animal viviparous.

A. NON-PLACENTAL MAMMALS.—The young not provided with a placenta.

Order 1. *Monotremata.*—*Ex.* Duck-mole (Ornithorhynchus), Spiny Ant-eater (Echidna).

Order 2. *Marsupialia.*—*Ex.* Kangaroos (Macropodidæ), Kangaroo-bear (Phascolarctos), Phalangers (Phalangistida), Opossums (Didelphidæ), Tasmanian Devil (Dasyurus).

B. PLACENTAL MAMMALS.—The young provided with a placenta.

Order 3. *Edentata.* — Sloths (Bradypodidæ), Armadillos (Dasypodidæ), Hairy Ant-eaters (Myrmecophagidæ), Scaly Ant-eaters (Manis).

Order 4. *Sirenia.*—*Ex.* Manatee (Manatus), Dugong (Halicore).

Order 5. *Cetacea.*—*Ex.* Whalebone-whales (Balænidæ), Sperm-whales (Physeteridæ), Dolphins and Porpoises (Delphinidæ).

Order 6. *Ungulata* (Hoofed Quadrupeds).—*Ex.* Rhinoceros; Tapir; Horse, Ass, and Zebra (Equidæ); Hippopotamus; Hogs, and Peccaries (Suida); Camels and Llamas (Camelidæ); Giraffe; Stags, Elk, Rein-deer (Cervidæ); Antelopes (Antilopidæ); Sheep and Goats (Ovidæ); Oxen and Buffaloes (Bovidæ).

Order 7. *Hyracoidea.*—*Ex.* Hyrax.

Order 8. *Proboscidea.*—*Ex.* Elephants (Elephas).

Order 9. *Carnivora.*—*Ex.* Seals (Phocidæ), Bears (Ursidæ), Raccoons (Procyon), Badgers (Melidæ), Weasels and Otters (Mustelidæ), Civets and Genettes (Viverridæ),

Dogs, Wolves, and Foxes (Canidæ); Hyænas (Hyænidæ), Cats, Lynxes, Leopards, Tigers, Lions (Felidæ).

Order 10. *Rodentia.*—*Ex.* Hares and Rabbits (Leporidæ), Porcupines (Hystricidæ), Beavers (Castoridæ), Mice and Rats (Muridæ), Dormice (Myoxidæ), Squirrels and Marmots (Sciuridæ).

Order 11. *Cheiroptera.* — *Ex.* Common Bats (Vespertilionidæ), Horseshoe Bats (Rhinolophidæ), Vampire bats (Phyllostomidæ), Fox-bats (Pteropidæ).

Order 12. *Insectivora.*—*Ex.* Moles (Talpidæ), Shrew-mice (Soricidæ), Hedgehogs (Erinaceidæ).

Order 13. *Quadrumana.*—*Ex.* Aye-aye (Cheiromys), Lemurs (Lemuridæ), Spider-monkeys (Ateles), Howlers (Mycetes), Macaques (Macacus), Baboons (Cynocephalus), Gibbons (Hylobates), Orang (Simia), Gorilla and Chimpanzee (Troglodytes).

Order 14. *Bimana.*—Man (Homo Sapiens).

INVERTEBRATE ANIMALS.

SUB-KINGDOM I.—PROTOZOA.

CHAPTER I.

1. General Characters of the Protozoa. 2. Classification. 3. Gregarinidæ.

The sub-kingdom *Protozoa* (Gr. *protos*, first; and *zoön*, an animal), as the name implies, is the lowest division of the animal kingdom, and its limits are therefore necessarily not yet strictly defined. The *Protozoa* comprise an enormous number of animals, almost all of which are so small as to be invisible to the naked eye, and can only be satisfactorily examined under pretty high powers of the microscope. For this reason, and because they are almost universally found in water, these creatures, often popularly called "animalcules," are almost unknown to the majority of people. Some few, however, attain a large size, and of these the sponges are familiar examples. The microscopical forms of the *Protozoa* swarm in most stagnant pools, and in all waters charged with organic matter so as to afford them food. Every worker with the microscope is familiarly acquainted with them, and they exhibit phenomena which in many cases render them objects of the highest interest. From their low position in the animal scale, it arises that the *Protozoa* are mainly characterized by the absence of organs and structures which occur in higher beings, and they possess few positive characters by which they can be distinguished.

The *Protozoa* may be defined as *animals, generally of very minute size, composed of a nearly structureless, jelly-like substance (called "sarcode"), showing no composition out of distinct segments, having no distinct internal cavity, no nervous system, and either no organs devoted to digestion, or at best a very rudimentary alimentary apparatus.*

Of all the points enumerated in this definition as characteristic of the *Protozoa*, none is more important than the nature of the body-substance. The body in all known *Protozoa* is composed of a substance which is generally known by the name of "protoplasm"—or, better, "sarcode" (Gr. *sarx*, flesh; *eidos*, form). This sarcode is a gelatinous substance, very like white-of-egg to look at, and really of nearly the same chemical constitution, consisting mainly of albumen, or of some body allied to albumen. Generally, however, it contains numerous oil-globules scattered through it. The sarcode shows the physiological property of "contractility"—that is to say, under appropriate stimuli, or at the will of the animal, it may be made to contract or shorten its dimensions, thus giving rise to movements. As a rule, no other structures appear in the sarcode except minute rounded particles, or granules and molecules, but in some cases larger definite structures are formed out of it. Of this nature is the so-called "nucleus" found in many *Protozoa.*

As regards their internal *structure*, some *Protozoa* exhibit nothing worthy of the name of structure at all, the entire body being simply composed of sarcode, containing scattered granules (for example, the *Foraminifera*). In other cases there are found certain definite bodies which are known as the "nucleus" and "nucleolus," and which are usually, if not always, connected with reproduction. Very often, too, there are found certain minute cavities or chambers which close and expand at definite intervals, and which are known as the "contractile vesicles." These are, doubtless, rudimentary organs of circulation. In one division of the *Protozoa* (the *Infusoria*) there is a permanent mouth and a short gullet, but in all the others there are no definite organs connected with the process of digestion. In no *Protozoön*, however, without exception, have any traces of a nervous system been hitherto detected; and in none, even in those which possess a mouth, is there any distinct and definite cavity or chamber within the body in which the particles of food are received. No organs of sense exist in any of the *Protozoa*—that is to say, there are no distinct organs fitted for the reception of impressions produced by light or sound; but the general surface of the body appears capable of receiving the impressions produced by contact with foreign bodies, and therefore acts as an organ of touch. The power of active locomotion is enjoyed by most of *Protozoa;* but in some cases this is very limited, and in other cases the animal is permanently fixed (as in the sponges). The apparatus of locomotion in the *Protozoa* is of a varied nature. In many cases, especially in the higher forms, movements are effected by means of little hair-like processes, which are called "cilia" (Lat. *cilium*, an eyelash), and which have the power of vibrating or lashing to and fro with great rapidity. In other cases the cilia are accompanied or replaced by one or more long whip-like bristles, which act in the same fashion, and are known as "flagella." Among the lower *Protozoa* the most characteristic organs of locomotion are the so-called "pseudopodia" (Gr. *pseudos*, falsity; *podes*, feet). These consist of variously-shaped filaments, threads, or finger-like processes of sarcode, which the animal can thrust out from any or every part of its body. They are not, however, definite and permanent organs like the cilia, for they can be produced at will, and when they are again withdrawn they simply melt into the sarcode of the body, and leave no traces of their existence.

As regards the classification of the *Protozoa*, a rough and useful division is into mouth-bearing or "stomatode" *Protozoa*, in which there is a distinct mouth; and mouthless or "astomatous" *Protozoa*, in which there is no mouth. It is somewhat doubtful, however, if the mouth-bearing forms (namely, the *Infusoria*) can properly be kept in the *Protozoa*, so that this arrangement is not a very good one. More scientifically, the *Protozoa* are divided into three great divisions or "classes," known by the names *Gregarinidæ*, *Rhizopoda* and *Infusoria*, all of which require special examination.

Class I. Gregarinidæ.—The *Gregarinidæ* may be defined as *parasitic Protozoa which have no mouth, and have no power of giving out pseudopodia.* They are usually looked upon as forming the lowest class of the *Protozoa;* but in all probability much of their degraded character, as we shall see in other cases, is due to the fact that they are internal parasites, and are therefore not dependent on their own exertions for food. They vary in size from less than the head of a small pin up to nearly half an inch in length, when they look something like small worms; and they are found inhabiting the intestines of various animals, especially the cockroach and the earthworm.

In anatomical structure a *Gregarina* usually presents the appearance of a single cell, consisting of an ill-defined membranous envelope, filled with a more or less granular sarcode containing fatty granules, and having in it a little central bladder or vesicle—the "nucleus"—which in turn encloses a solid particle or "nucleolus" (Fig. 2, *a*). The outer covering

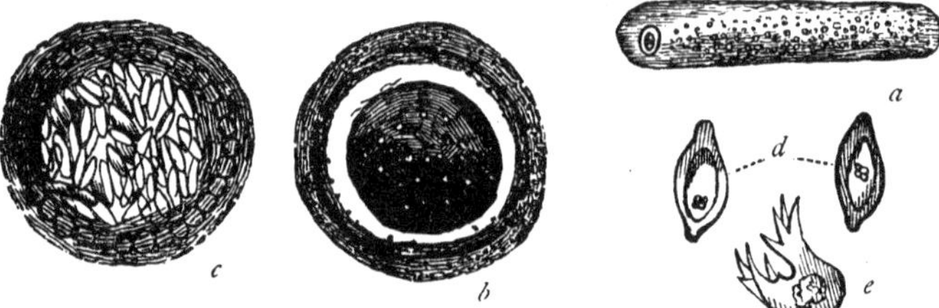

FIG. 2.—Anatomy and reproduction of the Gregarina of the earthworm (after Lieberkühn) *a* Adult Gregarina; *b* The same "encysted;" *c* With the contents broken up into pseudonavicellæ; *d* Free pseudonavicellæ; *e* Contents of the pseudonavicellæ when liberated.

or cuticle with which the protoplasmic body is enclosed may be quite smooth, or it may be furnished with bristles or spines, and in some cases even cilia have been observed. Beyond the nucleus and nucleolus (which are probably connected with reproduction), no definite organs have been detected in the *Gregarinæ;* and all the processes of assimilating food and getting rid of waste or injurious products must be effected by the general surface of the body. As we shall see, however, this is common in internal parasites, which are not necessitated to live upon solid food, but which are enabled to subsist simply by imbibing the nutritive juices of their hosts.

The following is a brief outline of the process of reproduction as it has been observed in the *Gregarinæ*, sometimes in a single individual, sometimes in two individuals which have come together and completely coalesced and melted into one another. The *Gregarina* becomes completely motionless, assumes a globular form, and develops round itself a thick structureless coat or envelope, when it is said to be "encysted" (Fig. 2, *b*). The nucleus then disappears, and the sarcode of the body breaks up into little masses, which are at first rounded, but afterward become pointed at both ends, when they are called "pseudonavicellæ" (Fig. 2, *c*). The cyst then breaks and the pseudonavicellæ escape, when they give origin to little masses of sarcode, which have the power of active movement and of throwing out pseudopodia, thus coming closely to resemble the animalcule which will be directly described as the *Amœba* (Fig. 2, *e*). These little amœba-like masses, if they find a suitable locality, are finally developed into new *Gregarinæ*.

CHAPTER II.

RHIZOPODA.

THE next class of the *Protozoa* which we have to consider comprises the most characteristic and typical forms of the whole sub-kingdom. The name of *Rhizopoda*, or "root-footed" animalcules (from the Greek, *rhiza*, root; and *podes*, feet), is derived from the fact that they all possess the power of throwing out at will from various parts of the body the processes of sarcode which have been already spoken of as pseudopodia, and by which they both move and obtain food. In fact, the *Rhizopoda* may be shortly defined as *Protozoa which have no mouth and possess the power of giving out pseudopodia.* The pseudopodia vary a good deal in shape and in other characters in different orders of the *Rhizopoda*, but they have invariably the character of being nothing more than temporary threads or finger-like processes of sarcode, which can be thrust out at will, and which melt again into the substance of the body when they are withdrawn.

Five distinct types of structure are known in the *Rhizopoda*, and these constitute as many distinct orders, which are known by the names of the *Monera*, *Amœbea*, *Foraminifera*, *Radiolaria*, and *Spongida*.

ORDER I. MONERA.—This name has been proposed for a small group of organisms which merely require to be mentioned. They are all microscopic in size, and inhabit the sea. Their sarcode-body is entirely structureless and devoid of definite organs of any kind. They have the power, however, of throwing out innumerable processes of the body-substance or "pseudopodia," and these agree in their characters with those which will be afterward described as characterizing the *Foraminifera*. They are, namely, very long and delicate filaments of sarcode, which unite in various directions so as to form a net-work, in which the particles of food are entangled. The body is completely naked, and the *Monera* differ from the

Foraminifera, chiefly if not entirely, in this absence of any hard covering or shell.

ORDER II. AMŒBEA.—This order is characterized by the fact that *the pseudopodia are mostly blunt and finger-like in shape, and that the sarcode of the body contains the structures known as the "nucleus" and "contractile vesicle."*

As the *type* of the order may be taken the *Amœba* or Proteus-animalcule, so called because of the incessant and illimitable changes of form which it exhibits (Gr. *amoibos*, changing). The *Amœba* is a little microscopical creature which may commonly be detected in stagnant water, especially where there is decaying vegetable matter. When examined under the microscope, all that would probably be seen at first would be a shapeless or irregularly-spherical mass of gelatinous, jelly-like sarcode, containing scattered granules. Soon the creature might be observed to push out a finger-shaped prolongation of its own substance; and it would soon be found that similar processes or pseudopodia could be pushed out at will from almost any point of the body and again retracted within it without leaving any trace behind. As a result of this, the form of the animal is constantly changing, and hence its common name of Proteus-animalcule (Fig. 3, *a*). By means of these temporary processes of sarcode, the *Amœba*

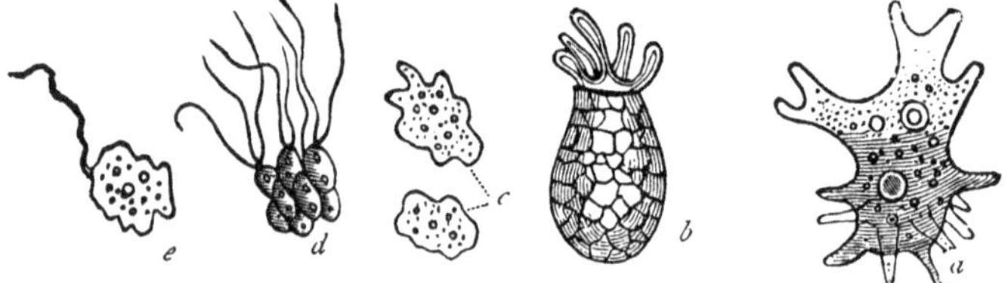

FIG. 3.—Morphology of Rhizopoda. *a Amœba radiosa*, showing the pseudopodia, the contractile vesicle, nucleus, and vacuoles; *b Difflugia*, with the pseudopodia protruded from the anterior end of the carapace; *c* Detached sponge-particles or "sarcoids;" *d* Ciliated sponge-particles of *Grantia; e* Sponge-particle of the fresh-water sponge (*Spongilla*) with a single cilium.

both moves and obtains food. Locomotion is effected in a kind of creeping manner, the animal pushing out the pseudopodia in one direction and then pulling the body in the same direction. In the same way, when any minute particle of food, such as a microscopic plant, comes within its reach, the *Amœba* wraps a pseudopodium round it, and then withdrawing the pseudopodium, lodges the nutrient particle securely in the substance of the body. It follows from this that the *Amœba* has no permanent mouth—no aperture, that is, which is especially employed in the admission of food. Any part of the surface can be pushed out into a pseudopodium, and therefore any part of the surface can be extemporized into a mouth. The process of taking food, in fact, in the *Amœba*, has been aptly compared to thrusting a stone or any other solid body into a lump of dough. The central portion of the body of the animal is softer and more fluid than the outer layers, and the particles of food, on reaching this point, undergo a sort of digestion, and are subjected to a species of movement or rotation in the interior of the animal. Each particle of food, in the process of being taken into the body, usually carries with it a little drop of water; and in this way a number of clear spaces are formed, which are usually quite round, and look like distinct cavities. These spaces are called "vacuoles;" but they are not distinct organs of any kind, though formerly regarded as distinct stomachs. Having undergone digestion, any portions of food which may be indigestible or insoluble are simply thrust out again through the walls of the body. This appears to be effected at one particular part of the body; but there is no permanent aperture for the purpose. There are no distinct vessels which serve to convey the nutritive fluid derived from the digestion; but there does appear to be a rudimentary organ by which this fluid is driven through the body. If we watch an *Amœba* carefully, there is usually no difficulty in observing that every now and again there appears at one particular place a clear spot, "like a window," which slowly expands to its full extent, and then usually contracts slowly till it disappears altogether. This process of gradual expansion and contraction is what is called "rhythmical"—that is to say, it is repeated at tolerably regular intervals, perhaps twice a minute. In some cases the vesicle, when contracted, remains so for a long time, but it always reappears in the same place. It is known as the contractile vesicle; and there can be little doubt that it is a permanent organ. It is, in fact, a little clear space or cavity in the substance of the body, filled probably with the nutritive fluid derived from the digestion, and no doubt serving by its contraction to drive this fluid to various parts of the body. In its function, then, the contractile vesicle of the *Amœba* is to be looked upon as the first indication which we have in the entire animal kingdom of that most important organ, the heart.

The *Amœba* possesses no breathing-organs, of any kind, and no excretory organs, so that these functions must be performed by the general surface of the body in a manner somewhat the same as the exhalation from the skin which takes place in the higher animals. There are also no traces of a nervous system, and no organs of sense, and the only other structure of any kind is what is known as the nucleus. The nucleus is simply a small rounded or oval granular mass, and there may be more than one in the same individual. Its function, however, is quite unknown, though it is probably connected with reproduction. The means employed by the *Amœba* to perpetuate the species are various, but the only one which need be mentioned is the process by self-division. This is what is technically called "fission" (Lat *findo*, I cleave), and it consists in a gradual division or cleavage of the body into two parts, each of which then becomes a separate and independent individual. In some cases this process is slightly varied, a single pseudopodium alone being cast off and becoming a fresh *Amœba*, but this does not differ essentially from the former.

Regarding the *Amœba* from a physiological point of view, we see that, though the animal nourishes itself and maintains its existence perfectly, the process of nutrition is carried on in the simplest possible manner and with the simplest possible apparatus. There is no permanent mouth, no stomach or alimentary canal of any kind, no respiratory or excretory organs, and even no distinct aperture for the extrusion of indigestible food. The only distinct structure which is at all concerned in nutrition is a rudimentary contractile cavity, the first foreshadowing of the heart in the higher animals. As regards the functions of relation, it is questionable how far the *Amœba* can be said to have distinct perceptions or sensations of any kind. It has no nervous system or organs of sight or hearing, and in all probability it has nothing more than a general sensibility to light. It appears, however, to be fully aware when any object comes in contact with a pseudopodium, and even to have some idea whether this is fit for food. Locomotion, as we have seen, is entirely effected by the temporary processes of sarcode or pseudopodia, and there are no permanent organs set aside either for locomotion or for prehension—that is, for seizing external objects.

The only other member of the *Amœbea* which deserves notice is the *Difflugia* (Fig. 3, *b*), which is not uncommonly found in fresh water. *Difflugia* in its essential structure does not differ from the *Amœba*, but the greater part of the body is enclosed in a sort of case or carapace, mostly composed of grains of sand, within which the animal can retire completely. The carapace is open at one end, and the pseudopodia are protruded from this aperture. The animal generally creeps about head-downward, so to speak; that is to say, with the closed end of the carapace elevated above the surface on which it is moving.

ORDER III. FORAMINIFERA (Lat. *foramen*, an aperture; *fero*, I carry).—The next order of the *Rhizopoda* is that of the *Foraminifera*, comprising animals which at first sight appear to be highly complex, but which are really much less highly organized than the *Amœba*. The *Foraminifera* may be defined as *Rhizopoda in which the body is protected by a shell or "test;" there is no nucleus or contractile vesicle; and the pseudopodia are extremely long and threadlike, and interlace with one another so as to form a net-work.*

The most obvious and striking character of the *Foraminifera* is the possession of an outer case or shell, and for a long time they were known to naturalists by their shells alone. As the shell or test is usually very beautiful and often very complex, the *Foraminifera* were consequently placed at first among the true shell-fish (*Mollusca*), very much in advance of their true position. When, however, the anatomical structure of the group came to be investigated, it was soon found that they were really referable to the *Protozoa*, and that in point of fact they even occupy a low position in this sub-kingdom. However elaborate and complicated the shell may be, the body of the contained animal is composed simply of granular gelatinous sarcode, highly elastic and contractile, and usually reddish or yellowish in color. This sarcode not only fills the shell, but also in many cases gains the exterior by means of little perforations in its walls, and forms a thin film over its outer surface. Wherever the sarcode is exposed, whether this be only at the mouth of the shell, as in *Miliola* (Fig. 4, *b*), or whether it be over the whole surface, as in *Discorbina* (Fig. 4, *c*), it has the power of giving off pseudopodia. The pseudopodia, however, differ greatly from those of the *Amœba*, and they show some remarkable characters. They are extremely long, threadlike processes, instead of being blunt and finger-shaped, and they have the curious property that they run into one another and interlace toward their extremities, so as to form a net-work which has been aptly compared to an "animated spider's web." Lastly, the microscope reveals in the pseudopodia a very curious circulation of minute solid particles or granules, which travel in all directions through the pseudopodial net-work. Internally, the sarcode-body of the *Foraminifera* exhibits absolutely no structures or definite organs of any kind. Even the nucleus and contractile vesicle which occur in the *Amœba* are here absent, and the only traces of structure are to be found in the existence of scattered granules.

Simple as is the sarcode-body of the *Foraminifera*, it has in all cases the power of secreting a skeleton or shell, which is technically called the "test" (Lat. *testa*, a shell). The shell is usually "calcareous"—that is to say, composed of carbonate of lime; but it is sometimes "arenaceous," or composed of particles of sand united together firmly by an unknown animal cement. In either case, the shell may exhibit one or other of two very distinct types of structure. In the one type (as in *Miliola*, Fig. 4, *b*), the shell-walls are not per-

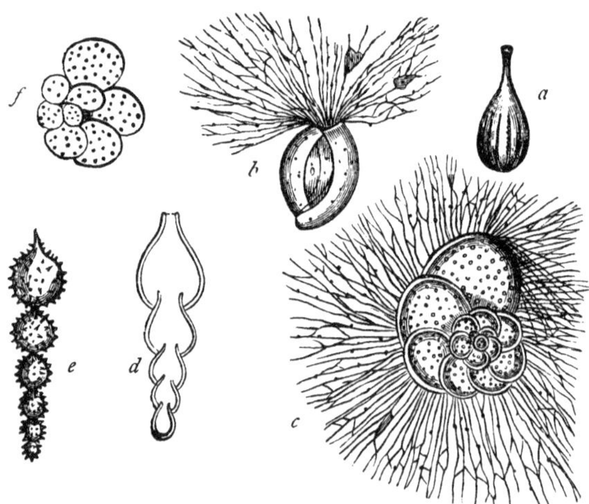

FIG. 4.—Morphology of Foraminifera. *a Lagena vulgaris*, a monothalamous Foraminifer; *b Miliola* (after Schultze), showing the pseudopodia protruded from the oral aperture of the shell; *c Discorbina* (after Schultze), showing the nautiloid shell with foramina in the shell-walls, giving exit to pseudopodia; *d* Section of *Nodosaria* (after Carpenter); *e Nodosaria hispida*; *f Globigerina bulloides*.

forated with holes, and the pseudopodia are therefore all emitted from the mouth of the shell. In the other type (as in

Discorbina, Fig. 4, *c*) the shell-walls are perforated with a number of little apertures or "foramina," from which the order derives its name. These foramina are the mouths of tubes which pierce the walls of the shell, and thus establish a free communication between the interior and exterior. In this way the sarcode which fills the inside of the shell is enabled to reach the outer surface, so as to form a film, from any part of which the pseudopodia may be given off. The presence or absence of foramina is believed to constitute a true structural distinction, and the *Foraminifera* may be thereby divided into two great groups (*Perforata* and *Imperforata*).

According to the form of the shell, also, the *Foraminifera* may be conveniently, though arbitrarily, divided into two great sections. The simplest form of shell is seen in such an example as *Lagena* (Fig. 4, *a*), where the shell consists of but a single chamber; and the animal, in fact, is nothing more than a little mass of sarcode, surrounded by a calcareous envelope. *Lagena*, then, may be taken as the type of what are called the "monothalamous" *Foraminifera* (Gr. *monos*, single; *thalamos*, a chamber)—that is to say, of those forms in which the animal consists of a single segment, and the shell of a single chamber. All the *Foraminifera* without exception commence life as "simple" or "monothalamous" forms, like *Lagena*, but it is comparatively seldom that they retain this simplicity throughout life. In the great majority of cases the primitive mass of sarcode commences a process of budding, or "gemmation," by which it becomes converted from a "simple" into a "compound" form. The original sarcode-mass, that is to say, begins to throw out buds in some determinate direction; all the buds thus produced remaining connected with one another, and all surrounding themselves with a calcareous covering. In this way we get ultimately a compound organism, composed of a number of little masses of sarcode, all permanently united to one another, and all enclosed in a common shell. We get then, ultimately, such a form as *Nodosaria* (Fig. 4, *d*, *e*), which may be regarded as a good example of these so-called "compound" or "polythalamous" *Foraminifera* (Gr. *polus*, many; *thalamos*, a chamber). The exact form of shell which is produced by this process of budding will depend upon the direction in which the buds are given off by the primordial segment. If the buds are given off in a line, we get such a form as *Nodosaria*. If they are given off in a spiral direction, each succeeding segment being a little larger than the one before it, and the coils of the spiral all lying in one plane, then we get such a shell as *Discorbina* (Fig. 4, *c*). This is one of the commonest forms of shell among the *Foraminifera*, and it is often called the "nautiloid" shell, from the close resemblance which it bears in shape to the well-known shell of the pearly Nautilus. It was, in fact, this external similarity which induced the older naturalists to place the *Foraminifera* among the *Mollusca* in the neighborhood of the cuttle-fishes. There are numerous other types of shell, all of which can be referred to the manner in which gemmation is carried on by the primordial segment; but the two forms above mentioned may be taken as sufficient examples. It may be mentioned, however, that there are forms in which the new segments are added in a very irregular manner, and the resulting colony has no very definite shape, as in *Globigerina* (Fig. 4, *f*).

Affinities of the Foraminifera.—In spite of their beautiful, and in many cases complex, shells, the anatomical structure of the *Foraminifera* is so simple that it may fairly be questioned whether in a systematic arrangement they should not be placed at the bottom of the whole sub-kingdom *Protozoa*. Perhaps the nearest relatives of the *Foraminifera* are the *Polycystina*, a group of organisms which we have yet to consider. These differ from the *Foraminifera* in little or nothing, except that the shell is composed of flint. The *Foraminifera* are also clearly related to those forms of the *Amœbea* which possess shells, such as *Difflugia*. The sarcode-body of *Difflugia*, however, contains a nucleus and a contractile vesicle, and the pseudopodia are thick and blunt, so that the differences are sufficiently weighty. There are also very interesting points of relationship between the *Foraminifera* and the sponges, which cannot be touched upon here. A few words, however, may be said on the physiological deductions which may be drawn from the study of the *Foraminifera*. Regarded from a physiological point of view, the structural simplicity of the *Foraminifera* renders them all the more wonderful. We have in them the great equation of life presented to us in perhaps its simplest form. They are composed of an organic substance, but cannot be said to possess "organization," being "structureless, and without permanent distinction or separation of parts." * Nevertheless they perform all the physiological functions; they assimilate food—they live, grow, and maintain their integrity in the face of the destructive forces constantly at work upon them—they reproduce their like—and they have certain relations with the external world, being at any rate capable of independent locomotion. All these vital actions they effect without possessing a single organ permanently set apart for the performance of any one of them. Lastly, they have the power of building up an outer envelope or shell, which is always beautiful, and is often of the most complex character, and constructed upon a regular mathematical plan. The *Foraminifera*, then, of all known animals, offer the most convincing illustration of two laws: firstly, that there is something in the action and nature of vital forces altogether distinct from any thing hitherto observed in the physical forces; and sec-

* Huxley.

ondly, that life is the *cause* of organization, and not the *result* of it: in other words, an animal is organized, or possesses structure, because it is alive; it does not live because it is organized.

DISTRIBUTION OF FORAMINIFERA IN SPACE.*—The *Foraminifera* are exclusively marine or inhabitants of the ocean, and have a world-wide distribution. They are mostly very minute, but some of the extinct forms attained a size of as much as three inches in circumference (e. g., the Nummulite, Fig. 5). Some forms may be obtained adhering to the roots of tangle (*Laminaria*) at or near low-water mark, but they are mostly to be dredged from tolerably deep water. In the deepest parts of the ocean which have yet been examined by the dredge—at a depth, namely, of nearly three miles—*Foraminifera* have been obtained in abundance. There is also no doubt that in many parts of the deep ocean, especially where warm currents exist, there are now forming deposits of the shells of *Foraminifera*, which may well be compared to the great masses of white chalk with which the geologist is familiar. *Foraminifera* may generally be obtained for examination from the shakings of sponges or from the sand of the sea-shore, especially in warm climates. To give some idea of their abundance, it may be stated that Plancus found about 6,000 specimens in an ounce of sand from the Adriatic; but D'Orbigny calculated that no fewer than between three and four millions were present in an ounce of sand from the Caribbean Sea.

DISTRIBUTION OF FORAMINIFERA IN TIME.—It is not the object of the present work to enter into the consideration of the past existence of different groups of animals, since this presupposes some knowledge of geology, but the *Foraminifera* present some points of special interest which may be very shortly noticed. In the first place, as far as is yet known, the *Foraminifera* were the earliest and oldest of created beings. The oldest fossil which has hitherto been exhumed by the labors of geologists is believed to have been a *Foraminifer*, † of large size, and with some decided affinities to existing forms. In the second place, it is only by an examination of the distribution of the *Foraminifera* in past time that we can arrive at any adequate notion of the importance of these microscopic creatures when looked at in the aggregate. The great geological formation known as the white chalk—a formation which forms the well-known chalk-cliffs of the south of England, and which stretches over a great part of the continent of Europe, attaining sometimes a thickness of not less than 600 feet—is almost wholly composed of the shells of *Foraminifera*, visible only to the microscope. The smallest fragment of the common chalk, with which every one is familiar, contains numbers of these minute shells; and it is a singular fact that some of the species in the chalk are indistinguishable from forms which now occur in the ooze which forms the bed of the Atlantic at great depths. The stone of which Paris is built is to a very great extent composed of the shells of *Foraminifera*, especially of the *Miliola*; and it is hardly an exaggeration to say that Paris is mainly built up out of these minute organisms. Another remarkable formation is that known as the "Nummulitic limestone," from the presence in it of a large coin-shaped *Foraminifer*, the Nummulite (Fig. 5), generally about as large as a shilling.

The Nummulitic limestone stretches from France on the west to the frontiers of China on the east, and is almost everywhere readily recognizable as

* Under the term "Distribution in Space" come all the facts relating to the *present* occurrence of any animal or group of animals upon the globe. Under the term "Distribution in Time" come all the facts relating to the *past* occurrence of any animal or group of animals upon the globe.

† The *Eozoön Canadense* of the Laurentian Rocks of Canada.

a distinct formation. It attains in places a thickness of several thousand feet, and is especially largely developed in the Alps. It has an historic interest from the fact that the Pyramids are built of it, and that the Nummulites in it were noticed by Herodotus, the "father of history."

FIG. 5.—*Nummulites lævigatus.*

ORDER IV. RADIOLARIA (Lat. *radius*, a ray).—The third order of the *Rhizopoda* is that of the *Radiolaria*, essentially distinguished by the fact that the sarcode-body has the power of secreting a "siliceous" or flinty skeleton, either in the form of a shell, or of detached spicules or needles; while the pseudopodia are long and thread-like, and stand out from the body like *rays*. In this last character the *Radiolaria* approach very closely to the *Foraminifera;* and the resemblance between the two groups is still further increased by the fact that the pseudopodia often run into one another so as to form a net-work, and sometimes show a circulation of granules along their edges. Three groups of organisms have been described as belonging to the *Radiolaria*, and we may briefly notice an example of each of these.

In the first family we have organisms like *Acanthometra* (Fig. 6, *a*), in which the body is composed of sarcode, supported by a framework of siliceous or flinty rods, which all meet in a common centre. The spines or rods are all perforated by canals, and each conveys a pseudopodium, which is protruded from an aperture at its apex. Many pseudopodia, however, are given off from the surface of the body directly, and are not enclosed in the spines. The *Acanthometræ* are all minute, and are found floating near the surface in the open ocean, sometimes in great numbers.

In the second family (*Polycystina*, Fig. 6, *b*) we have a number of beautiful little organisms closely allied to the *Foraminifera*, but differing in the fact that the body is enclosed in a glassy shell composed of flint. The shell is perforated by numerous holes through which the pseudopodia are emitted, and it is usually of extreme beauty, being sculptured in various ways, and often adorned with spines. The sarcode of the body is usually olive brown in color, and often does not quite fill the shell.

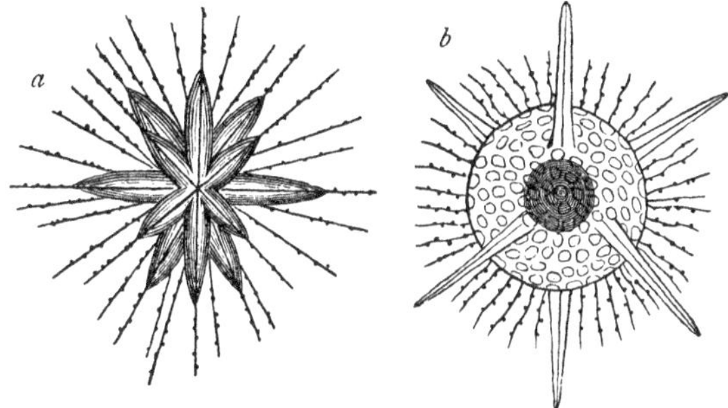

Fig. 6.—*a Acanthometra lanceolata; b Haliomma hexacanthum*, one of the *Polycystina* (after Müller).

The pseudopodia are filamentous, and exhibit a slow circulation of granules along their borders, but they do not run into one another. All the *Polycystina* are microscopic, and they are all inhabitants of the sea. They are best known to students of the microscope as the "Fossil Infusoria of Barbadoes," as they occur in incalculable numbers in a sandstone in that island.

In the third family (*Thalassicollida*, Fig. 7) are included a number of singular gelatinous organisms which may be as large as an ordinary marble, but are often hardly visible to the unassisted eye. They are found floating passively at the surface of most seas.

The body in all the *Thalassicollida* is composed of sarcode, and has the power of giving off thread-like radiating pseudopodia, which sometimes run into one another and form networks. In all cases the sarcode-body appears to have the power of secreting flint in some form or other. In *Collosphæra* (Fig. 7, *a*), the flint is secreted in the form of a shell or test, perforated by large apertures. In *Thalassicolla* (Fig. 7, *b*), the silica forms groups of needles or "spicula," scattered here and there in the jelly-like sarcode.

Order IV. Spongida.—The last order of the *Rhizopoda* is that of the *Spongida*, the exact nature and position of which have only recently been determined. For a long time sponges were pretty generally regarded as being vegetables, and it is only since the microscope has been employed in their elucidation that their true nature has been made out. Most naturalists are now agreed as to the propriety of placing the sponges in the animal kingdom, and they are generally referred to the *Rhizopoda*, though they are sometimes looked upon as constituting a distinct and separate class of the *Protozoa*. The apparent complexity of structure which the sponges exhibit is due to the fact that what we ordinarily

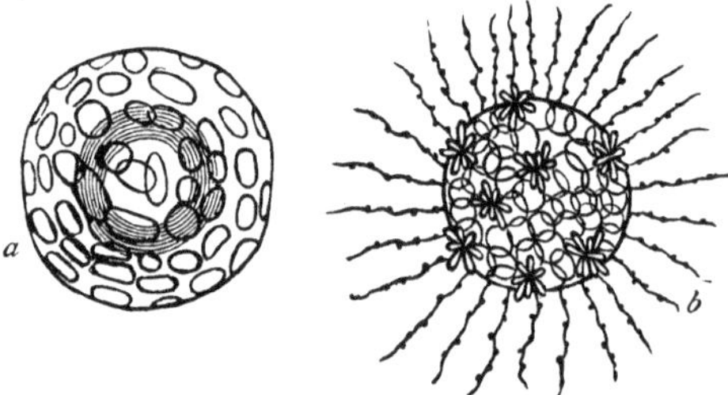

FIG. 7.—*a* Siliceous shell of *Collosphæra*; *b* *Thalassicolla*, showing the radiating pseudopodia and groups of siliceous spicula (after Müller).

term a sponge is really a *colony* or aggregation of separate masses of sarcode, greatly resembling *Amœbæ* in structure, and having the power of secreting a skeleton or supporting framework common to the whole assemblage. Sponges, in fact, may be defined as *compound Rhizopoda, forming masses which are traversed by canals opening on the surface, and supported by a framework of horny fibres or of calcareous or flinty needles.*

There are, then, two essential elements in the structure of a sponge—namely, the sarcode-bodies which constitute the animal itself, and which are collectively termed the "sponge-flesh," and the hard framework or "skeleton" upon which the flesh is supported. To understand the nature of these fully, we may take an ordinary horny sponge, such as we are constantly in the habit of using. As we see the sponge in this country, we are only acquainted with the skeleton, which is composed of an enormous number of horny fibres, all interlaced and interwoven with one another, but leaving numerous holes and canals between their bundles (Fig. 9, *d*). In its living condition, however, the whole of this skeleton is covered inside and outside—saturated, in fact—with a kind of slimy material very like white-of-egg to look at. This is the so-called sponge-flesh, and, upon examining this with a microscope, it is found to be composed of an enormous number of minute masses of sarcode, all more or less completely independent of each other, and each very closely resembling an *Amœba*. These separate "sponge-particles," or "sarcoids," as they are called, consist, in fact, of granular sarcode, capable of pushing out little processes or threads of sarcode in the form of pseudopodia, and sometimes furnished with an internal solid mass or nucleus (Fig. 3, *c*). In some cases each sarcoid carries a single lash-like vibrating filament or cilium (Fig. 3, *d*, *e*). Each sarcoid has the power, if detached, of independent movement, and each can obtain food for itself. As the sponge, however, is a fixed animal, some provision is necessary by which food shall be conveyed to the sarcoids in the interior of the mass. This is effected by a remarkable water-carrying or "aquiferous" system in the following manner: The entire sponge is riddled in every direction by an immense number of canals, all opening on the surface, and communicating freely

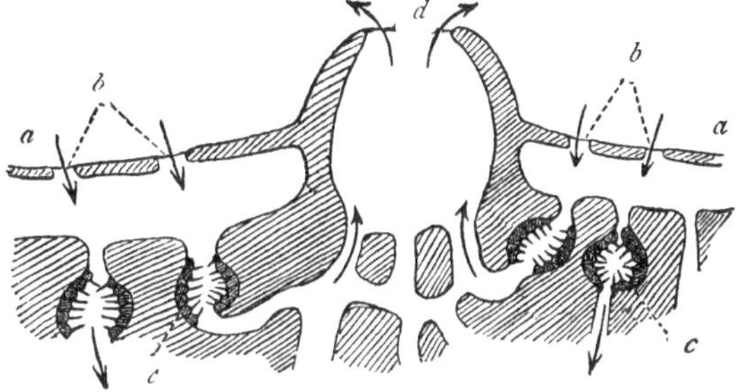

FIG. 8.—Diagrammatic section of *Spongilla* (after Huxley). *a a* Outer or superficial layer of the sponge; *b b* Inhalant apertures, or "pores;" *c c* Ciliated chambers; *d* An exhalant aperture, or "osculum." The arrows indicate the direction of the currents.

with one another in the interior of the mass. The canals are of different sizes, and, as can readily be observed in an ordinary sponge, their external openings are also of different sizes. A few of the holes are of much larger size than the others, and these, for reasons which will be seen directly, are called the "exhalant apertures," or "oscula." The great majority of the holes are very minute, and these are known as the "inhalant apertures," or "pores." In a living sponge a more or less constant circulation of water is carried on by means of this canal system. The water is admitted by means of the pores (Fig. 8, *b b*), is driven into the interior of the sponge, and is finally expelled in steady streams from the oscula (Fig. 8, *d*). The mechanism by which this circulation of water is effected, was long unknown, but it is now known to consist in aggregations of sponge-particles provided with cilia which all work toward the interior of the sponge (Fig. 8, *c c*). The circulation of water in this manner can be readily observed in many of our common marine sponges, and it is under the control of the animal to a great extent. The large apertures or oscula are permanent, but they can be closed and opened at will; while the smaller apertures or pores appear to be formed afresh, wherever they are wanted, at any point of the surface. By means of the currents of water each individual sarcoid or sponge-particle is enabled to obtain food, so that the whole sponge, as remarked by Huxley, "represents a kind of subaqueous city, where the people are arranged about the streets and roads in such a manner that each can easily appropriate his food from the water as it passes along." It is also not improbable that the process is at the same time a rudimentary form of respiration.

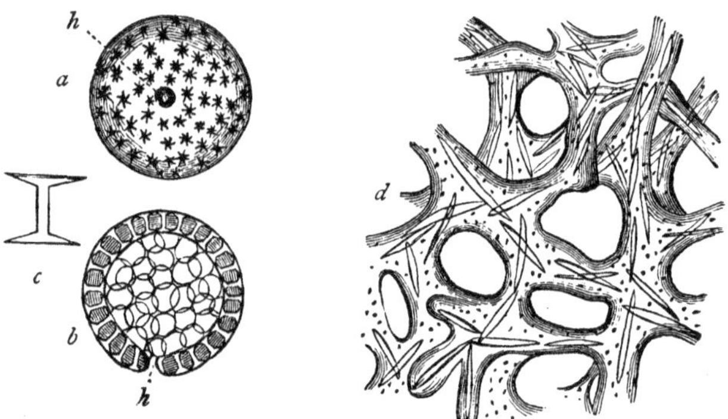

FIG. 9.—*a* Gemmule of *Spongilla*; *h* Hilum; *b* Diagrammatic section of the gemmule, showing the outer layer of spicules or amphidiscs, and the inner mass of cells; *h* Hilum; *c* One of the amphidiscs seen in profile; *d* Fragment of the skeleton of a horny sponge (after Bowerbank), showing the interlacing horny fibres with spicula. All much magnified.

Such, then, are the general phenomena exhibited by any sponge, and the chief variations which occur among the sponges are to be found in the nature of the skeleton. In the sponges of commerce the skeleton consists of matted fibres composed of a substance nearly allied to horn. In other forms the skeleton is calcareous, or composed of lime; and in other cases, again, it is siliceous, or composed of flint. The Venus's flower-basket (Euplectella), which looks like a goblet woven of spun glass, is a familiar example of the flinty sponges. In most cases, the skeleton, and often the flesh as well, is furnished with more or less numerous needles or spicula, generally of flint, but sometimes of lime, which assume a great variety of shapes, and appear to exercise different functions (Fig. 9, *c*, *d*).

As regards the reproductive process in the sponges, it will be sufficient to state very briefly the leading phenomena which have been observed in the fresh-water sponge (*Spongilla fluviatilis*). If a specimen of *Spongilla* be observed toward the approach of winter, its deeper portions will be found to be filled with numerous small, rounded bodies, like seeds, which have been called "gemmules." Each gemmule (Fig. 9, *a*, *b*) exhibits at one point a small aperture, and is found to be composed of a leathery membrane, surrounded by a layer of sarcode, in which are imbedded a number of spicula. These spicula consist each of a central rod or axle carrying a toothed wheel or disk at each end (Fig. 9, *c*). In the interior of the capsule thus formed is a mass of cells, of which the central ones contain numerous reproductive germs. When the spring comes, these masses are discharged into the water through the aperture of the gemmule, and become developed into fresh *Spongillæ*. In addition to this method of reproduction, the fresh-water sponge during the summer months has the power of producing true eggs or ova, and sperm-cells. The impregnated ova develop themselves into embryos, which are provided with numerous cilia or vibrating hairs, by means of which they swim about freely. Finally, upon finding a suitable locality, they fix themselves to some solid object, lose their cilia, and grow up into *Spongillæ*. Indeed, as a general if not universal rule, the embryos of the sponges are provided with cilia, and are thus capable of active locomotion. In this way is secured the extension over a wide area of these otherwise fixed and plant-like organisms.

Distribution of Sponges in Space.—It remains only to add a few words on the distribution of sponges in space. With the single exception of *Spongilla*, all known sponges are inhabitants of the sea; but the former is to be found in lakes and rivers in most parts of the world. The marine sponges are found mostly attached to stones and other foreign objects between tide-marks and in deep water. The sponges of commerce are mostly obtained from the Grecian Archipelago, but inferior kinds are imported from the Bahama Islands. One common sponge (*Cliona*), instead of incrusting other objects, inhabits branching cavities in shells, which it excavates for itself. It apparently lives upon the animal matter contained in the shell, and few oyster-shells can be picked up upon our shores which do not exhibit the perforations and mines of some species or other of this genus. Fossil shells, also, often occur, which show that these mining sponges have enjoyed a vast antiquity.

CHAPTER III.

INFUSORIA.

THE last class of the *Protozoa* is that of the *Infusoria*, so called because of their being frequently developed in organic infusions under the following singular circumstances: If some water be taken, and any animal or vegetable substance be soaked or boiled in it, a solution is formed containing organic matter, or, in other words, an "organic infusion." It is unnecessary to say that if this infusion be examined under the microscope, after boiling, nothing will be detected in it—nothing living, at any rate. If examined, however, at the end of a few days' time—if the circumstances have been favorable—a vast number of living forms will now be found in it. Among these will be found several of the members of the present class, and hence the name applied to them of Infusorian animalcules, or *Infusoria*. It is unnecessary to enter here into the question how these living beings are produced, since the subject is one of great obscurity, and opinions are still divided upon it. It is sufficient to remark that there are eminent observers who hold that the appearance of the *Infusoria* in this fashion is to be explained upon the theory that they have been spontaneously produced out of the inorganic materials of the infusion, in opposition to the general view that they are derived from preëxistent germs.

The position of the *Infusoria* is somewhat doubtful, and it appears probable that they will ultimately have to be regarded as a separate sub-kingdom, or as a branch of a higher sub-kingdom (*Annuloida*). In the mean while it is most convenient to retain them in their present place, at the head of the sub-kingdom *Protozoa*. Regarded in this light, the *Infusoria* present a great advance in structure over all the forms which we have hitherto studied—an advance which is especially seen in the constant presence of a permanent mouth. The *Infusoria* may be defined as *Protozoa which are provided with a mouth, and generally a rudimentary digestive canal. They do not possess the power of emitting pseudopodia, but are furnished with vibratile cilia or contractile filaments. They are mostly microscopic in size, and their bodies usually consist of three distinct layers.* They are mostly simple free-swimming organisms, but they sometimes form colonies by budding, and are fixed to some solid object in their adult condition. As types of these two sections of the *Infusoria*, we may take respectively *Paramœcium* and *Epistylis.*

Paramœcium (Fig. 10) is a beautiful slipper-shaped little

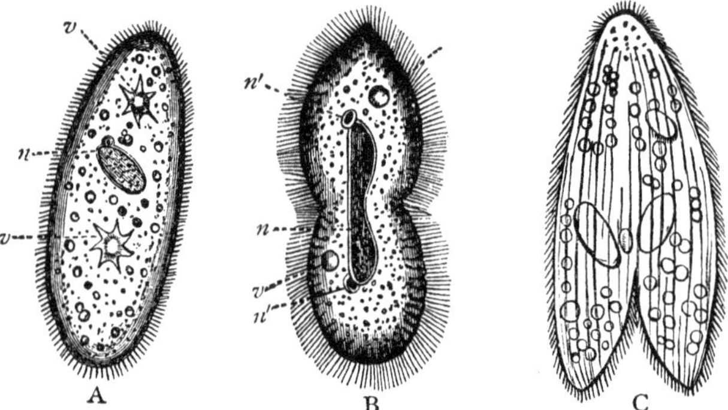

Fig. 10.—Ciliated Infusoria. A. *Paramœcium*, showing the nucleus (*n*) and two contractile vesicles (*v*); B. *Paramœcium bursaria* (after Stein), dividing transversely, *n* Nucleus; *n'* nucleolus; *v* Contractile vesicle; C. *Paramœcium aurelia* (after Ehrenberg), dividing longitudinally.

creature, which may be found commonly in stagnant waters or in artificially-prepared infusions. The body is nearly quite transparent, and consists of three layers—1. A structureless, transparent, external film or pellicle; 2. A central mass of soft semi-fluid sarcode; and 3. An intermediate layer of firm and consistent sarcode. The external membrane or cuticle is richly covered with minute vibrating hairs or cilia, which appear, however, to be really derived from the middle layer. The cuticle is also perforated by the aperture of the mouth, which is continued into a short, funnel-shaped gullet. The gullet, however, is not continued into any distinct stomach, but opens directly into the soft, semi-fluid sarcode which constitutes the central abdominal cavity. The particles of food on passing through the gullet are directly received into the central mass of diffluent sarcode, where they undergo a kind of slow circulation or rotation. As in the case of the *Amœba*, each particle of food generally carries with it a little water, so that the appearance is produced of a number of little clear spaces in the central sarcode. These are now called vacuoles, or food-vacuoles; but they were originally described by Ehrenberg, the famous Prussian microscopist, as so many distinct stomachs, in consequence of which he named the *Infusoria* the *Polygastrica* (Gr. *polus*, many; and *gaster*, stomach). The vibrating cilia which clothe the surface of *Paramœcium* serve partly to drive the animal rapidly through the water, and partly to set up currents by means of which food is conveyed to the mouth. All the nutrient particles obtained in this way undergo the circulation in the central sarcode above spoken of, where they are partially or completely digested. The indigestible portions of the food appear to be got rid of by a second aperture (*anus*) placed near the mouth. The only other organs possessed by *Paramœcium* are the so-called nucleus and nucleolus, and the contractile vesicle (or vesicles), all of which appear to be situated in the cortical layer of the body. The nucleus (Fig. 10, *n*) is a little solid body, composed of an external membrane, with granular contents, and having the nucleolus (*n'*) firmly attached to its exterior in the form of a little spherical particle. Both appear to be organs of reproduction, the nucleus being an ovary, and the nucleolus a spermarium. The names, therefore, of nucleus and nucleolus are extremely inappropriate, as they lead to confusion with the wholly distinct structures which receive these names in an ordinary animal or vegetable cell. The contractile vesicle (*v*) has exactly the same structure as in the *Amœba*. It is simply a little contractile cavity filled with a fluid apparently derived from the digestion, and contracting and dilating at regular intervals. There is usually only a single vesicle present, but there may be two or more.

Reproduction in *Paramœcium* may be effected by fission—that is to say, by a simple splitting of the body of a single individual into two portions, each of which becomes a fresh being. The process of fission may commence at the surface, or it may begin at the nucleus. In other cases, two *Paramœcia* come together and adhere closely to one another. The nucleus and nucleolus enlarge, and the nucleolus of each is transferred to the other, apparently through the mouth. As the result of this, numerous germs are produced, which, after their liberation from the body of the parent, are developed into fresh individuals.

Epistylis, which is a good example of the fixed *Infusoria*, may be regarded as essentially similar to *Paramœcium* in its anatomical structure. In place, however, of a single free-swimming organism, we have now a colony of more or less closely related beings, the whole assuming a plant-like form, and being rooted to some solid object. The colonies of *Epistylis* may not uncommonly be found adhering to the stems of water-plants or to the backs of our common water-beetles, and the trained eye readily recognizes them as a grayish-white down or nap. On placing a portion of this under the microscope, we see a number of little oval cups or "calyces" supported upon a branched stem. Each cup contains a sarcode-body, essentially the same as *Paramœcium* in structure, consisting of granular sarcode, with vacuoles, a nucleus, and a contractile vesicle. The end of the cup farthest from the stalk terminates in a rounded aperture, through which there can be protruded a ciliated disk. On one side of this disk is the aperture of the mouth, leading into a kind of gullet, which is also furnished with large vibrating cilia. This, in turn, opens directly into the soft, granular sarcode of the abdominal cavity, which exhibits a constant though slow rotation.

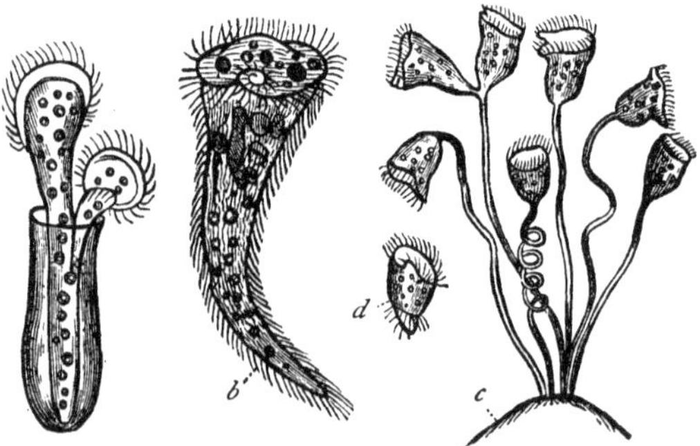

FIG. 11.—Ciliated Infusoria. *a Vaginicola*; *b Stentor Mülleri*, the Trumpet Animalcule; *c* Group of *Vorticellæ*; *d* Detached bud of *Vorticella*.

A still commoner and equally beautiful example of the Stalked Infusoria is the so-called Bell-animalcule (*Vorticella*, Fig. 11, *c*), which may be found in any stagnant pool attached to the stems of aquatic plants. The body in *Vorticella* forms a kind of cup or "calyx" supported upon a long stalk, which is in turn fixed to some solid object. The stem contains a contractile fibre in its interior, and the animal can by this means push itself out or coil itself up with the utmost rapidity. The vibrating filaments or cilia are not scattered over the whole surface of the bell-shaped body, but are collected to form a kind of fringe or circle round the mouth of the calyx. Nearly in the centre of this ring, or on one side, is placed the aperture of the mouth, which leads by a short gullet straight into the central soft sarcode of the interior of the body. A nucleus and contractile vesicle are also present, so that in the essential points of its anatomy *Vorticella* does not differ from a free-swimming Infusorian such as *Paramœcium*. Indeed, a transition between the two forms is found in the so-called Trumpet animalcule or *Stentor* (Fig. 11, *b*), which can detach itself and swim about at will, at the same time that it is ordinarily fixed by its thinner extremity to some solid object. In *Vaginicola* (Fig. 11, *a*), again, we have an animalcule closely related to *Stentor*, but having the body protected by a horny or membranous sheath.

All the *Infusoria* we have been hitherto considering belong to a section of the class in which the surface is furnished with more or less numerous cilia. There are other forms, however, in which there are no cilia, but the body is furnished with a number of radiating filamentous tubes, the extremities of which form little sucking-disks. Finally, there is another section in which the organs of locomotion are in the form of long, contractile filaments, termed "flagella," which may be combined with cilia, or may be the only locomotive organs present. In accordance with these differences, the *Infusoria* are divided into the three orders of the *Ciliata*, *Suctoria*, and *Flagellata*, of which the ciliated forms are by far the most numerous and most important.

Distribution of Infusoria in Space.—As regards the distribution of *Infusoria* in space, there is little to say, except that they are of universal occurrence in fresh water over the whole globe, and that they occur also in the sea. In fact, the only conditions which appear to be necessary for their existence are a certain quantity of water holding organic matter in solution. Wherever these conditions are fulfilled, *Infusoria* are certain to make their appearance. The attached forms of *Infusoria*, however (such as *Vorticella*, *Epistylis*, *Stentor*, and others), do not appear to be ever developed in artificial infusions, and they are to be sought for on the stems of water-plants, and in other similar localities. It seems hardly necessary to remark that, as before defined, the occurrence of fossil *Infusoria* is not to be looked for, as they possess no hard structures which are capable of permanent preservation. It is only to be added in this connection that, if the animalcule known as *Noctiluca* be rightly referred to this class, the *Infusoria* take a very decided share in producing the diffused phosphorescence or luminosity of the sea, which is occasionally such a beautiful spectacle even in our own climate.

SUB-KINGDOM II.—CŒLENTERATA.

CHAPTER IV.

1. Characters of the Sub-Kingdom. 2. Divisions. 3. General Characters of the Hydrozoa. 4. Explanation of Technical Terms.

In the sub-kingdom *Cœlenterata* are included the sea-anemones, corals, sea-jellies, sea-firs, and other allied animals, and the whole division may be looked upon as forming the most typical section of the animals formerly called by Cuvier *Radiata.* In addition, however, to the above-mentioned animals, Cuvier included in his *Radiata* all the members of the modern sub-kingdom *Protozoa*, together with the sea-mats or lowest class of the *Mollusca*, and the sea-urchins, star-fishes, wheel-animalcules, internal parasites, and others which are now placed in a separate sub-kingdom by themselves (*Annuloida*). The old *Radiata*, therefore, was an extremely heterogeneous assemblage, and there is no advantage to be derived from its employment even in works such as this present. The division *Cœlenterata*, or "hollow-entrailed" animals (Gr. *koilos*, hollow; and *enteron*, intestine), includes all those radiate animals which are more or less closely allied to the sea-anemones on the one hand, and to the sea-firs on the other. Most of the *Cœlenterata* come under the conveniently loose term of "zoophytes," or plant-animals, from the external resemblance which many of them show to plants.

The *Cœlenterata* may be defined as *animals whose alimentary canal communicates freely with the general cavity of the body ("somatic cavity"). The body is essentially composed of two layers or membranes, an outer layer or "ectoderm," and an inner layer or "endoderm." No circulatory organs exist, and in most there are no traces of a nervous system. Peculiar stinging organs, or "thread-cells," are usually if not always present, and in most cases there is a radiate or star-like arrangement of the organs, which is especially perceptible in the tentacles, which are in most instances placed round the mouth. Distinct reproductive organs exist in all.*

The leading feature which distinguishes the *Cœlenterata*, and the one from which the name of the sub-kingdom is derived, is the peculiar arrangement of the digestive system. In the *Protozoa*, as we have seen, a mouth is only very rarely present, and in no case is there any definite internal cavity bounded by the walls of the body, to which the name of "body-cavity" or "somatic cavity" could be properly applied. In most of the higher animals, on the other hand, not only is a permanent mouth present, but the walls of the body enclose a distinct and permanent chamber or body-cavity. Further, in most cases the mouth opens into an alimentary or digestive tube, which is always distinct from the body-cavity, and never opens into it, usually passing through it to open on the surface by another distinct aperture (the anus). In most cases, therefore, the alimentary canal is a tube which communicates with the outer world by two apertures—a mouth and an anus—but which simply passes through the body-cavity without in any way communicating with it. In the *Cœlenterata* (Fig. 12) the condition of parts is intermediate in its arrange-

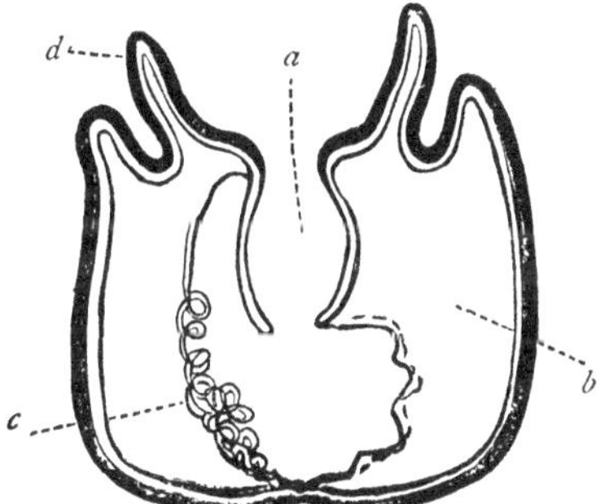

FIG. 12.—Diagrammatic vertical section of a Sea-anemone (*Actinia*). *a* Stomach; *b* Mesentery; *c* Convoluted cord or 'craspedum;" *d* Tentacle. The dark line indicates the "ectoderm," the fine line and clear space adjacent mark the "endoderm."

ment. There is a distinct and permanent mouth, and there is a distinct and permanent body-cavity, but the mouth opens into, and thus communicates freely with, the body-cavity. In some cases the mouth opens straight into the general body-cavity, which then serves as a digestive cavity as well (Fig. 13, *a*). In other cases there intervenes between the mouth and the body-cavity a short alimentary tube, which communicates externally with the outer world through the mouth, and opens below by a wide aperture into the general cavity of the body. In no case is there a distinct intestinal tube which runs through the body-cavity and opens on the surface by a mouth at one end and an excretory aperture or anus at the other. Another leading character of the *Cœlenterata* is the composition of the body out of two fundamental membranes (Fig. 12), which are usually of a very simple structure, but which may be more or less complicated by the development of muscular fibres and other tissues. The outer of these layers or membranes is known as the "ectoderm," and it forms the whole of the outer surface of the body, terminating at the margins of the mouth. The inner layer is known as the "endoderm," and it lines the whole of the interior of the body, being prolonged into the tubular tentacles round the mouth. Both of these membranes, but especially the endoderm, are usually more or less richly furnished with vibrating cilia. The peculiar microscopic organs called "thread-cells," or "nettle-cells," which communicate to many of the *Cœlenterata* (such as the sea-jellies) their peculiar power of stinging, are structures found in the integument of almost all the members of this sub-kingdom, and sometimes in internal parts as well. They are very beautiful objects of microscopical examination, and differ very considerably in the details of their structure. They are, however, in most respects essentially the same as in the common *Hydra* or fresh-water polype, in which the thread-cells (Fig. 13, *d*) are "oval elastic sacs, containing a long, coiled filament, barbed at its base and serrated along its edges. When fully developed the sacs are tensely filled with fluid, and the slightest touch is sufficient to cause the retroversion of the filament, which then projects beyond the sac for a distance, which is not uncommonly equal to many times the length of the latter" (Huxley).

In accordance with the above-mentioned differences in the arrangement of the digestive system, the *Cœlenterata* are divided into two great classes, termed respectively the *Hydrozoa* and the *Actinozoa*. In the *Hydrozoa*, there is no body-cavity distinct from the digestive cavity—or, in other words, the body-cavity *is* the digestive cavity. In the *Acti-*

nozoa, on the other hand, there is a distinct digestive cavity, but this opens directly into the general body-cavity, so that the two form distinct but freely-communicating divisions of the same chamber.

Class I.—Hydrozoa.

The *Hydrozoa* are defined as *Cœlenterata in which the walls of the digestive sac are not separated from those of the general cavity of the body, the two coinciding with one another. The reproductive organs are external, in the form of outward processes of the body-wall* (Fig. 13, *a*, *b*).

The *Hydrozoa* are all aquatic in their habits, and, with the exception of two genera, all are inhabitants of salt water. The class includes both simple and composite organisms, of which the most familiar are the sea-firs and their allies (Hydroid zoophytes), the fresh-water polype or *Hydra*, the sea-jellies (*Medusæ*), and the Portuguese man-of-war (*Physalia*). Owing to the extremely complicated nature of many of the *Hydrozoa*, it appears advisable to preface their description by an explanation of some of the more important terms which are employed in connection with various members of the class.

General Terminology of the Hydrozoa.

Individual.—In order to understand fully the meaning which is attached to the term "individual" in zoological language, it is necessary to glance briefly at the general features of reproduction as displayed in different sections of the animal kingdom. Reproduction is the process by means of which new individuals are produced and the perpetuation of the species insured. This end may be attained in various ways, but these all come under the two heads of "sexual" and "non-sexual" reproduction. In *sexual* reproduction, by which alone can fresh beings be produced among the higher animals, the essential element of the process consists in the formation of two distinct structures, a germ-cell or ovum, and a sperm-cell or spermatozoid. By the union of these distinct reproductive elements fresh beings can be produced. As a general rule, the germ-cell is produced by one individual (female), and the sperm-cell by another (male); but among the lower animals it is not uncommon for the same individual to produce both of these elements, in which case the individual is said to be "hermaphrodite." Among the lower animals, however, fresh beings may be produced without the contact of a sperm-cell and an ovum—that is to say, without any genuine act of reproduction. The processes by which this can be effected in different animals vary considerably, but they are all spoken of as forms of "non-sexual" reproduction. The only varieties, however, of the process which require consideration, are those in which fresh beings are produced by what is called "gemmation" or "fission."

Gemmation (Lat. *gemma*, a bud) consists in the production of a bud or buds, usually from the outside, but sometimes from the inside, of an animal; which buds become developed into more or less completely independent beings. The fresh beings thus produced by budding are all known as *zoöids*, and are not spoken of as distinct animals for reasons which will be immediately evident. When the zoöids produced by budding remain permanently attached to one another and to the parent organism which produced them, the case is said to be one of "continuous" gemmation, and the ultimate result of this is to produce a colony or composite structure, composed of a number of similar and partially independent beings, all produced by budding, but all remaining in organic connection. This is seen very well in the sponges, in the compound *Foraminifera*, and in a great number of the *Hydrozoa*. When, on the other hand, the zoöids produced by budding become finally detached from the parent organism, we have a case of what is called "discontinuous" gemmation. In this case, the detached zoöids become completely independent beings; and they are often wholly unlike the original zoöid in structure and in habits, so much so that they have in various cases been described as altogether distinct animals. Discontinuous gemmation is very well seen in many of the *Hydrozoa*, and in them the case is still further complicated by the coexistence of discontinuous gemmation with the continuous form of the process. Thus, it is not an uncommon thing among the *Hydrozoa* to find a composite organism or colony produced from a primordial zoöid by continuous gemmation, and having at the same time the power of giving rise to detached and completely independent beings by a process of discontinuous gemmation.

In what is called "fission" (Lat. *findo*, I cleave), exactly the same results are attained as in gemmation, but in a slightly different manner. In gemmation the new beings are produced by means of buds thrown out by a primitive zoöid. In fission the new beings are produced by a cleavage or division of a primitive zoöid into two or more parts, each of which becomes finally developed into a new and more or less completely independent being. In fission, as in gemmation, the new beings or zoöids may remain permanently in connection with one another, when the process is a continuous one, and a composite organism is produced, as in many corals. Or, in other cases, the new zoöids produced by fission are detached to lead an independent existence, as in some of the *Hydrozoa*, the process thus becoming a discontinuous one.

We are now able to understand what is meant, in strict zoological language, by the term "individual," as applied to animals. Zoologically speaking, an *individual* is defined as "*equal to the total result of the development of a single ovum.*" In the higher animals there is no sort of difficulty about this, for each ovum gives rise to no more than one single animal, which cannot produce fresh beings in any other way than by producing another ovum. In this case, therefore, each animal is an individual. In the lower animals, however, the being produced by an ovum has often the power of giving rise to fresh beings by a process of gemmation or fission, and these beings may either remain attached to one another so as to form a colony, or may become detached to lead independent lives. In either case, the term "individual" can only be properly applied to the whole assemblage of beings produced in this way, however much they may differ from one another in appearance, structure, or mode of life. In these cases, therefore, the individual may be, *firstly*, a single independent being—as, for instance, an *Amœba*, or an Infusorian such as *Paramœcium*; *secondly*, a colony or composite organism composed of a number of more or less nearly similar beings or zoöids, produced by budding from a primitive zoöid—as, for instance, a sponge, or such an Infusorian as *Epistylis*; and *thirdly*, an assemblage of zoöids produced by budding or fission from a primitive being, but not necessarily remaining connected with one another or exhibiting any common features of likeness, as we shall see is the case in many of the *Hydrozoa*. Lastly, cases may occur in which the individual consists partly of similar zoöids which remain permanently connected with one another, and partly of dissimilar zoöids which are detached to lead an independent life, all alike being the result of the development of a single ovum.

Zoöid (Gr. *zoön*, animal; *eidos*, form).—The term "zoöid" is indifferently applied to all the more or less completely independent beings which are produced by budding, or by cleavage from a primitive organism. It does not matter, therefore, for the purposes of this definition, whether these beings remain permanently attached to the original organism, or whether they are finally separated to enjoy an independent existence.

Hydrosoma (Gr. *hudra*, a water-serpent; *soma*, body).—The term "hydrosoma" is one which is very conveniently applied to the entire organism in any *Hydrozoön*, whether this be simple, or whether it be composite and made up of a number of connected zoöids.

Polypite.—That portion of any *Hydrozoön* which is concerned with the process of digestion, or, in other words, the "alimentary region," is termed the "polypite"—the more generally current term of "polype" being now restricted in meaning to the same region in the higher *Cœlenterata* (*Actinozoa*). In such of the *Hydrozoa* as the fresh-water polype or *Hydra*, in which the hydrosoma is simple, the whole organism is termed a polypite; but the term is more generally employed to indicate the nutritive zoöids of any compound *Hydrozoön.*

Cœnosarc.—The term "cœnosarc" (Gr. *koinos*, common; *sarx*, flesh) is employed to designate the common trunk or flesh by which the separate polypites of any compound *Hydrozoön* are united into a single organic whole.

Polypary.—The term "polypary" or "polypidom" is applied to the horny or chitinous outer covering or envelope with which many of the *Hydrozoa* are furnished. These terms have also been not uncommonly employed to designate the very similar structures produced by the much more highly organized sea-mats and their allies (*Polyzoa*), but it is better to restrict their use entirely to the *Hydrozoa.*

CHAPTER V.

DIVISIONS OF THE HYDROZOA.

THE *Hydrozoa* are divided into four great divisions, each of which requires some notice, as presenting points of special interest. These divisions or sub-classes are known by the names of *Hydroida*, *Siphonophora*, *Discophora*, and *Lucernarida*.

SUB-CLASS HYDROIDA.

This sub-class comprises all the sea-firs and their allies, commonly known to naturalists as the "Hydroid zoophytes," from their resemblance to the fresh-water polype (*Hydra*), which is also a member of this division. The *Hydroida* are defined by the fact that *they consist of an alimentary region or "polypite," which is furnished with a mouth and prehensile tentacles at one end, and with an adherent disk at its other extremity.* In some few cases the hydrosoma consists of but one such polypite (as in the *Hydrida* and some of the *Corynida*); but generally the hydrosoma is composed of a greater or less number of similar polypites all united by a cœnosarc or common trunk (as in the majority of the *Corynida*, and in the *Sertularida* and *Campanularida*). In the great majority of cases, also, the hydrosoma is not unattached, but is fixed to some solid object by one extremity. The Hydroid zoophytes exhibit three principal types of structure, which constitute so many orders.

ORDER I. HYDRIDA.—In the first order we have only the well-known fresh-water polypes or *Hydræ*, of which we may take the common green *Hydra* (*H. viridis*) as the type. This singular little creature may be found with a little trouble in most of our streams and pools, and it is quite visible to the naked eye, though it can only be satisfactorily examined by the help of the microscope. When uncontracted, the body of the *Hydra* is in the form of a cylindrical tube (Fig. 13, *a*, *b*), composed of the two fundamental layers, the ectoderm and endoderm, of which the former contains many thread-cells, and is likewise furnished with numerous green granules, stated to be identical with "chlorophyll," or the green coloring-matter of plants. At the base of the cylindrical body is a kind of disk-shaped sucker, by means of which the animal can attach itself at will to any foreign body. Its favorite position appears to be that of hanging head-downward, suspended from the stem of some water-plant. It is not, however, permanently fixed, but it can detach itself and change its place at will. At the opposite extremity of the body is placed the aperture of the mouth, surrounded by a circle of from five to fifteen small tubular filaments, which are termed the "tenta-

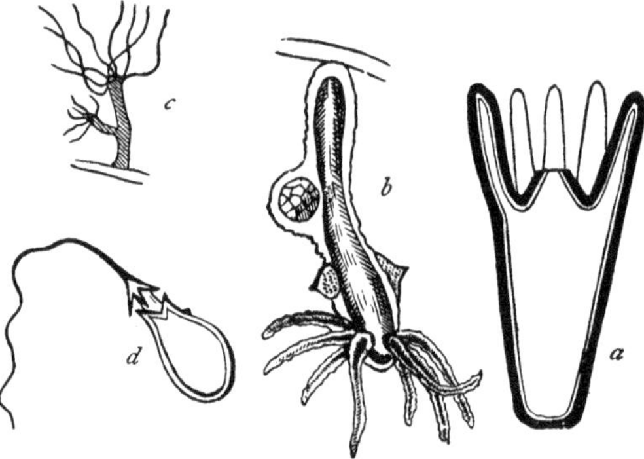

FIG. 13.—Morphology of Hydrozoa. *a* Diagrammatic section of *Hydra*: the dark line is the ectoderm, the fine line and clear space adjacent indicate the endoderm; *b* *Hydra viridis*, showing a single ovum contained in the body-wall near the lower extremity, and two conical elevations containing sperm-cells near the bases of the tentacles. *c* *Hydra vulgaris*, with an undetached bud—enlarged; *d* Thread-cell of the *Hydra*, greatly magnified.

cles" (Fig. 13, *b*). Each tentacle consists of a tubular prolongation of both ectoderm and endoderm, and encloses a canal which opens at its base into the general cavity of the body. The ectoderm of the tentacles is richly furnished with thread-cells, and they are also well supplied with muscular fibres. They exhibit the most extraordinary contractility, being capable of retraction till they appear as nothing more than so many little warts or tubercles, and of being extended to a length which is in some species many times longer than the body itself. (In *Hydra fusca* the tentacles can be protruded to a length of more than eight inches.) The tentacles are the organs by means of which the *Hydra* obtains its food, consisting chiefly of minute aquatic organisms, such as small worms, insects, *Crustacea* and *Rotifera*. These are seized by the tentacles and gradually drawn into the mouth; but in addition to this merely mechanical action, the tentacles appear to exercise a benumbing or even fatal influence upon the animals grasped by them—this being apparently due to the thread-cells with which they are furnished. The mouth in the *Hydra* opens directly into a capacious cylindrical cavity, which is excavated along the whole length of the body, and which is both the body-cavity and the stomach in one. This cavity (Fig. 13, *a*, *b*) is filled with water derived from the exterior, and also with the nutritive particles derived from the food. Indigestible fragments appear to be rejected by the mouth, though an anal aperture has been asserted to be present. There are no internal organs of any kind. Physiologically, therefore, the *Hydra* presents little advance upon the higher *Protozoa*, such as the *Infusoria*. There is a permanent mouth, surrounded by permanent and special organs adapted for the seizure of food. There is also a permanent internal cavity for the reception and digestion of the food, but this is not shut off from the general cavity of the body. There is no organ for the propulsion of the nutritive fluid through the body, no nervous system or organs of sense, and no special respiratory or excretory organs. Another and striking proof of the essentially low position of the *Hydra* in the animal scale is to be found in its extraordinary capacity of resisting mutilation, or, in fact, mechanical injury of any kind short of absolute annihilation. The briefest illustration of this fact is all that can here be given, but with that the name of Trembley of Geneva must be associated. This well-known observer, in a long series of experiments, most of which have been successfully repeated by subsequent naturalists, discovered that the *Hydra* could be mechanically divided with a knife into any number of fragments, with the sole result that each and all of these possessed the power of developing themselves into fresh and independent polypites. Further, the animal could even be turned inside out, with a necessary transposition of the ectoderm and endoderm, without any apparent inconvenience or interference with its health.

Reproduction in the *Hydra* is effected non-sexually by gemmation, and sexually by the production of ova and sperm-cells; the former process being followed in summer and the latter in autumn, few individuals appearing to survive the winter. In the first or non-sexual method, the *Hydra* throws out one or more buds, usually from near the fixed extremity (Fig. 13, *c*). These buds at first consist simply of a tubular prolongation of the ectoderm and endoderm, enclosing a cavity which communicates with the general cavity of the body. A new mouth and tentacles are soon developed at the free end of this bud, and after a longer or shorter period the new *Hydra*, thus produced, is detached to lead an independent life. Each *Hydra* can produce many such buds during the summer season, and the liberated buds can also repeat the same process, so that in this way reproduction is rapidly carried on. In the second or sexual method of reproduction, ova and sperm-cells are produced toward the winter in external processes of the body-wall. The spermatozoa are developed in little conical elevations, which are produced near the bases of the tentacles; and the ova are formed in much larger elevations, of which there is ordinarily but one, placed nearer to the fixed extremity of the animal (Fig. 13, *b*). When mature, the ovum is fertilized by the sperm-cells, both being set free into the water by the rupture of the body-wall. The embryo *Hydra* is at first covered with vibrating cilia, and swims freely about, until it meets with a suitable locality. It then fixes itself by one extremity, the cilia drop off, and a mouth and tentacles are developed at the free end of the body.

ORDER II. CORYNIDA.—In the second order of the Hydroid zoophytes, known as the *Corynida* or *Tubularida*, we have a number of organisms which in their essential structure are closely related to the *Hydra*, but which differ considerably in the nature of the reproductive process. All of them are marine, with the single exception of the genus *Cordylophora*, which inhabits fresh water. Some of the members of the order are simple, consisting of no more than a single polypite. In these cases there is an exceedingly close approach to the structure of the common *Hydra*, but the polypite is permanently fixed without the power of voluntarily changing its place, while the reproductive process is considerably different. In the majority of the *Corynida*, however, the hydrosoma is compound, consisting of a greater or less number of separate polypites or zoöids, all connected with one another by a common flesh or cœnosarc, and all forming parts of a plant-like rooted colony. In some of the *Corynida* the polypites are naked, but in most cases the cœnosarc is protected by a horny-looking chitinous* envelope or "polypary," as in *Tubularia indivisa* (Fig. 14). In no case, however, is this horny covering so prolonged as to form little cups in which each polypite is contained. It always stops short at the bases of the polypites, and in this way the *Corynida* can always be distinguished from their near allies, the sea-firs (*Sertularida*).

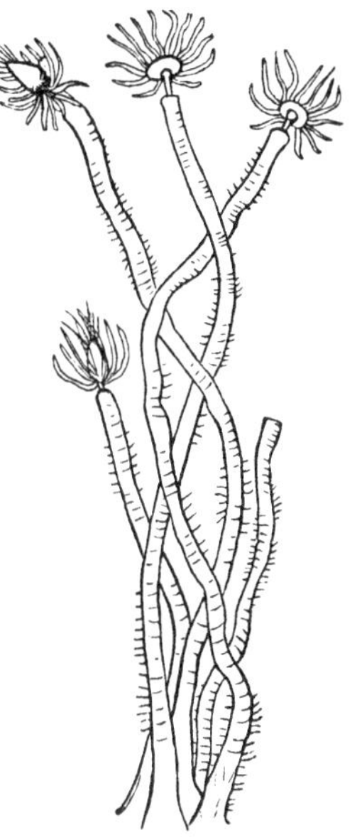

FIG. 14.—Fragment of *Tubularia indivisa*, natural size.

As a good example of the *Corynida*, the common pipe-coralline (*Tubularia indivisa*) may be taken. In this animal (Fig. 14) we have a gregarious zoophyte consisting of numerous clustered horny tubes, fixed by their bases to shells or stones, and inhabiting most seas. The tubes are usually unbranched, though often considerably interwoven together. Each tube is filled with a soft, semi-fluid, reddish cœnosarc, and gives exit at its free extremity to a single polypite. The polypites are bright red in color, and are not retractile within their tubes, the horny polypary extending only to their bases. The polypites are somewhat conical in shape, the mouth being placed at the apex of the cone, and they are furnished with two sets of tentacles. One set consists of numerous short tentacles placed directly round the mouth, the other is composed of from thirty to forty tentacula of much greater length arising from the polypite about its middle or near its base. Near the insertion of these tentacles the generative buds are produced at proper seasons. In *Eudendrium* (the branched pipe-coralline) the essential structure is much the same as in *Tubularia*, but the hydrosoma is now truly compound, consisting of a number of non-retractile reddish polypites, united by a cœno-

* *Chitine* is a substance which is nearly allied to horn, but is distinguished from it by the fact that it is not soluble in caustic potash.

sarc, which is furnished with a horny polypary, the whole colony assuming a singularly close resemblance to a plant. In *Cordylophora*—the only fresh-water member of the order—we find also a branched composite hydrosoma carrying numerous polypites, and having the cœnosarc defended by a horny sheath (Fig. 15, *a*, *b*). In *Coryomorpha*, finally, we have a type of the

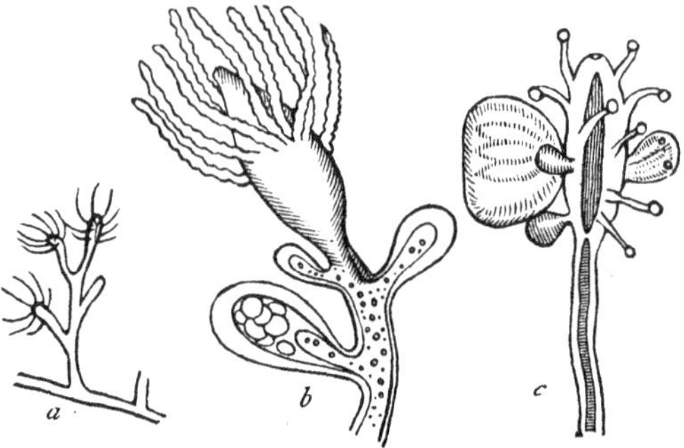

FIG. 15.—*a* Fragment of *Cordylophora lacustris*, slightly enlarged; *b* Fragment of the same, considerably enlarged, showing a polypite and three gonophores in different stages of growth; *c* Portion of *Syncoryne Sarsii*, with medusiform zoöids budding between the tentacles.

Corynida, in which the hydrosoma consists of no more than a single polypite, and there is no polypary. It is about four inches in length, and is fixed by filamentous roots to the bottom of the sea. It consists of a single whitish polypite, striped with pink, and terminating upward in a pear-shaped head, furnished with two sets of tentacles, the shortest of which form a circlet round the mouth.

As regards the generative process in the *Corynida*, it may be as well to consider the general phenomena of reproduction as carried on by all the Hydroid zoophytes, the general characters of the process being of a most remarkable nature. As has been already explained, the individual in the case of the compound *Hydrozoa* consists of an aggregation or colony of partially independent beings or zoöids, produced by gemmation or fission from a primordial organism. This is the case in all composite animals, such as sponges, sea-mats, corals, and many others. In many of the compound *Hydrozoa*, however, the case becomes still further complicated. In many of these organisms, namely, the zoöids differ very much from one another both in structure and in function. One set of zoöids is entirely devoted to the duty of providing food for the colony, and in these no reproductive organs are ever developed. These nutritive zoöids are all like each other in form, and the whole assemblage of them has been appropriately termed the "trophosome" (Allman), from the Greek *trepho*, I nourish; and *soma*, body. The colony or trophosome thus formed by the nutritive zoöids can go on increasing by the production of fresh zoöids for an almost indefinite period; but in all cases there ultimately comes a time when it becomes necessary to produce the essential elements of reproduction in order to secure the perpetuation of the species. The nutritive zoöids, as just stated, cannot produce the ova and sperm-cells, being destitute of reproductive organs, and the colony is therefore compelled to produce a second set of buds, which have the power of producing the essential elements of reproduction. These buds are collectively called the "gonosome" (Gr. *gonos*, offspring; and *soma*, body). The generative buds have the further peculiarity that not only can they produce the generative elements, but they are altogether unlike the nutritive zoöids in appearance. This difference in external appearance and in structure is sometimes so great as to lead to a most remarkable series of phenomena. In the simplest form in which these generative buds or "gonophores" appear, they have the form of mere protuberances of the ectoderm and endoderm (Fig. 16, *a*), enclosing a cavity derived from the body-cavity. In these buds the generative elements—ova and spermatozoa—are developed (Fig. 15, *b*). In other instances, the generative buds have a more complicated struct-

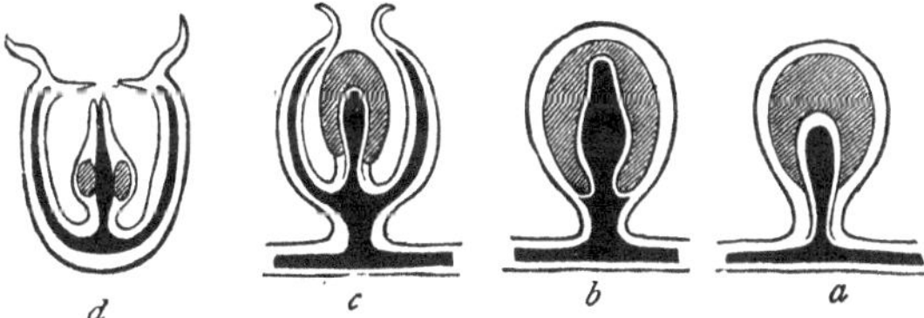

Fig. 16.—Generative buds or gonophores of the *Hydrozoa* diagrammatically represented. *a* Simple gonophore, consisting merely of a protuberance of the ectoderm and endoderm; *c* Gonophore which has the structure of a Medusa (medusoid), but is not detached; *d* Free medusiform gonophore.

ure. They consist now (Fig. 16, *c*) of a bell-shaped disk, which is attached by its base to the parent organism, and has its cavity turned outward (see also Fig. 15, *c*). From the roof of this disk there is suspended a kind of handle, which corresponds to the clapper of the bell, and is termed the "manubrium" (Lat. for handle). From the fixed or proximal extremity of the central process or manubrium proceed four canals, which extend to the margin of the bell, where they all open into a circular canal surrounding the mouth of the bell. This bell-shaped reproductive bud may attain no higher development than this, and may remain permanently attached to the parent organism from which it is produced In other cases, however, a higher state of development i reached (Fig. 16, *d*). The generative bud or gonophore becomes detached from its parent colony; the manubrium or central process develops a mouth at its free extremity; the mouth of the bell becomes partially closed by an inward prolongation or shelf, called the "veil;" and a series of tentacles are developed from its margin. The generative bud, thus liberated, leads a wholly independent existence. The manubrium, having developed a mouth, assumes the functions of a true polypite, and its cavity acts as a digestive sac. The whole organism swims about freely, and has the power of assimilating food, and thus of attaining to a comparatively gigantic size. This independent existence, however, only goes on till such time as the elements of reproduction can be produced. The ova and sperm-cells are developed in specialized portions of this generative bud, and then it ceases to exist. The ova, however, when fertilized, do not develop themselves into the free-swimming bell-shaped organisms in which they were actually produced, but into the plant-like, rooted, and compound zoophyte, from which the generative buds were originally given forth. These free-swimming bell-shaped reproductive buds or gonophores, as we shall see, are identical structurally with the smaller forms of the so-called sea-jellies or *Medusæ*; and it is now known that most if not all of these *Medusæ*, though originally described as distinct beings, are really nothing more than the free generative buds of the fixed *Hydrozoa*. We have here, then, an instance of what has been not quite appropriately called "alternation of generations." We have a compound fixed animal, in many respects comparable to a plant, producing a special series of buds which are devoted to the process of reproduction. These buds are cast off as independent beings to lead an independent life, and they are furnished with the necessary organs to preserve their existence till they are able to mature the reproductive elements. When once able to consummate this, they die; but the young to which they give origin are wholly unlike themselves. The young, namely, instead of being free-swimming "medusiform" beings, become developed into the fixed, plant-like colony from which the generative buds were originally produced. The term "alternation of generations" is not an altogether good one, and does not quite express the facts of the case. There is not any alternation of generations, but there is an alternation of generation with gemmation or budding. The only true generative act takes place in the reproductive zoöid or gonophore, in which the ova and sperm-cells are developed. The production of this gonophore from the parent organism (trophosome) is a process, not of generation, but of gemmation or budding. The whole process, therefore, is, properly speaking, not an "alternation of generations," but an alternation of generation with gemmation.

To recapitulate, then—the process of reproduction in the Hydroid zoophytes is carried on by means of reproductive buds or gonophores, which are produced at special seasons, and in which the reproductive elements are developed. These generative buds differ a good deal in their character, but three chief kinds may be distinguished: 1. Simple closed sacs or protuberances formed out of both ectoderm and endoderm, and having the special elements of generation developed in their interior. 2. Bell-shaped buds, attached to the parent colony by their bases, and having a central process or manubrium, which is furnished with a mouth and central cavity, from which there is given off a system of canals to ramify in the substance of the disk. The reproductive elements are developed either in the walls of these canals or between the ectoderm and endoderm of the manubrium. From the resemblance of these buds in anatomical structure to the so-called sea-jellies or *Medusæ*, they are usually spoken of as "medusiform gonophores," or simply as "medusoids." In this form, however, though highly organized, the buds never become detached from the parent colony. 3. Buds which become developed into bell-shaped medusiform bodies exactly similar in structure to the last, but detached to lead an independent existence. These free-swimming medusiform gonophores are anatomically indistinguishable from ordinary *Medusæ*; and it is now known that most, if not all, of the so-called "naked-eyed" *Medusæ*, are really the detached generative buds of other orders of *Hydrozoa*. The special elements of reproduction are developed in these detached buds, but the resulting embryos are not developed into *Medusæ*, such as produced the ova and sperm-cells, but straightway grow up into the plant-like, sexless colony, from which the medusiform gonophores were originally budded forth. In these cases, therefore, the individual Hydroid consists of a fixed, rooted colony or trophosome, producing fresh zoöids by a process of budding, but incapable of producing the essential elements of reproduction, together with a free and independent series of generative buds, or gonosome, in which the elements of reproduction are developed.

ORDER III. SERTULARIDA.—In this order of the *Hydroida* we have the most familiar and best known of all our zoophytes —namely, the sea-firs and their allies. The horny, plant-like polyparies of the *Sertularida* are familiar to every visitor at the sea-side, and by those unacquainted with their true nature they are almost universally set down as sea-weeds. The *Sertularida* are very closely allied to the compound forms of the *Corynida*, resembling them in being rooted, plant-like colonies, composed of a number of similar polypites or zoöids, produced by budding from a primitive zoöid. As in the Tubularians among the *Corynida*, the whole cœnosarc is enveloped in a horny or chitinous envelope or polypary (Fig. 17, *a*), and this is the structure which is most familiarly known to sea-side observers. The *Sertularida*, however, are distinguished from the *Corynida* by two points: Firstly, none of the *Sertularida* are simple, but are all compound, consisting of more or less numerous polypites, united by a branched cœnosarc. Secondly, the polypary of the *Sertularida* differs from that of the *Corynida* in not simply reaching to the bases of the polypites, but in being prolonged to form a number of little cups or "hydrothecæ" (Fig. 17, *a*, *b*) within which the polypites are lodged. Each polypite has a cup of its own, within which it can entirely withdraw, and from which it can protrude its free extremity.

The polypites of the *Sertularida* have essentially the same structure as in the *Corynida*, and each may be compared to a little *Hydra*. Each, namely, consists of a soft, contractile, and extensile body, which is furnished at its free extremity with a mouth and a circlet of prehensile tentacles, richly furnished with thread-cells. The mouth opens into a chamber which occupies the whole length of the polypite, and which is to be regarded as the combined body-cavity and digestive sac. At its lower end this chamber opens by a constricted aperture into a tubular cavity, which is everywhere excavated in the substance of the cœnosarc (Fig. 17, *b*). The nutrient particles obtained by each polypite thus serve for the support of

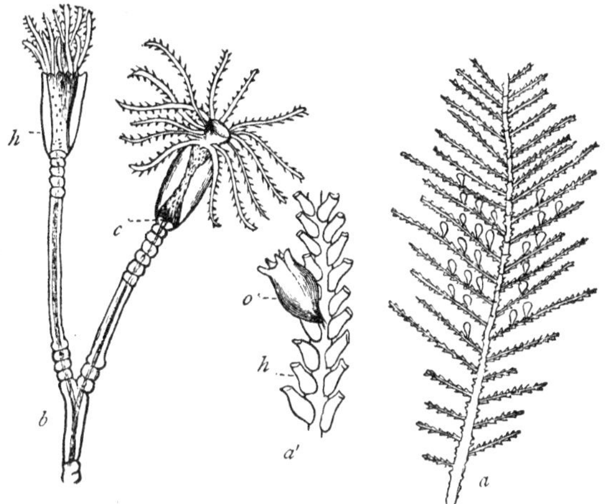

FIG. 17.—*a Sertularia* (*Diphasia*) *pinnata*, natural size; *a'* Fragment of the same enlarged, carrying a male capsule (*o*), and showing the hydrothecæ (*h*); *b* Fragment of *Campanularia neglecta* (after Hincks), showing the polypites contained in their hydrothecæ (*h*), and also the point at which the cœnosarc communicates with the stomach of the polypite (*c*).

the entire colony, and are distributed throughout the entire organism. The nutritive fluid prepared in the interior of each polypite gains access through the above-mentioned aperture to the cavity of the cœnosarc, which, by the combined exertions of the whole assemblage of polypites, thus becomes filled with a granular nutritive liquid. This cœnosarcal fluid is in constant movement, circulating through all parts of the colony, and thus maintaining its vitality—the cause of the movement being probably due, in part, at any rate, to the existence of vibrating cilia.

The process of reproduction varies somewhat in different members of the order. In all alike, however, the ordinary polypites are incapable of producing the essential elements of reproduction, and for this purpose special generative buds have to be developed. In the typical Sertularians the reproductive buds are developed at certain seasons in great numbers, and they constitute what used to be called the "ovarian vesicles" or "capsules." These reproductive buds are enclosed in horny cups or receptacles, often of a very beautiful shape, and much larger in size than the ordinary hydrothecæ (Fig. 17, *a*, *a'*). Each bud may be compared to a polypite destitute of a mouth and tentacles, being composed of a protuberance of the ectoderm and endoderm, containing a prolongation from the general cavity of the cœnosarc. The essential elements of reproduction are developed between the ectoderm and endoderm of the bud, and the resulting embryo is finally liberated as a little oval body covered with cilia, with which it swims freely about, until it meets with a suitable locality, when it fixes itself, loses its cilia, and by budding soon develops another colony.

In one division of this group—often described as a separate order, under the name of *Campanularida*—some points of difference are observable. In the typical Sertularians the little cups or hydrothecæ for the polypites are placed on the sides of the branches, and they are not stalked (Fig. 17, *a'*), while the reproductive elements are produced in fixed buds. In the *Campanularida*, on the other hand (Fig. 17, *b*), the hydrothecæ are supported upon stalks, and are placed at the ends of the branches, while the generative buds are usually detached to lead an independent existence. In these forms the reproductive zoöids or gonophores start as simple buds; but they become gradually developed into free-swimming medusoids, such as have been before alluded to. Each medusoid consists of a little transparent, glassy bell, from the under surface of which there is suspended a modified polypite, in the form of a manubrium (Fig. 18). The whole organism swims gayly through the water, propelled by the contractions of the bell or disk; and no one would suspect now that it was in any way related to the fixed, plant-like zoophyte from which it was originally budded off. The central polypite is furnished with a mouth at its distal end, and the mouth opens into a digestive sac. From the upper end of this stomach

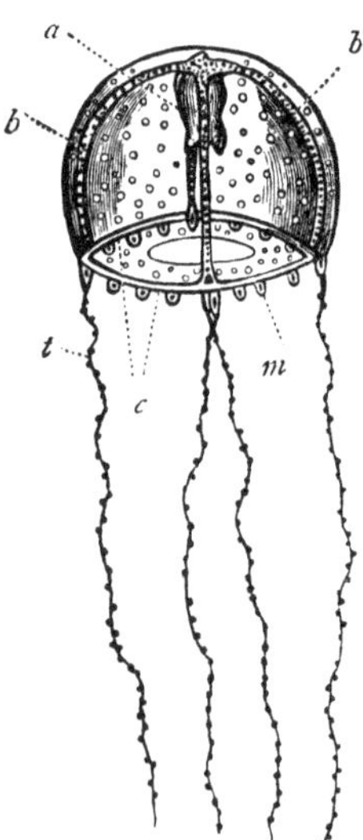

FIG. 18.—Gonophore of one of the *Campanularida*.

proceed four radiating canals which extend to the margins of the bell, where they all open into a circular vessel which runs round the mouth of the bell. From the circumference of the bell hang also a number of delicate extensile filaments or tentacles; and the margin is further adorned with a series of brightly-colored spots, which are probably rudimentary organs of vision and hearing. The mouth of the bell is partially closed by a delicate transparent membrane or shelf, the so-called "veil." Thus constituted, these beautiful little beings lead an independent and locomotive existence for a longer or shorter period. Ultimately ova and sperm-cells are produced in special organs, which are developed in the course of the radiating canals of the disk. The resulting embryos are minute free-swimming bodies, covered with cilia, which finally fix themselves, and develop into the plant-like colonies from which the medusoids were derived.

CHAPTER VI.

SUB-CLASS SIPHONOPHORA.

THE animals included under the name of *Siphonophora* are often known as the "oceanic Hydrozoa," as they are not fixed like the Hydroid zoophytes, but are found swimming at the surface of the open ocean, far from land. They are all singularly delicate and beautiful organisms, but they require little notice here. They are distinguished from the Hydroid zoophytes, which we have been just considering, by the fact that the hydrosoma consists of numerous polypites, united by a common trunk or cœnosarc, which is very rarely branched, and is never furnished with any hard outer covering or polypary, so that it remains permanently soft and flexible throughout life. As in the *Hydroida*, the reproductive organs are in the form of special buds, which have the power of developing the essential elements of generation, and which are often detached as free-swimming medusoids.

The entire sub-class is divided into two great groups or orders, and it will be sufficient to consider shortly a typical form of each. In the first order—that of the *Calycophoridæ*—the cœnosarc is thread-like, cylindrical, unbranched, and highly contractile. The cavity of the cœnosarc dilates at one end into a peculiar ciliated chamber, which is the distinguishing character of the order. The name of *Calycophoridæ* (Gr. *kalux*, a cup; and *phero*, I bear) is, however, derived from another circumstance—namely, that one end of the cœnosarc is always furnished with a series of bell-shaped disks, which are known as "swimming-bells" or "nectocalyces." Each nectocalyx consists of a bell-shaped cup (Fig. 19, *v*, *v'*), attached by its base to the cœnosarc, and having its cavity turned outward. In the substance of the disk run at least four canals, which communicate with the cavity of the cœnosarc, and proceed to the margin of the bell, where they all open into a circular vessel. The mouth of the bell is also furnished with a delicate ledge, which runs round its circumference, and is known as the "veil." The structure, therefore, of the nectocalyces is very similar to that of an ordinary medusiform gonophore, the chief difference being the absence in the former of the central polypite or manubrium. The nectocalyces are highly muscular, and have the power of alternately contracting and dilating, thus driving the whole organism through the water. In *Diphyes* (Fig. 19), which may be taken as the type of the group, there is a long thread-like trunk or "cœnosarc" (*c*) which bears at intervals minute polypites, each of which is protected by a delicate glassy overlapping plate, termed a "bract." At one extremity of the cœnosarc are two large mitre-shaped swimming-bells or nectocalyces (*v*, *v′*), by the contractions of which the entire organism is driven through the water. The cœnosarc with its polypites can be withdrawn, when necessary, into a kind of chamber between the two swimming-bells; but when unretracted the organism often attains a length of several inches. Its name is derived from the fact that the two nectocalyces can be separated from the cœnosarc by the least touch, and it was for this reason originally supposed to consist of two distinct animals loosely attached to one another. The tentacles are comparatively speaking of great length, and are furnished with lateral branches containing numerous thread-cells. The mouths of the polypites are not provided with a circlet of tentacles, but each has a single long tentacle arising from near its base. The reproductive organs of the *Calycophoridæ* are in the form of medusiform gonophores, which are budded from the stalks of the polypites, and which are mostly detached to lead an independent existence.

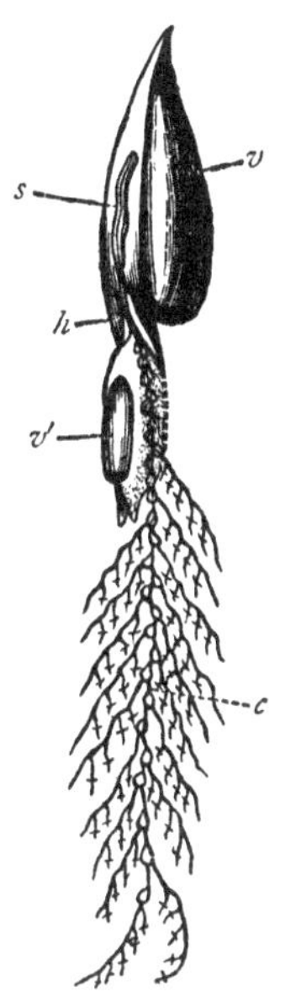

FIG. 19.—*Diphyes appendiculata*, one of the *Calycophoridæ* (after Kölliker). *v*, *v′*, Swimming-bells; *c* Cœnosarc, carrying the polypites, each with its tentacle and protected by a bract.

The second order of the oceanic Hydrozoa is that of the

Physophoridæ (Gr. *physa*, a bladder; and *phero*, I carry), of which the most familiar, though not the most typical, example is the Portuguese man-of-war, *Physalia utriculus* (Fig. 20, *a*). The *Physophoridæ* are distinguished from the organisms which we have been just considering by the fact that one extremity of the cœnosarc is developed into a structure which is known as the "float" or "pneumatophore" (Gr. *pneuma*, air; and *phero*, I carry). The float contains a larger or smaller sac, composed of some elastic, horny substance, probably chitine, often communicating with the exterior by one or more apertures, and always more or less completely filled with air. This sac is enclosed in a reflection of the ectoderm and endoderm, so that it is really outside the cavity of the cœnosarc. The function of the float is no doubt that of enabling the organism to maintain its position at the surface of the sea. As in the *Calycophoridæ*, the cœnosarc is always perfectly flexible, contractile, and soft, and is never furnished with any chitinous covering or polypary. There may or may not be swimming-bells, and the tentacles are very complicated in structure, and often attain a length of many inches. The polypites present no special points of interest, but are often furnished with the protective plates, which have been already spoken of as "bracts."

As a good example of the *Physophoridæ*, the Portuguese man-of-war may be taken (Fig. 20, *a*). It is composed of a large, spindle-shaped float, often of several inches in length, upon the under surface of which are arranged a number of polypites, together with highly-contractile tentacles of great length, and reproductive organs. The tentacles are richly furnished with thread-cells; and it has the power of stinging very severely. *Physalia* is commonly found floating at the surface of tropical and sub-tropical seas, and fleets of it are occasionally driven upon temperate shores.

Another very beautiful member of the *Physophoridæ* is the *Velella vulgaris*, which occurs abundantly in many seas. It is about two inches in length by one and a half in height. One end of the cœnosarc is greatly expanded and flattened out into an oval disk, which carries a vertical triangular crest, running obliquely across its upper surface (Fig. 20, *b*). The whole organism is semi-transparent and of a beautiful bluish color, and it floats at the surface of the sea with the vertical crest exposed to the influence of the wind, and thus officiating as a sail. From the under surface of the disk are suspended the various appendages of the organism, consisting of a single large central polypite, a number of processes, like polypites in shape, and carrying medusiform gonophores; and lastly, a single series of tentacles which arise from the cœnosarc quite independently of the polypites.

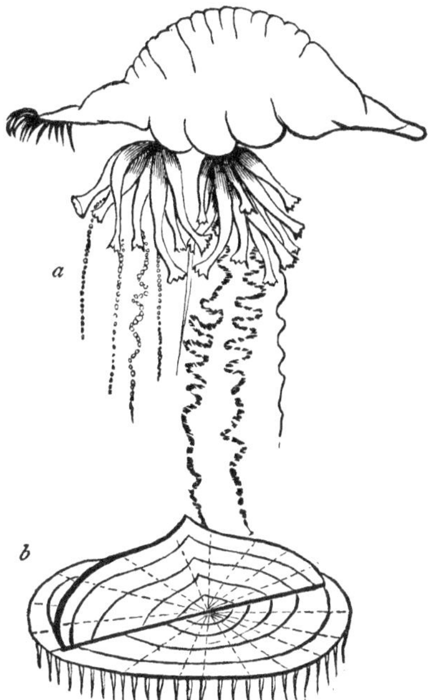

Fig. 20.—*a* Portuguese man-of-war (after Huxley); *b* *Velella vulgaris* (after Gosse).

CHAPTER VII.

SUB-CLASS DISCOPHORA.

THE group of *Hydrozoa* here spoken of as *Discophora* or *Medusidæ* comprises most of the familiar organisms known to visitors at the sea-side as sea-jellies, jelly-fishes, or sea-nettles; this last name being derived from the power possessed by some of them of stinging pretty severely in virtue of the possession of numerous thread-cells. Under the name, however, of sea-jellies are included a number of large organisms, extremely common at certain seasons in our seas, but now known to be properly referable to another group of the *Hydrozoa* (viz., *Lucernarida*). It is these large forms which alone possess any power of stinging man, and to these the term of "sea-nettles" ought properly to be restricted. They are better known under the name of "hidden-eyed" *Medusæ*, applied to them by the late Edward Forbes. Under the present group of the *Discophora* are included only a number of small jelly-fishes, found in great abundance at certain times, floating in the open sea, but nevertheless very little known to the general public in consequence of their very minute size. These delicate and diminutive organisms were originally described by Edward Forbes, for reasons to be immediately stated, as the "naked-eyed" *Medusæ*. It is now known, however, that most of these naked-eyed *Medusæ* are in reality nothing more than the free-swimming generative buds, or medusiform gonophores, produced by budding from so many of the other *Hydrozoa*, and then detached, as we have formerly seen, to lead an independent existence. That this is their true nature, is shown by the fact that the eggs which they produce develop themselves, not into fresh *Medusæ*, but into various other forms of *Hydrozoa*, which are fixed or oceanic. Under these circumstances, therefore, the naked-eyed *Medusæ* which can be shown to be of this nature, cannot, of course, be regarded as distinct animals at all. Still, there remains a considerable group of naked-eyed *Medusæ* to which this explanation has not hitherto been shown to apply. In most of the members of this group the course of development is quite unknown, and therefore their true nature is a matter of doubt. Two families, however, of this group are stated to produce eggs which develop directly into *Medusæ*, such as those which gave origin to the eggs; and, if this observation is confirmed, these, at any rate, must be regarded as true *Discophora*. In the mean while, therefore, it is best to regard the group of the *Discophora* or *Medusidæ* as of a questionable nature, and as including forms which may ultimately be shown to be nothing more than the detached zoöids of other *Hydrozoa*. Under these circumstances it will not be requisite to do more than very briefly to describe the anatomical structure of a typical *Medusid;* and this is the less necessary, since it will be seen at once that the structure is in all essential respects identical with what has been already described in speaking of the free medusiform gonophores of the Hydroid zoophytes.

In all the naked-eyed *Medusæ*, of which *Modeeria* (Fig. 21)

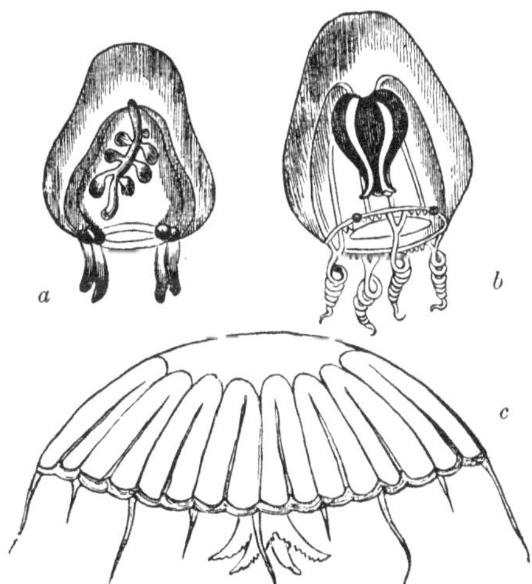

FIG. 21.—Naked-eyed Medusæ. *a Sarsia gemmifera; b Modeeria formosa; c Polyxenia Alderi* (after Gosse).

may be taken as a good example, the general structure is briefly as follows: The hydrosoma is perfectly free and is oceanic, being found swimming near the surface in the open ocean. The body is composed of a thick, transparent, gelatinous disk or swimming-bell (the nectocalyx), by the pulsations of which the animal is driven through the water. From the under surface or roof of this bell-shaped disk is suspended a single polypite (the manubrium), which bears to the disk the same relative position as the clapper does to an ordinary hand-bell. The distal end of the central polypite is furnished with a mouth, the lips of which are often prolonged into four longer or shorter lobes or processes. The mouth opens into a digestive sac, occupying the axis of the polypite; and from the upper end of this proceed four radiating canals, which run in the substance of the disk to its margin, where they are united by a single circular vessel, the whole system constituting the so-called "gastro-vascular" canals. The margin of the bell is narrowed by a kind of shelf, which runs round the whole circumference, leaving a central aperture, and which is known as the "veil." From the margin of the disk hang more or less numerous tentacles, which are hollow processes of the ectoderm and endoderm, and which communicate with the circular vessel of the canal-system. Also round the circumference of the swimming-bell are disposed certain "marginal bodies," which are doubtless organs of sense. Some of these marginal bodies consist of little rounded sacs or "vesicles," filled with a transparent fluid, and containing mineral particles, apparently of carbonate of lime. These are probably rudimentary organs of hearing. Others of the marginal bodies are in the form of little masses of coloring-matter or pigment, often of a strikingly bright color, enclosed in distinct cavities. These are known as the "pigment-spots" or "eye-specks," and they are believed to be rudimentary organs of vision. They are placed in a conspicuous and unprotected position on the margin of the disk, and hence these organisms were termed "naked-eyed" *Medusæ* by Edward Forbes. The reproductive organs are mostly developed in the course of the radiating gastro-vascular canals, but are sometimes situated in the walls of the central polypite. The above is the essential structure of any of the ordinary naked-eyed *Medusæ*; and it is hardly necessary to remark that it is exactly similar to what has been formerly described as distinguishing the undoubted free-swimming reproductive buds of the fixed *Hydrozoa*. The probabilities, therefore, as before said, are in favor of the belief that the entire group of the *Discophora* will have to be ultimately done away with.

The naked-eyed *Medusæ* are all exceedingly elegant and attractive, when examined in a living condition, resembling little bells of the most transparent glass, adorned here and there with the most brilliant colors. They occur, in their proper localities and at proper seasons, in enormous numbers, and they constitute one of the staple articles of diet to the Greenland whale. They are mostly phosphorescent, or capable of giving out light at night, and they appear to be one of the principal sources of the luminosity of the sea. It does not seem, however, that they phosphoresce unless disturbed or irritated in some way.

CHAPTER VIII.

SUB-CLASSES LUCERNARIDA AND GRAPTOLITIDÆ.

THE last remaining group of the living *Hydrozoa* is that of the *Lucernarida* (Lat. *lucerna*, a lamp), under which name are included a considerable number of forms, differing from one another to a great extent in external appearance. It will be sufficient here to describe one or two typical forms.

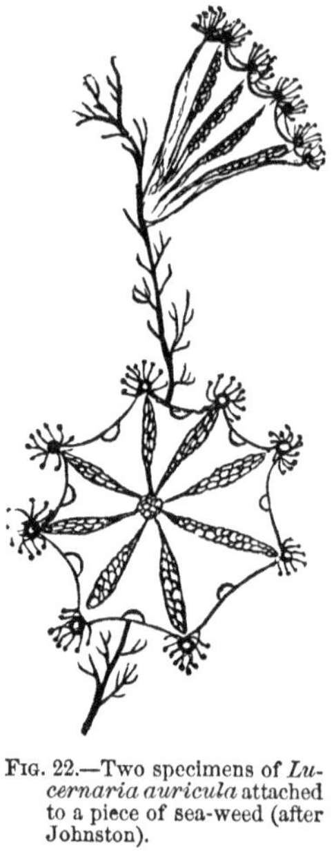

FIG. 22.—Two specimens of *Lucernaria auricula* attached to a piece of sea-weed (after Johnston).

One group of the *Lucernarida* is represented by *Lucernaria* itself (Fig. 22), which occurs not uncommonly in temperate seas. In *Lucernaria* we have a cup-shaped body, of a more or less gelatinous consistence, usually found attached by its smaller extremity to sea-weeds, this end of the body being developed into a small sucker. Like the *Hydra*, however, *Lucernaria* is not fixed, but can detach itself at will, and can even swim freely by means of the alternate contraction and expansion of the cup-shaped body (or "umbrella," as it is termed). Round the margin of the cup are tufts of short tentacular processes, and in its centre is fixed a single polypite, furnished with a four-lobed mouth. The essential elements of reproduction are developed within the body of *Lucernaria* itself, and it does not give off any generative buds, as so commonly occurs in other forms.

Another type of the *Lucernarida* is represented by the organisms formerly termed "hidden-eyed" *Medusæ*, and familiarly known as sea-nettles or sea-blubbers. Every sea-side visitor is familiar with the great circular disks of jelly which are left upon the sands by the retreating tide during the summer months; and many must have noticed on a calm day the large, transparent disks of these same creatures slowly flapping their way through the water. Not a few, too, must have learned by painful experience that some of these singular organisms have the power of stinging most severely, if incautiously handled. The forms included under the old name of "covered-eyed" *Medusæ* differ considerably from one another in their nature, and even in their structure, though they all present, in spite of their much greater size, a decided resemblance to the naked-eyed *Medusæ* already described. Some of the covered-eyed *Medusæ* produce eggs which are developed into organisms resembling themselves; but most of them are now known to be nothing more than the free-swimming reproductive buds of minute rooted *Hydrozoa*. It will be sufficient here to describe shortly the life-history of one of the more remarkable forms of this section.

If we commence with the young form of one of these sin-

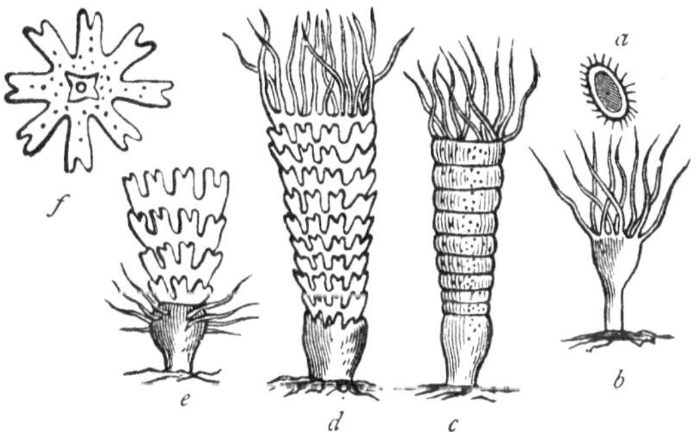

FIG. 23.—Development of Lucernarida (*Chrysaora*). *a* Ciliated embryo; *b* Hydra-tuba; *c* Hydra-tuba beginning to divide by transverse cleavage; *d* The cleavage still further advanced; *e* A form in which the cleavage has proceeded still further, and a fresh circle of tentacles has been produced near the base; *f* Free-swimming, generative zoöid, produced by fission from the Hydra-tuba.

gular animals, we find that the egg gives origin to a little microscopic ciliated body, which swims about freely by means of the cilia with which its surface is covered (Fig. 23, *a*).

This little body, on finding a suitable locality, fixes itself by one end, and develops a mouth and tentacles at the other, when it is known as a "Hydra-tuba" (Fig. 23, *b*), from its resemblance in shape to the fresh-water polype or *Hydra*. The Hydra-tuba is only about half an inch in height, and it possesses the power of forming large colonies by gemmation, while it is incapable of developing the essential elements of reproduction. Under certain circumstances, however, reproductive zoöids are produced by the following singular process: The Hydra-tuba becomes elongated, and exhibits a number of transverse grooves (Fig. 23, *c*). These grooves go on getting deeper and deeper, and become lobed at their margins, till the whole organism assumes the aspect of a pile of saucers placed one above the other (Fig. 23, *d*). The tentacles now disappear, and a fresh circle is formed close to the base of the Hydra-tuba (Fig. 23, *e*). Finally, all the saucer-like segments above the new circle of tentacles drop off one by one, and present themselves in the form of independent, free-swimming *Medusæ* (Fig. 23, *f*). These reproductive zoöids or *Medusæ* eat voraciously, and increase rapidly in size, becoming not only comparatively, but often actually, gigantic. Thus, in one case the reproductive zoöid has been known to attain a size of seven feet across, with tentacles fifty feet in length, though the fixed organism from which it was produced, was no more than half an inch in height. These gigantic reproductive bodies live an independent life until they are able to produce ova and sperm-cells, when they die. The fertilized egg, however, develops itself, not into the monstrous organism by which it was produced, but into the little fixed sexless Hydra-tuba, from which the generative bud was detached. We have, then, here another instance of the so-called "alternation of generations."

It is now known, then, that most of the great sea-blubbers which abound around our coasts in summer are really the detached reproductive buds of minute fixed *Hydrozoa*; and it may be as well to mention the leading features in their structure, and the points by which they may be distinguished from the smaller or naked-eyed *Medusæ*, to which they have a decided superficial likeness. In the commonest forms of these zoöids (such as the familiar sea-blubbers, *Aurelia* and *Cyanea*), the body consists of a great bell-shaped gelatinous disk or "umbrella," from the roof of which is suspended a single polypite, the lips of which are extended into lobed processes, often extending far below the margin of the disk (Fig. 24).

The digestive cavity of the polypite gives out from its upper extremity a series of radiating gastro-vascular canals, which proceed toward the margin of the umbrella. These radiating canals are never less than eight in number, and on their way to the margin of the disk they break up into a great number of smaller vessels, which unite with one another to form a complicated net-work. At the margin of the bell they all

FIG. 24.—Generative zoöid of one of the *Lucernarida* (*Chrysaora hysoscella*). (After Gosse.)

open into a circular vessel, which in turn sends processes into a series of marginal tentacles, which are often of extraordinary length. Besides the tentacles, the margin of the umbrella is provided with a number of marginal bodies, each of which consists of a little collection of pigment or "eye-speck," and a little sac filled with fluid and containing mineral particles. Each of these marginal bodies is covered and concealed from view by a kind of hood derived from the ectoderm. Hence the name of "hidden-eyed" *Medusæ* applied to these forms, in contradistinction to the "naked-eyed" *Medusæ*, in which the eye-specks are exposed to view. The reproductive organs are usually of some bright color, and "form a conspicuous cross shining through the thickness of the disk."

From the above description it will be evident that there is considerable resemblance between the so-called "hidden-eyed" *Medusæ*, or the reproductive zoöids of many of the *Lucernarida*, and the medusiform gonophores of so many of the *Hydrozoa*, as well as the true *Discophora* or naked-eyed *Medusæ*. The differences, however, between them are these: The swimming-disk of the naked-eyed *Medusæ* and of any medusiform gonophore is furnished at its mouth with an internal shelf or veil; the radiating gastro-vascular canals are very rarely more than four in number, and, should they subdivide (as in rare cases they do), they do not form an intricate net-work; lastly, the marginal bodies are simply placed in an uncovered situation on the margin of the disk. In the reproductive zoöids of the *Lucernarida* or hidden-eyed *Medusæ*, on the other hand, the swimming-disk or umbrella is destitute of any marginal shelf or veil; the radiating gastro-vascular canals are never less than eight in number, and they split up into numerous branches, which unite to form an intricate network; lastly, the marginal bodies are concealed from view by a kind of hood.

There still remains another family of the *Lucernarida* (viz., *Rhizostomidæ*) in which the reproductive process is carried on in the same way as in the forms we have just described, but the structure of the reproductive zoöids is somewhat different. In these, as in *Rhizostoma*, the generative zoöid is much like those just mentioned; but the umbrella is destitute of marginal tentacles; and, in place of a single central polypite, there hangs from the under surface of the umbrella a complex tree-like mass, the branches of which end in, and are covered by, small polypites and club-shaped tentacles. The umbrella itself does not exhibit any difference as compared with those already described, but the ova are produced in a genital cavity which is placed on the under surface of the umbrella.

SUB-CLASS GRAPTOLITIDÆ.—Before leaving the *Hydrozoa*, it will be as well to notice very briefly a group of extinct organisms which certainly belong to this class, and which probably find their nearest allies in the *Sertularians*. The *Graptolitidæ* are without a single living representative, and their antiquity is, indeed, very high, since it is doubtful if they ever pass above the group of rocks known to geologists as the Silurian formation. The most typical forms of the group agree with the living Sertularians in having a horny polypary, and in having the polypites protected by little horny cups or hydrothecæ, all springing from a common flesh or cœnosarc. The typical Graptolites, however, differ from all known Sertularians in the fact that the hydrosoma was not fixed to any solid object, but was permanently free.

Most of them, also, exhibit a very anomalous and remarkable structure, termed the "solid axis." This is a peculiar fibrous rod, which no doubt served to strengthen the polypary, and which is often prolonged beyond one or both ends of the polypary in a naked state. There is also good evidence that the reproductive process in the Graptolites was carried on in a manner somewhat similar to what is seen in the living Sertularians—namely, by means of reproductive buds enclosed in horny capsules. Graptolites most usually present themselves as beautiful silvery impressions, covering the surface of the black shales of various parts of the Silurian system.

CHAPTER IX.

ACTINOZOA.

THE second great class of the *Cœlenterata* is that of the *Actinozoa*, comprising the sea-anemones and their allies, the corals, the sea-pens, the sea-shrubs, and various other organisms. They are all defined as Cœlenterate animals in which there is *a distinct digestive sac which opens below into the general cavity of the body, but is nevertheless separated from the body-walls by an intervening space, which is divided into a number of vertical compartments by a series of partitions or "mesenteries," to the faces of which the reproductive organs are attached.* The *Actinozoa* (Fig. 12), therefore, differ fundamentally from the *Hydrozoa* in this, that whereas in the latter the digestive cavity is identical with the body-cavity, in the former there is a distinct digestive sac, which opens truly into the body-cavity, but is nevertheless separated from it by an intervening space. The result of this is, that while the body of a *Hydrozoön* exhibits on transverse section a single tube only, formed by the walls of the combined digestive and somatic cavity, the body of an *Actinozoön* exhibits *two* concentric tubes, one formed by the digestive sac and the other by the general walls of the body (Fig. 25, A). Further, in the *Actinozoa* the reproductive organs are always internal, and are never in the form of external processes of the body-wall as in the *Hydrozoa.*

In their minute structure the tissues in the *Actinozoa* differ little from those of the *Hydrozoa.* The body is essentially composed of two fundamental layers—an ectoderm and endoderm; but there are often well-developed layers of muscular fibres, somewhat obscuring this simplicity of structure. Thread-cells are most commonly present in abundance. Cilia are very generally developed, especially in the endoderm lining the body-cavity, where they serve to maintain a circulation of the contained fluids. The only digestive apparatus consists of a tubular or sac-like stomach, which opens inferiorly directly into the body-cavity (Fig. 12, *a*), and communicates

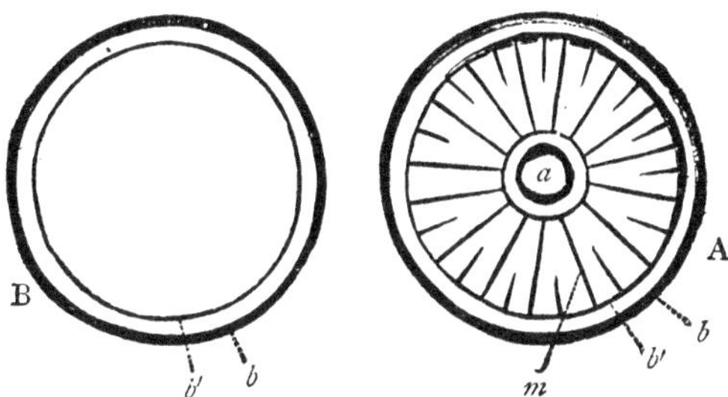

FIG. 25.—A. Transverse section of an *Actinozoön*. *a* Digestive sac; *b* Outer-wall of the body or ectoderm; *b'* Endoderm; *m* Mesenteries, connecting the stomach with the body-walls, and dividing the space between the two into a number of vertical chambers. B. Transverse section of the body of a *Hydrozoön*, showing the single tube formed by the walls of the body.

with the outer world through the mouth. A nervous system has not been shown to exist in any of the *Actinozoa* except the *Ctenophora*, and in none are there any traces of a circulatory system. Distinct reproductive organs are always present, and true sexual reproduction occurs in all the members of the class. In a great many forms, however, of the *Actinozoa* we have composite organisms or colonies, produced by a process of "continuous" gemmation or fission, the zoöids thus originated remaining attached to one another. In these cases—as in most of the corals—the separate beings or zoöids thus produced are termed "polypes," the term "polypite" being restricted to the *Hydrozoa*. In the simple *Actinozoa*, however, such as the sea-anemones, the term "polype" is applied to the entire organism, as consisting of no more than a single alimentary region. It follows from this, that the entire body of any *Actinozoön* may be composed of a single polype, or of several such produced by budding or cleavage, and united to one another by a common connecting structure or cœnosarc. Most of the *Actinozoa* are permanently fixed, like the corals; some, like the sea-anemones, possess a limited amount of locomotive power; and one order, the *Ctenophora*, is composed of highly-active free-swimming organisms. Some of them are unprovided with hard structures or supports of any kind, as the sea-anemones and *Ctenophora*; but a great many

5

secrete a calcareous or horny skeleton or framework which is known as the "coral" or "corallum."

The *Actinozoa* are divided into four orders—viz., the *Zoantharia*, the *Alcyonaria*, the *Rugosa*, and the *Ctenophora*.

ORDER I. ZOANTHARIA (Gr. *zoön*, animal; *anthos*, flower).—The *Zoantharia* comprise those *Actinozoa* in which the polypes are furnished with *smooth*, *simple*, *usually numerous tentacles*, which, like the mesenteries, are in multiples of *five* or *six*. The *Zoantharia* are divided into three groups, distinguished from one another by the presence or absence of a coral, and by its structure when present.

The first of these groups is termed *Zoantharia malacodermata*, or "soft-skinned" *Zoantharia*, because the polypes are either wholly destitute of a coral, or, if there is one, it consists merely of little scattered needles or spicules of carbonate of lime. Generally, too, the organism is simple, and consists of

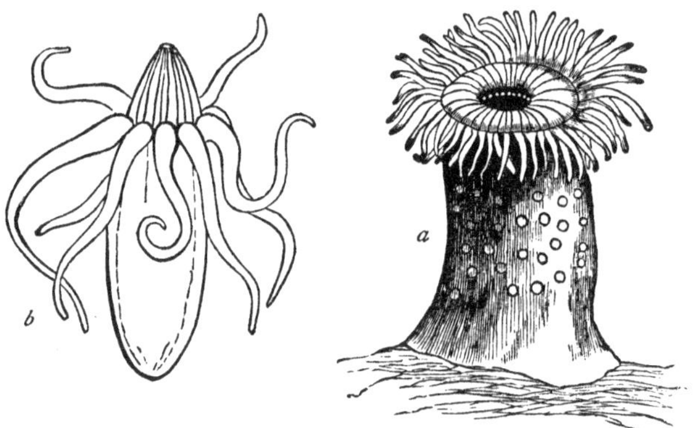

FIG. 26.—Morphology of *Actinidæ*. *a Actinia rosea*; *b Arachnactis albida* (after Gosse).

no more than a single polype. The best known of the members of this group are the beautiful sea-anemones or *animal-flowers* (*Actinidæ*), which occur so plentifully on every coast (Fig. 26, *a*). It will be as well to describe the structure of a sea-anemone somewhat in detail, as in this way a clear notion may be obtained of the general anatomy of the *Actinozoa*. The body of an ordinary sea-anemone (Fig. 26, *a*) is a truncated cone or short cylinder, termed the "column," and is of a soft, leathery consistence. The two ends of the column are termed respectively the "base" and the "disk," the former constituting a kind of sucker, by means of which the animal can attach itself at will, while the mouth is placed in the centre of the latter. The mouth is surrounded by a flat space, destitute of appendages, and the circumference of the disk is in turn surrounded by numerous simple tubular tentacles, arranged in alternating rows. The tentacles consist of both ectoderm and endoderm, enclosing a tube which communicates with the body-cavity. By the muscular contraction of the walls of the column, the fluid contained in the body-chambers can be forced into the tentacles, which can be thus protruded a great length, while they can also be usually retracted. In some cases the tentacles are furnished with perforations at their extremities. The mouth (see Fig. 12, *a*) leads directly into the stomach, which is a wide, membranous tube, opening by a wide aperture into the body-cavity below, and extending about half-way between the mouth and the base. The wide space between the stomach and body-walls is subdivided into a number of separate compartments by radiating vertical plates, which are called the "mesenteries," and to the faces of which the reproductive organs are attached, in the form of reddish bands, containing either ova or sperm-cells. Below the stomach, attached to the free edges of the mesenteries, are a series of singularly twisted threads or cords (Fig. 12, *c*), which are filled with thread-cells, and are termed "craspeda." The function of these is not well understood; but it is believed that in some cases they can be emitted through apertures, which are occasionally found in the walls of the column. The sea-anemones are mostly to be found between tide-marks, in rock-pools, or on ledges of stone, adhering by means of the expanded base. They are not, however, permanently fixed, but can change their place at will. In the nearly allied *Ilyanthus* and *Arachnactis* (Fig. 26, *b*) the base is tapering, and it appears that the animal spends the greater part of its existence in an unattached, free condition. The true sea-anemones, as already said, are all simple, each consisting of a single polype; but there are closely-related forms (such as *Zoanthus*) in which the organism is compound, consisting of numerous polypes united by a creeping, fleshy trunk or cœnosarc.

The second group of the *Zoantharia* is termed that of the *Zoantharia sclerodermata*, from the nature of the skeleton or coral. In this group are all the so-called "reef-building" corals, which are the makers of the well-known "coral-reefs."

The members of this group all possess the power of secreting carbonate of lime within their tissues, so as to form a more or less continuous skeleton or corallum. From the fact that this corallum is secreted by the inner layer of the polypes, and is therefore truly *within* the body, it is said to be "sclerodermic," in opposition to the kind of coral produced by other forms (such as the red coral), where the skeleton is secreted by the outer layer of the polypes, and is therefore *outside* them. In this latter case the coral is said to be "sclerobasic." (For illustrations of these different kinds of corals, *see* Fig. 29.) In the typical form of sclerodermic coral, the skeleton is in the form of a conical cup, the upper part of which is hollow. The lower part is divided into a series of compartments by vertical plates, which are called the "septa," and which correspond to the mesenteries of the living animal. Sometimes the space contained within the walls of the cup or "corallite" is broken up by horizontal plates called "tabulæ;" but, when these are present, there are generally no septa. In the form of coral just described we have a single corallite, produced by one polype, and this simple condition may be maintained throughout life. In the great majority of cases, however, the polypes bud, so as to form a colony, all bound together by a common flesh or cœnosarc. When such a colony, therefore, produces a sclerodermic coral, in place of a single corallite, we have a composite skeleton composed of a number of little cups or corallites, each of which was produced by one polype, and all of which are united by means of a common calcareous basis secreted by the cœnosarc (Fig. 29, *a*).

In accordance with their mode of formation, an ordinary compound sclerodermic coral may be distinguished from a sclerobasic coral by the fact that it would show a number of little cups in which the polypes were contained, whereas these cups would be absent in the latter. In accordance, also, with the fundamental character of the order *Zoantharia*, the corals of the present group always show septa which are some multiple of *five* or *six*.

When it is understood that compound corals, such as we have been speaking of, are produced by the combined efforts of a number of polypes, essentially the same in structure as our ordinary sea-anemones, it is readily intelligible that under favorable circumstances large masses of coral may be produced in this way. When these masses attain such a size as to be of geographical importance, they are spoken of as "coral-reefs," and the phenomena exhibited by these are of such interest as to demand some notice. The coral-producing polypes require for their existence that the average temperature of the sea shall not be less during winter than 66°; and coral-reefs are, therefore, not found in temperate seas. Reefs, however, abound in all the seas not far removed from the equator, being found chiefly on the east coast of Africa and the shores of Madagascar, in the Red Sea and Persian Gulf, throughout the Indian Ocean and the whole of the Pacific Archipelago, around the West-Indian Islands, and on the coast of Florida. The headquarters, however, of the reef-building corals may be said to be around the islands and continents of the Pacific Ocean, where they often form masses of coral many hundreds of miles in length. According to Darwin, coal-reefs may be divided into three principal forms—viz., Fringing-reefs, Barrier-reefs, and Atolls, distinguished by the following characters:

1. *Fringing-reefs* (Fig. 27, 1).—These are reefs, usually of a moderate size, which may either surround islands or skirt the shores of continents. These shore-reefs are not separated from the land by any very deep channel, and the sea on their outward margins is not of any great depth.

2. *Barrier-reefs* (Fig. 27, 2). — These, like the preceding, may either encircle islands or skirt continents. They are distinguished from fringing-reefs by the fact that they usually occur at much greater distances from the land, that there intervenes a channel of deep water between them and the shore, and soundings taken close to their seaward margin indicate great depths.

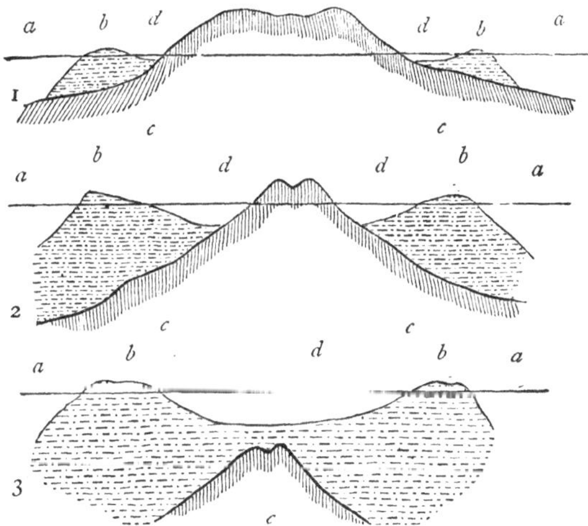

FIG. 27.—Structure of Coral-reefs. 1. Fringing-reef; 2. Barrier-reef; 3. Atoll; *a* Sea-level; *b* Coral-reef; *c* Primitive land; *d* Portion of sea within the reef, forming a channel or lagoon.

As an example of this class of reefs may be taken the great barrier-reef on the northeast coast of Australia, the structure of which is on a gigantic scale. This reef runs, with a few trifling interruptions, for a distance of more than a thousand miles, with an average breadth of thirty miles, and an area of thirty-three thousand square miles. Its average distance from the shore is between twenty and thirty miles, the depth of the inner channel is from ten to sixty fathoms, and the sea outside is "profoundly deep" (in some places over eighteen hundred feet).

3. *Atolls* (Fig. 27, 3).—These are oval or circular reefs of coral enclosing a central expanse of water or lagoon. They seldom form complete rings, the reef being usually breached by one or more openings. They agree in all particulars with those barrier-reefs which surround islands, except that there is no central island in the lagoon which they enclose.

The last group of the *Zoantharia* comprises composite organisms in which the cœnosarc is supported upon a central axis or sclerobasic skeleton. These *Zoantharia sclerobasica* require no notice, except simply to remark that they are distinguished from other sclerobasic corals (such as the *Gorgonidæ*) by the fact that each polype possesses tentacles which are a multiple of *six* in number.

ORDER II. ALCYONARIA.—The second great order of living *Actinozoa* is distinguished by the fact that the polypes are furnished with *fringed* tentacles, and that these, as well as the mesenteries and somatic chambers, are always some multiple of *four*. With one doubtful exception, all the *Alcyonaria* are composite, their polypes being connected together by a cœnosarc. The body-cavities of the polypes are connected with a system of canals which are excavated in the cœnosarc, and communicate freely with one another, so that a free circulation of nutrient fluids is thus kept up. The structure of the polypes of the *Alcyonaria* is, in all essential anatomical features, the same as in the sea-anemones, the number of the mesenteries and tentacles being the chief distinction.

Of the various different organisms included under this order, one of the best known is the "Dead-men's-fingers," or *Alcyonium*, which occurs commonly in most seas. It forms spongy-looking masses of a yellow or orange color, attached to shells and other marine objects. The whole mass is covered with little star-shaped apertures, through which the delicate polypes can be protruded and retracted at will. Another well-known member of this order—the type of another family—is the "sea-rod" (*Virgularia mirabilis*), which occurs not very rarely in shallow seas. *Virgularia* occurs in the form of a long rod-shaped body of a light flesh-color, supported upon a calcareous rod, somewhat like a knitting-needle, which is covered by the cœnosarc. From the cœnosarc are given out lateral processes, each of which bears numerous polypes. Closely allied to *Virgularia* is the "Cock's-comb" *Pennatula* (Fig. 28); but in this the lower end of the cœnosarc is naked and fleshy, and the polype-bearing fringes are considerably longer, giving the whole organism very much the appearance of a feather.

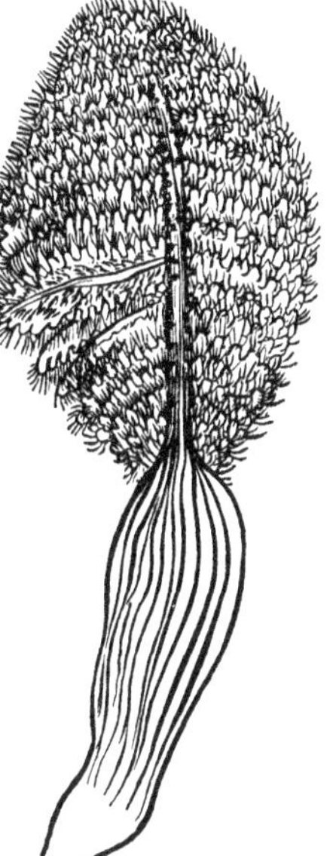

Fig. 28.—Pennatulidæ. The Cock's-comb (*Pennatula phosphorea*). (After Johnson.)

Another family of the *Alcyonaria* is represented by the so-called "Organ-pipe corals," of which *Tubipora musica* is a well-known example. In this there is a well-developed sclerodermic coral consisting of numerous cylindrical tubes, which are not divided by vertical partitions (septa), but which are connected by strong transverse plates. The coral is bright red in color, and the polypes are usually bright green.

The best known, however, of the *Alcyonaria* is the family *Gorgonidæ*, represented by the sea-shrubs, fan-corals, and the red coral of commerce. A few of the members of this family live in temperate waters, but they attain their maximum in point of size and numbers in the seas of the tropics. In all the *Gorgonidæ* the organism consists of a composite structure made up of numerous polypes united by a common flesh or cœnosarc (Fig. 29, *b*), the whole supported by a central branched axis or coral. The coral varies in composition, being sometimes calcareous—as in red coral—sometimes horny, and sometimes partly horny and partly calcareous, as in *Isis* (Fig. 29). In all cases, however, the corallum differs altogether from the sclerodermic corallum, which has been described as so characteristic of the reef-building corals. The coral in the present instance is always what is called "sclerobasic"—that is to say, it always forms an internal axis, covered by the cœnosarc with the polypes produced therefrom. It is, therefore, *outside* the polypes, and bears to the cœnosarc the same relation that the trunk of a tree bears to its investing bark. This is well shown in Fig. 29, *b*, where there is represented one of these sclerobasic corals in which the corallum consists of alternate horny and calcareous joints. The polypes of all the *Gorgonidæ* agree, of course, with their order in having eight tentacles each, and by this they are distinguished from the few *Zoantharia* in which there is a sclerobasic coral.

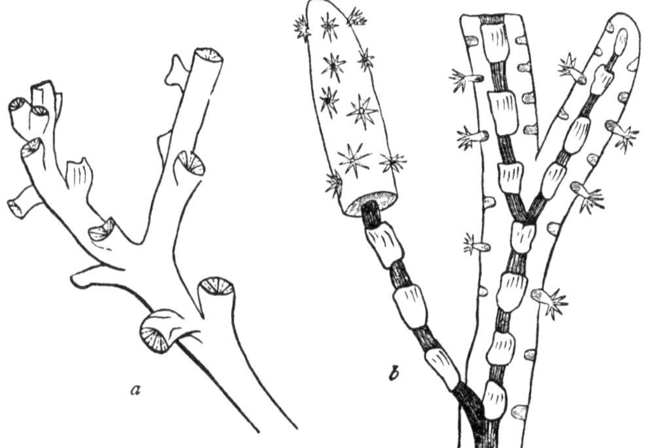

Fig. 29.—Sclerodermic and Sclerobasic Corals. *a* Portion of branch of *Dendrophyllia nigrescens*, a sclerodermic coral (after Dana); *b* Longitudinal section of *Isis hippuris*, a sclerobasic coral, exhibiting the external bark or cœnosarc, with its imbedded polypes, supported by the internal axis or skeleton (after Jones).

The best known of the *Gorgonidæ* is the *Corallium rubrum*, or "red coral" of commerce, which is largely imported from the Mediterranean. In this species there is a bright-red, finely-grooved, calcareous coral, usually more or less repeatedly branched. The coral is invested by a bright-red cœnosarc or bark, which is studded with numerous little apertures. The polypes can be protruded from these openings at will, and are milk-white in color, with eight fringed tentacles each. The entire cœnosarc is excavated into a number of communicating canals, with which the cavities of the polypes are connected, the whole system being filled with a nutritive fluid known as the "milk."

Order III. Rugosa (Lat. *rugosus*, wrinkled).—This order merely requires mention, as all its members are extinct, and are therefore only known to us by their hard parts or skeletons. They agree with the *Zoantharia sclerodermata* in having a well-developed sclerodermic corallum, but differ from them in the fact that the septa are always some multiple of *four;* and there are generally transverse plates or tabulæ combined with the vertical plates or septa. On the other hand, they agree with the *Alcyonaria* in having their parts in multiples of four, but differ from them in having a well-developed sclerodermic corrallum in which septa are present.

ORDER IV. CTENOPHORA (Gr. *kteis*, a comb; *phero*, I carry).—The fourth and last order of the *Actinozoa* is that of the *Ctenophora*, comprising a number of free-swimming oceanic creatures, very different in appearance from any of the forms which we have hitherto been considering. They are all transparent, gelatinous, glassy-looking creatures, which are found near the surface in the open ocean, swimming rapidly by means of bands of cilia. The cilia are arranged in a series of transverse ridges, which are disposed in longitudinal bands, the whole constituting locomotive organs which are known as "ctenophores." In none are there any traces of a corallum or skeleton, and thread-cells are asserted to be universally present.

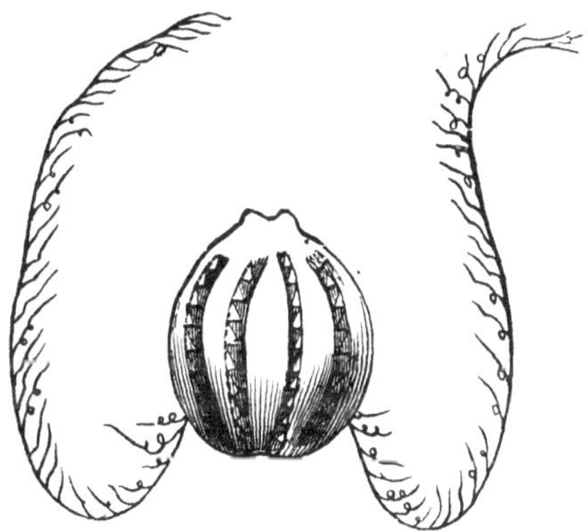

FIG. 30.—Ctenophora. *Pleurobrachia pileus.*

As the type of the order, we may take one of the commoner forms, which is known by the name of *Pleurobrachia* or *Cydippe* (Fig. 30). The body of *Pleurobrachia* is transparent, colorless, gelatinous, and melon-shaped, and exhibits two poles, at one of which is placed the mouth. The globe-like body is divided into a number of crescentic lobes by eight ciliated bands or ctenophores, which proceed from near the mouth to near the opposite pole of the body. Besides the cilia there are two very long and flexible tentacular processes, which are fringed on one side by smaller secondary branches. The tentacles arise each from a kind of sac, one placed on each side of the body, and they can be instantaneously and completely retracted within these sacs at the will of the animal. The mouth of *Pleurobrachia* opens into a spindle-shaped digestive sac or stomach, which in turn opens below into a wider and shorter cavity termed the "funnel;" from this there proceed in the axis of the body two small canals, which open at the opposite pole of the body. The funnel communicates with a complicated system of canals, which are ciliated internally, and are filled with a nutrient fluid. In the angle between the two canals which run from the base of the funnel to the surface is a little vesicle or sac, believed to be a rudimentary organ of hearing, and placed upon this is a little mass which is generally believed to be of a nervous nature. If this is correct, this is the first indication which we have hitherto encountered of a genuine nervous system. The reproductive organs are developed in the walls of the canal-system.

The only other form of the *Ctenophora* which deserves mention is the "Venus's girdle" (*Cestum Veneris*), which agrees in essentials with *Pleurobrachia*, but is greatly enlongated in a direction at right angles to the alimentary canal, till we have a ribbon-shaped body produced, four or five feet in length and two or three inches high. *Cestum* is not uncommon in the Mediterranean, and has the power of phosphorescence, appearing at night as a moving and twisting band of flame.

SUB-KINGDOM III.—ANNULOIDA.

CHAPTER X.

ECHINODERMATA.

THE third primary division of the animal kingdom is known by the name of *Annuloida*, and includes two groups of organisms which are extremely unlike one another in appearance, and are termed respectively the *Echinodermata* and the *Scolecida*. In the former we have the sea-urchins, star-fishes, and their allies, formerly classed in the old sub-kingdom *Radiata;* in the latter are a number of internal parasites, with some minute aquatic creatures, all formerly referred elsewhere. Different as are these two groups in appearance and habits, they are nevertheless united by the following peculiarities:* *They possess a distinct alimentary canal, usually communicating with the outer world by two apertures (a mouth and a vent), but in any case completely shut off from the general cavity of the body. In all there is a distinct nervous system; and in all there is a peculiar system of canals termed the "water-vascular" or "aquiferous" vessels, which usually communicate with the exterior of the body.* It should be mentioned that many naturalists dissent from this grouping together of the *Echinodermata* and *Scolecida* into a single sub-kingdom, *Annuloida.* Many other arrangements have been proposed, most of which present some special advantages and some disadvantages. In the mean while, in the confessedly uncertain state of this department of Natural History, it has been thought well to adhere to the arrangement proposed by Prof. Huxley, an arrangement with many obvious drawbacks, and at best but provisional.

* Some of the internal parasites of this sub-kingdom have no alimentary canal at all, but this does not affect the value of the above definition.

CLASS I.—ECHINODERMATA.

The members of this class are popularly known as sea-urchins, star-fishes, brittle-stars, feather-stars, sea-cucumbers, etc., and derive their name of *Echinodermata* (Gr. *echinos*, a hedgehog; and *derma*, skin) from the generally prickly nature of their integuments. In all, the skin is possessed of the power of secreting carbonate of lime, but in very different degrees. In the sea-urchins this goes so far that the body becomes enclosed in an immovable box, composed of numerous calcareous plates firmly jointed together. In the star-fishes and their allies the skin is rendered prickly by grains, tubercles, or spines of calcareous matter, and the body is either destitute of regular plates or is only partially enclosed by them. In the sea-cucumbers, again, the calcareous matter is mostly only present in the form of minute grains scattered in the skin. When adult, they all show a more or less distinctly radiate structure, which is most conspicuous in the star-shaped star-fishes and sand-stars, but can be detected in all the members of the class. When young, however, they almost always exhibit what is called "bilateral symmetry"—that is to say, they show similar parts on the two sides of the body. In all Echinoderms there is a water-vascular system of tubes, which is termed the "ambulacral system," which generally communicates with the exterior, and which in most cases is used in locomotion. An alimentary canal is always present, and is always completely shut off from the general cavity of the body. A vascular or circulatory system is sometimes present. There are always distinct organs of reproduction, which are almost always placed in different individuals, so that the sexes are distinct. The nervous system is in the form of a ring surrounding the gullet and sending branches in a radiating manner to different parts of the body.

The *Echinodermata* are divided into seven orders, as follows:

1. *Echinoidea* (Sea-urchins).
2. *Asteroidea* (Star-fishes).
3. *Ophiuroidea* (Sand-stars and Brittle-stars).
4. *Crinoidea* (Feather-stars).
5. *Cystoidea* (extinct).
6. *Blastoidea* (extinct).
7. *Holothuroidea* (Sea-cucumbers).

This is by no means a true arrangement of these orders, but it is convenient to consider them in this sequence.

ORDER I. ECHINOIDEA.—The animals included in this order vary from the shape of a sphere or globe to that of a disk, and they are all commonly known as "sea-urchins" or "sea-eggs." They are all characterized by the fact that the body is encased in a "test" or "shell" (Fig. 31, 2) composed of numerous calcareous plates mostly immovably jointed together so as to form a kind of box. The intestine is convoluted, and there is a distinct vent, or anal aperture.

The test of a sea-urchin, as just said, consists of many calcareous plates accurately fitted together, and united by their edges. In all living forms the test is composed of ten zones of plates, each zone consisting of a double row. In five of these zones (1 *a*, 2 *a*) the plates are of large size, and are per-

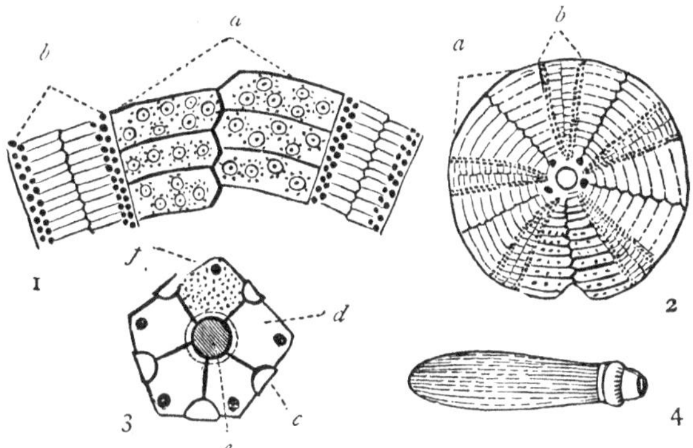

FIG. 31.—Morphology of Echinoidea. 1. Portion of the test of a sea-urchin (*Galerites*), enlarged, showing the ambulacral areas (*b*) and interambulacral areas (*a*). 2. Test of the same, viewed from above: *a* Interambulacra; *b* Ambulacra. 3. Genital disk of a sea-urchin (*Hemicidaris*) enlarged: *c* Ocular plate; *d* Genital plate; *e* Anal aperture; Madreporiform tubercle. 4. Spine of the same. (After Forbes.)

forated by no apertures. These are termed the "interambulacral areas." In the other five zones (1 *b*, 2 *b*) the plates are of small size, and are perforated by little apertures for the emission of delicate locomotive suctorial tubes (the so-called "ambularcal tube-feet"). These zones are therefore called the "ambulacral areas." Besides these main rows of plates which collectively make up the greater part of the test, there are other plates placed in the leathery skin round the mouth and vent. The most important of these form a kind of disk, which is placed at the summit of the shell. This disk (Fig. 31, 3) is composed of two sets of plates—one called the "genital plates," perforated for the ducts of the reproductive organs; the other set smaller, and each carrying a little "eye," hence their name of "ocular plates." One of the genital plates is also larger than the others, and carries a spongy mass which is called the "madreporiform tubercle," and which protects the entrance of the water-vascular or ambulacral system. The whole of the test is covered with numerous tubercles of different sizes, which carry longer or shorter spines (Fig. 32). The spines are jointed to the tubercles by a sort of "ball-and-

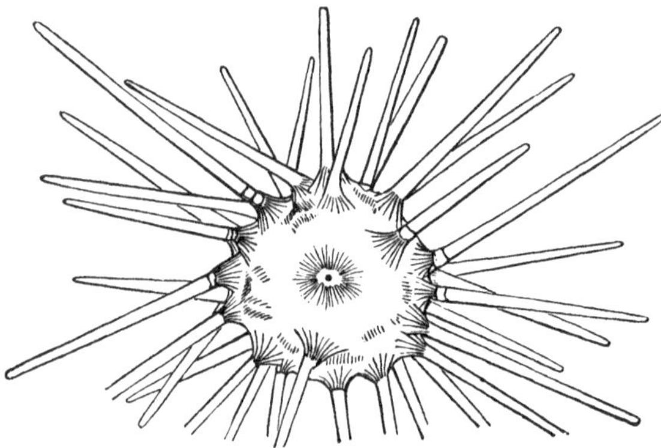

FIG. 32.—*Cidaris papillata* (after Gosse).

socket" or "universal" joint, and they are completely under the control of the animal, so as to be used both in locomotion and apparently as defensive weapons. In most common species the spines are short, but in many tropical forms they attain a very great length. Besides the spines, the outer surface of the test is furnished with curious little bodies called "pedicellariæ," which were long believed to be parasitic. They consist of two or three blades mounted upon a flexible stalk and constantly employed in snapping together like the beak of a bird. They occur in many other *Echinodermata*, and their use is obscure.

Locomotion is effected in the sea-urchins by a curious system of contractile tubes which are known as the "ambulacral tubes" or "tube-feet," and which are appendages of the water-vascular system. The following is essentially the arrangement of the whole aquiferous system. From the madreporiform tubercle on the largest of the genital plates there proceeds a membranous canal by which the outer water is conducted to a central tube, which forms a ring round the gullet. The tubercle is spongy, and is perforated with little holes, and its function is probably to act as a filter, and prevent foreign particles gaining access to the interior. From the "circular canal" round the gullet proceed five "radiating canals" which take their course toward the summit of the shell, underneath the ambulacral areas. In its course each radiating canal gives off numerous short lateral tubes—the ambulacral tubes or tube-feet—which gain the exterior of the shell by passing through the apertures in the ambulacral plates of the shell, and which terminate in little sucking-disks. The tube-feet can be distended with water by means of a series of little muscular bladders placed at their bases, and they can thus be thrust far out beyond the shell, into which they can be again withdrawn at the will of the animal. However long the spines may be, the animal can protrude the tube-feet to a still greater length; and by the combined action of the little suckers at their extremities locomotion is effected with moderate rapidity, considering the bulk of the body.

The digestive system in the *Echinus* consists of a mouth armed with a curious apparatus of calcareous teeth, which opens into a gullet, which in turn conducts to a distinct stomach. From the stomach there proceeds a long and convoluted intestine, which is attached to the interior of the shell by a delicate membrane or "mesentery," and terminates in a distinct vent. The surface of the mesentery, as well as that of the lining membrane of the shell, is richly ciliated, and thus serves to distribute the fluids of the body-cavity to all parts of the body. In this way, also, respiration is subserved, though it is probable that the chief agent in this function is to be found in certain specialized portions of the ambulacral system. The circulatory system consists in its central portion of two rings placed round the opposite ends of the alimentary canal, and united by an intermediate muscular cavity or heart. The nervous system consists of a gangliated cord placed round the gullet, and sending five radiating branches along the ambulacral areas. The sexes are distinct, but in both the reproductive organs are in the form of five membranous sacs placed in a radiating manner in the interambulacral areas, and opening at the genital plates. The embryo of the *Echinus* is at first a little free-swimming ciliated organism, and it passes through an extraordinary development, which can only be alluded to here. In its later stages it was originally described as a distinct animal under the name of "*Pluteus*." In this state the larva is a curious, easel-shaped body, with a distinct alimentary canal and an internal calcareous skeleton, and exhibiting distinct bilateral symmetry. The remarkable point, however, about its further development is, that the young *Echinus* is developed out of only a *portion* of the *Pluteus*, and the greater part of the latter, including the skeleton, is cast away as useless.

The majority of the sea-urchins are found at moderate depths in the sea, especially in the neighborhood of oyster-banks. Others spend their existence buried in the sand; and one species excavates holes for itself in the solid rock, apparently by some mechanical action.

ORDER II. ASTEROIDEA (Gr. *aster*, star; *eidos*, form).—As the structure of the sea-urchins may be taken as embodying the most important anatomical peculiarities of the *Echinodermata*, and as this has been described at some length, it will not be necessary to do more than briefly indicate the more important characteristics of the remaining orders. In the present order are included all the true star-fishes, the sand-stars and brittle-stars being generally regarded as a distinct

FIG. 33.—*Cribella oculata* (after Forbes).

group. The body in all the *Asteroidea* is more or less obviously *star-shaped* (Fig. 33), consisting of a central disk surrounded by five or more lobes or arms, which radiate from the body, are hollow, and contain prolongations from the stomach. The body is not enclosed in an immovable box or test, as in the sea-urchins, but the integument is of a leathery nature, and is richly furnished with calcareous plates, tubercles, and spines. The true star-fishes are distinguished from the nearly allied brittle-stars (*Ophiuroidea*) by the fact that the arms are direct prolongations of the body, that they contain prolongations of the stomach, and that they are deeply grooved on their under surfaces for the radiating vessels of the water-vascular system, which are further protected by a sort of internal skeleton. The upper surface of the body and arms is richly furnished with calcareous matter, in the form of prickles, tubercles, spines, and pedicellariæ, these last being peculiarly-modified spines. The upper surface, also, exhibits the madreporiform tubercle in the form of a concentrically-striated disk placed at the angle between two of the rays, and also the aperture of the anus, when this is present. The mouth is placed in the centre of the lower surface, and is not furnished with teeth. It leads by a short gullet into a stomach which usually terminates on the upper surface by an anal aperture; but this is occasionally wanting. From the stomach in all the *Asteroidea* proceeds a series of much-branched membranous sacs, two of which are prolonged into each ray. The water-vascular or ambulacral system is in most essential respects identical in structure with that of the sea-urchins, making due allowance for the different shape of the body. The nervous system consists of a gangliated ring surrounding the mouth and sending branches along each of the arms. The reproductive organs, like the nervous system, exhibit a radiate condition, being arranged in pairs in each ray.

The star-fishes are found on all shores, but many forms are properly inhabitants of deep water. They differ much in the general shape of the body. In the common cross-fish (*Uraster rubens*) the disk is small, and is furnished with long, finger-like rays, which are properly five in number. In the *Cribellæ* (Fig. 33) the general shape is much the same. In the sun-stars (*Solaster*) the disk is large and well marked, the rays are from twelve to fifteen in number, and they are shorter than the diameter of the disk. In the cushion-stars (*Goniaster*) the body is in the form of a five-angled disk, more or less flattened on both sides, the rays being only marked out by the ambulacral grooves upon the lower surface.

ORDER III. OPHIUROIDEA (Gr. *ophis*, snake; *oura*, tail; *ei*-

dos, form).—In this order we have only the common sand-stars (*Ophiura*) and brittle-stars (*Ophiocoma*), all closely allied to the true star-fishes in external appearance, especially in their strikingly radiate form. The body in the *Ophiuridæ* consists of a circular central disk covered with small calcareous plates, and giving off five long, slender, *snake-like* arms (Fig. 34, *a*, *b*), which may be simple or branched, but which do not contain any prolongations from the stomach, nor have their under surfaces excavated into grooves for the protrusion of ambulacral tube-feet. The arms, in fact, are not prolongations or lobes derived from the body itself, but are special appendages added for purposes of locomotion and prehension. The arms

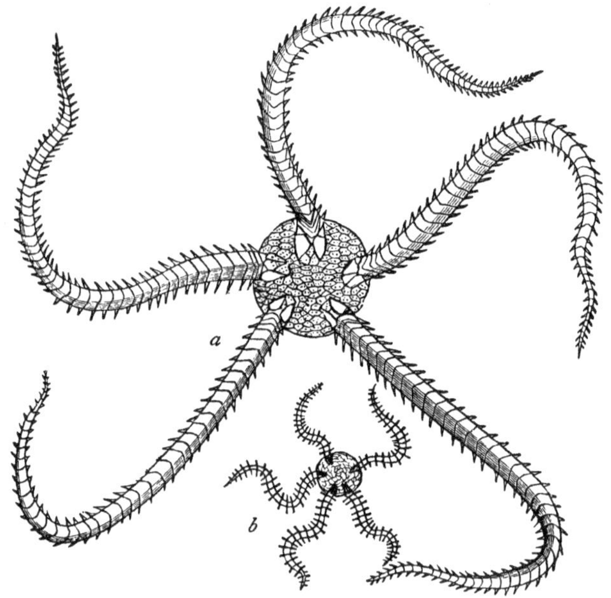

FIG. 34.—Ophiuroidea. *a Ophiura texturata*, the common sand-star; *b Ophiocoma neglecta*, the gray brittle-star (after Forbes).

are very much longer than the diameter of the disk, and are protected by four rows of calcareous plates—one above, one below, and one on each side. In the centre of each arm is a row of calcareous pieces which form a kind of internal axis or skeleton, below which is placed the radiating ambulacral vessel. All the internal organs are contained within the disk, and none of them pass into the arms except the nerve-cords and ambulacral vessels. The mouth is placed in the centre of the under surface of the disk, and opens into a globular, simple stomach, which is not furnished with an anal aperture, all indigestible particles being got rid of through the mouth. In various points of their anatomy the *Ophiuroidea* differ considerably from the true star-fishes, to which they are most nearly related, but these differences do not require further notice.

The habits of the brittle-stars and sand-stars are various, but many of them may be found in rock-pools or under stones at low water on most shores.

Fig. 35.—*Comatula rosacea.* *a* Free adult; *b* Fixed young (after Forbes).

Order IV. Crinoidea (Gr. *krinos*, a lily; *eidos*, form).—In this order are comprised *Echinodermata*, in which the body is fixed, during the whole or a portion of the existence of the animal, to submarine objects by means of a jointed flexible stalk or column. The *Crinoidea* were formerly very numerous, both individually and in types, but they are represented at the present day by no more than three or four living forms, of which one only (the feather-star) is at all of common occurrence. The body in the Crinoids consists of a central disk or cup formed of calcareous plates, and protecting the body of the animal. From the margins of this cup spring five or more arms which are arranged in a radiating manner, so as to form a more or less feathery crown. In one of our living forms, the animal, when full grown, is free; but in all other living genera, and in the great majority of fossil forms, the body was attached throughout life to the sea-bottom by means of a jointed stalk attached to the lower surface of the cup (Fig. 36), thus somewhat resembling a *lily*.

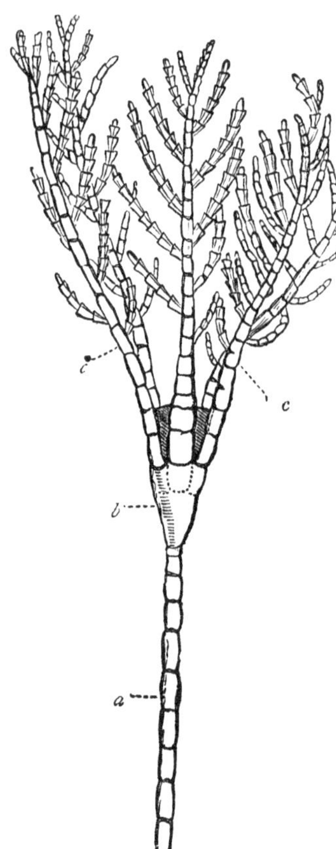

Fig. 36.—*Rhizocrinus lofotensis*, a living stalked Crinoid (after Wyville Thomson), four times the natural size. *a* Stem; *b* Cup; *c c* Arms.

The commonest living species is the rosy feather-star (*Comatula rosacea*), which occurs not very rarely on European coasts (Fig. 35). This beautiful animal consists of a central body or disk, from which proceed five radiating arms, which divide almost directly after their origin into two secondary branches, so that ultimately there are produced ten long and slender rays. Each arm is furnished on both sides with a number of little jointed lateral processes or "pinnæ," so as to assume a feather-like appearance, from which its popular name is derived. The digestive system is furnished with both a mouth and a vent;

the water-vascular or ambulacral system appears to take no part in locomotion, and the reproductive organs are lodged in the lateral processes of the arms. The most remarkable point, however, about the *Comatula* is the manner in which it develops itself. When fully grown (Fig. 35, *a*) it presents no small superficial resemblance to some of the *Ophiuroidea*. When young (Fig. 35, *b*) the *Comatula* is so different in appearance from the adult, that it was originally described as a distinct animal. It consists now of a little cup-shaped disk with ten radiating arms above, produced by the splitting into two of five primary rays, and furnished inferiorly with a little flexible column or stalk composed of a number of calcareous joints. By this jointed stem the body is at this period of life fixed to sea-weeds or other submarine objects. When sufficiently mature, however, the body drops off its stalk, and then only requires to grow in size to become a fully-developed *Comatula*.

The stalked condition which we have just seen to constitute a merely temporary stage in the life-history of the *Comatula* is, on the other hand, the permanent state of parts in almost all the "stone-lilies" and other fossil *Crinoidea*, and in two or three living forms. Of these recent species, one of the most remarkable is one which has been recently discovered in the Atlantic and North Seas, and which has been described under the name of *Rhizocrinus lofotensis*. This curious species (Fig. 36) consists of a little thread-like, jointed stem supporting a calcareous cup, from which proceed five branched and jointed arms; and the stalked condition is here permanently retained during life.

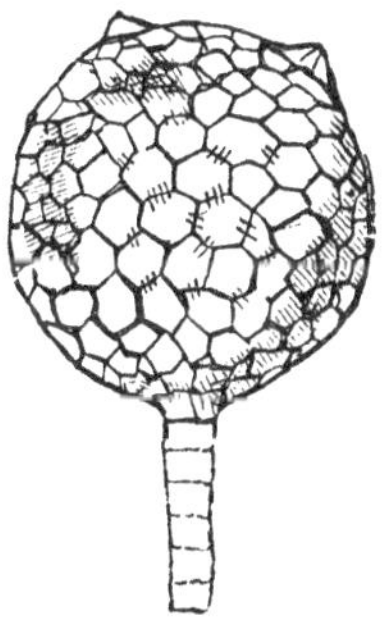
FIG. 37.—Cystoidea. *Echinosphærites.*

ORDERS V. AND VI. CYSTOIDEA AND BLASTOIDEA.—These orders merely require to be mentioned here, as all the forms included in them are extinct, and are unrepresented at the present day by living species. In both, the body is enclosed in a kind of box formed by jointed calcareous plates (Fig. 37), and it was in most cases permanently fixed to the sea-bottom by a jointed stalk or column. The arms, which form so conspicuous a feature in the true *Crinoidea*, were either absent or very rudimentary. Both orders are most closely allied to the *Crinoidea*, and they constitute probably the least highly-developed sections of the whole class of the *Echinodermata.*

ORDER VII. HOLOTHUROIDEA.—In this order are comprised the highest of the *Echinodermata*, all very different in outward appearance from any of the forms we have hitherto considered. They are commonly known as sea-cucumbers, or trepangs, but they are mostly rare and inconspicuous animals at the best. They are all more or less worm-shaped or snail-like in form, and they are either altogether destitute of calcareous matter in the skin, or with rare exceptions have only scattered grains and spines of this material. As a rule, the skin is simply leathery, and is endowed with wonderful contractility by means of powerful longitudinal and transverse muscles. In consequence of this, they can, in many cases, eject all or almost all their internal organs, and can sometimes divide their bodies into several parts when injured or alarmed. Locomotion is effected by alternate extension or contraction of their worm-like bodies, by anchor-shaped spicules of lime contained in the skin, or by rows of ambulacral tube-feet, like those of the sea-urchins, protruded through the integument. Sometimes the tube-feet are scattered over the whole surface of the body, and sometimes they are altogether absent. There is always a mouth at one extremity of the body, and a distinct vent at the other. The mouth is situated anteriorly, and is surrounded by a circlet of feathery tentacles (Fig. 38), which

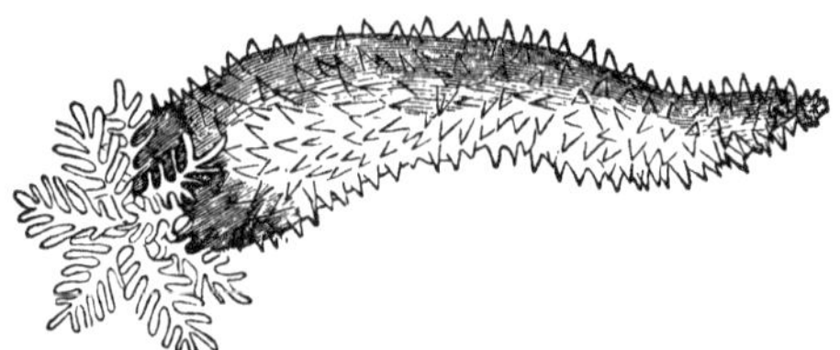

FIG. 38.—Holothuroidea. *Thyone papillosa* (after Forbes).

are believed to be modified tube-feet. The water-vascular or ambulacral system is sometimes quite rudimentary, but in other cases it much resembles that of the sea-urchins, except that the madreporiform tubercle is not placed on the outside of the body, but hangs down freely in the interior of the body. In most of the *Holothuroidea* there are appended to the termination of the intestinal canal two much-branched tubes, which are filled with sea-water from without, and are believed to exercise a respiratory function, hence the name of "respiratory tree" often applied to them.

The ordinary species of Holothurians, as already said, are all rare, and are mostly only to be obtained by dredging in tolerably deep water. Some of the tropical forms attain a large size, and some are largely searched after to be sold in the Chinese market, being regarded in that country as a delicacy.

CHAPTER XI.

CLASS II.—SCOLECIDA.

IN the second class of the sub-kingdom *Annuloida* are included a number of organisms which are, in many cases, very unlike one another in external appearance, but which, nevertheless, agree in one or two structural points of importance. The most important of these are the possession of a system of water-vascular vessels, the absence of a vascular system, and the possession of a nervous system composed of no more than one or two nervous masses or ganglia. The points by which the *Scolecida* are distinguished from the *Echinodermata* are, the absence of calcareous matter in the skin, the absence of any traces of a radiate arrangement of their parts, especially of the nervous system, the constant absence of any blood-circulatory apparatus, and the course of their development. The *Scolecida* (Gr. *skolex*, a worm) are often *vermiform* in shape, but many of them exhibit no worm-like characters, and one whole order is entirely microscopic. A great many of the *Scolecida* are internal parasites in other animals, and these are often collectively spoken of as *Entozoa* (Gr. *entos*, within; *zoön*, an animal). These parasitic forms subsist by an imbibition of the juices of their host through their delicate integument. They have, therefore, no necessity for acquiring food for themselves; and we find, in consequence, that many of them are wholly destitute of an alimentary canal, and that in all the organs of "relation" are very rudimentary. The *Scolecida* are divided into the following seven groups or orders:

1. *Tæniada* (Tape-worms).
2. *Trematoda* (Flukes).
3. *Turbellaria* (Ribbon-worms and Planarians).

4. *Acanthocephala* (Thorn-headed worms).
5. *Gordiacea* (Hair-worms).
6. *Nematoda* (Round-worms and Thread-worms).
7. *Rotifera* (Wheel-animalcules).

ORDER I. TÆNIADA (Gr. *tainia*, a ribbon).—In this order are comprised the *ribbon-shaped* Tape-worms (Fig. 39, 5) and

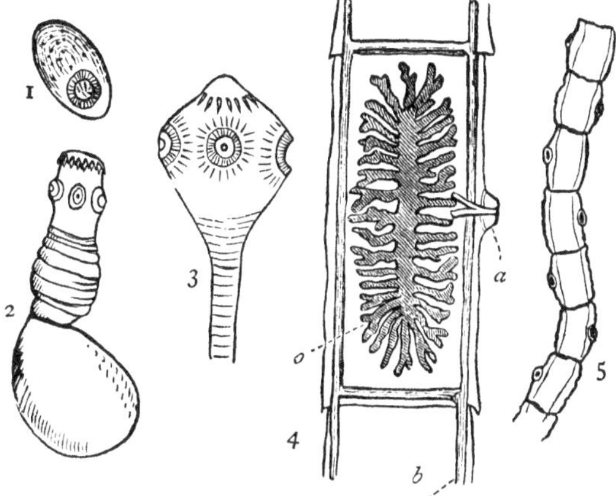

FIG. 39.—Morphology of Tæniada. 1. Ovum containing the embryo in its leathery case; 2. A bladder-worm (*Cysticercus longicollis*), magnified; 3. Head of the adult *Tænia solium*, enlarged, showing the suckers and crown of hooklets; 4. A single generative joint, enlarged to show the branched ovary (*o*), the generative pore (*a*), and the water-vascular canals (*b*); 5. Fragment of *Tænia solium*, showing the generative joints and the alternate arrangement of the generative pores.

the bladder-worms or cystic worms (Fig. 39, 2). These were formerly described as distinct groups; but it is now known that the latter are merely the young forms of the former. The peculiarity which distinguishes the development of the *Tæniada*, and which led to the cystic worms being described as distinct animals, is that the different stages of growth are always found inhabiting different animals or "hosts." If the fully-grown tape-worm is found in one animal, then its young form or cystic worm will always be found in another. Many animals are infested by tape-worms; but all the leading points of interest in the order will be brought out by a consideration of the commonest of the three tape-worms to which man is subject—namely, the common tape-worm, or *Tænia solium*.

The common tape-worm is found inhabiting the intestines of man, one only being generally present in the same individual. In shape (Fig. 39, 5) it is an extremely elongated, flattened, tape-like body, many feet in length, and composed of a number of flattened joints (Fig. 39, 4) all loosely united to one another. At one extremity the joints become much smaller and narrower, till ultimately a point is reached where the organism is firmly fixed to the mucous membrane of the intestine by means of a minute rounded head (Fig. 39, 3). The organs by which attachment is effected are, in this species, a crown of recurved hooks and four suckers. The head is in reality the true animal, and all the long, jointed, tape-like body which follows this, is really produced by a process of budding from the head. The head contains no reproductive organs, and is not furnished with a mouth or digestive organs of any kind, its nutrition being entirely effected by imbibition of the nutritive fluids elaborated by its host. A nervous system, in the form of one or two ganglia, sending filaments backward, is said to be present; but there is some doubt on this point. The water-vascular system (Fig. 39, 4) consists of two long vessels which run down each side of the body and communicate at each articulation by a transverse vessel, the whole opening in the last joint into a contractile vesicle. Each joint is sexually perfect, or hermaphrodite, containing both male and female reproductive organs (Fig. 39, 4), which open on the surface by a small raised aperture, the "generative pore." Almost the whole of each of the mature joints is filled up by a much-branched ovary. As the head is the true animal, and the numerous joints are only produced by budding, it follows that the entire organism is to be regarded as a kind of colony, constituted by a single sexless zoöid or "nurse," and numerous sexual zoöids, produced by budding from the former.

The process of development—that is to say, the process by which this composite organism, commonly known as the tape-worm, is produced—is a very remarkable one, and is briefly as follows: Each generative segment or joint, as already said, is hermaphrodite, and contains innumerable ova. These eggs, however, cannot be developed within the body of the animal infested by the tape-worm itself, but they are compelled to gain access to the body of some different species of animal, if development is to proceed. To secure this end, the mature joints of the colony break off, and are expelled from the alimentary canal of the host. The joints thus expelled die and decompose, and their contained eggs are thus set free. Each egg (Fig. 39, 1) is covered with a little leathery capsule which protects it from injury, and contains a minute embryo in its interior. If this microscopically small egg be swallowed—as in many ways it easily may be—by another warm-blooded animal (in this particular case by the *pig*), then a fresh series of changes ensues. The leathery case of the ovum is dissolved in the stomach of the new host, and the embryo is set free, when it bores its way through the walls of the stomach by means of little siliceous hooks with which it is provided. Having reached a suitable locality, the young tape-worm proceeds to surround itself with a kind of cyst, and it develops from its hinder end a kind of bladder filled with fluid (Fig. 39, 2). It is now a bladder-worm, or cystic worm, and as such would formerly have been regarded as a distinct animal. In the particular case of the *Tænia solium* which we are now considering, the cystic worm is found imbedded in the muscles of the pig, and it constitutes in that animal the disease known as the measles. In this cystic stage the young tape-worm may remain for an apparently indefinite period, being quite incapable of developing eggs, though sometimes fresh bladder-worms may be produced by a process of budding. For its further development it is necessary that it should now be introduced into the alimentary canal of man. If a portion of measly pork be eaten with these cystic worms imbedded in it, then the young tape-worm is liberated from its cyst: it fixes itself by means of its suckers and hooklets to the mucous membrane of the intestine, and its caudal bladder drops off. It is now converted into the head of the adult tape-worm. It finally commences to throw out buds from its hinder extremity, and in these buds or joints the reproductive elements are produced, so that ultimately we get the long, flattened jointed colony with which we started.

This extraordinary series of phenomena is now known to occur in other cases, but space will not admit our dwelling upon these. Another of the tape-worms of man (the *Tænia mediocanellata*) is developed in the same way from the measles of the ox. The tape-worm of the cat is the mature form of the bladder-worm of mice, and the tape-worm of the fox is derived from the cystic worm of hares and rabbits. Lastly, man is not only liable to be infested with the tape-worms derived from the cystic worms of other animals, but may be attacked by the cystic or immature forms of the tape-worms of other animals. Thus the disease known as "hydatids" in the human subject is caused by the presence in his tissues of the cystic worms which are ultimately developed into the tape-worm of the dog.

ORDER II. TREMATODA (Gr. *trema*, a pore or sucker).—The "suctorial" worms, or "flukes," as the members of this order are commonly called, are all internal parasites, inhabiting various situations in different animals, but especially affecting birds and fishes. They are all more or less flattened and rounded in shape, and are furnished with one or more *suckers*, by which they adhere. They are distinguished from the *Tæniada* by always possessing an alimentary canal, which is often much branched (Fig. 40, 1), is simply hollowed out

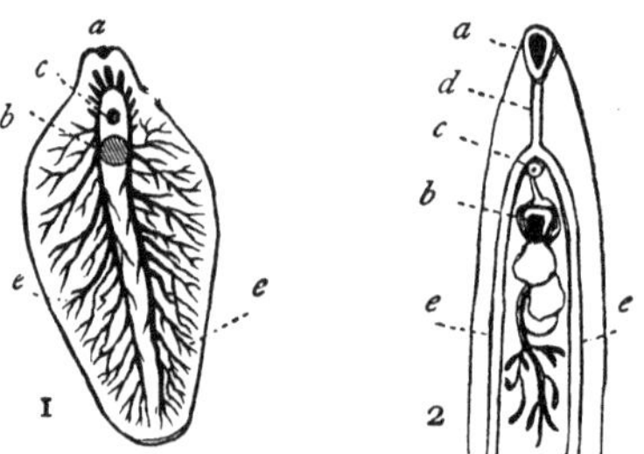

FIG. 40.—Trematoda. 1. *Distoma hepaticum*, the "liver-fluke," showing the branched alimentary canal: 2. Anterior extremity of *Distoma lanceolatum*, enlarged; *a* Anterior sucker; *b* Posterior sucker; *c* Generative pore; *d* Gullet; *e e* Bifurcating alimentary canal (after Owen).

of the tissues of the body, and is never provided with a distinct anus. The best known of the *Trematoda* is the common liver-fluke (*Distoma hepaticum*, Fig. 40, 1), which inhabits the gall-bladder or ducts of the liver in sheep, and is the cause of the disease known as the rot. In form it is ovate, flattened on the two sides, and presenting two suckers, of which the anterior is perforated by the aperture of the mouth. A branched water-vascular system is present, and opens posteriorly by a small aperture. The alimentary canal bifurcates shortly behind the mouth, the two divisions thus produced being much branched, and terminating posteriorly in blind extremities. In *Distoma lanceolatum* (Fig. 40, 2) the intestine is divided into two branches, but these are simple tubes, and are not branched.

ORDER III. TURBELLARIA.—The animals included in this order differ altogether from the *Trematoda* and *Tæniada* in being almost all aquatic in their habits and being all non-parasitic. They never possess sucking-disks or hooklets, and their integument is always furnished with vibrating cilia. A water-vascular system is always present, but it appears sometimes not to communicate with the exterior. The alimentary canal is sometimes simply hollowed out of the tissues and destitute of an anus, as in the *Trematoda*, or at other times suspended in a free space (body-cavity) and furnished with an anus. It may be simple or much branched.

The best known of the members of this order are certain little jelly-like, soft-bodied, ovate, or elliptical creatures, which are commonly found in fresh water or on the sea-shore, and are known as *Planarians*. The skin in these curious little animals (Fig. 41, 1, 2) is richly furnished with cilia, and also contains numerous cells which have been compared to the

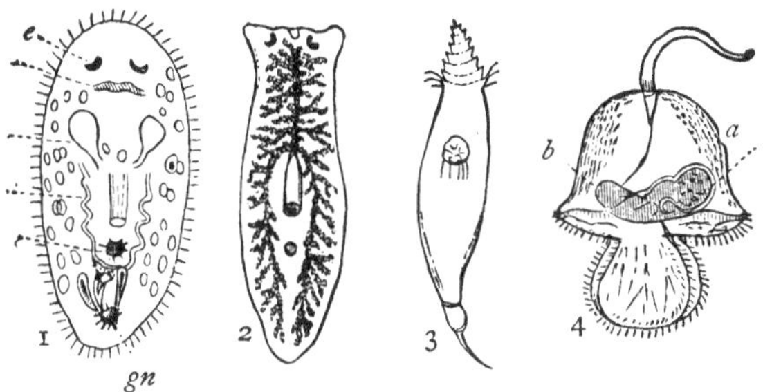

FIG. 41.—Turbellaria. 1. *Planaria torva:* *m* Mouth; *g* Nerve-ganglion; *e* Eyes; *o v* Ovary; *t* Spermarium; *g n* Genital opening: 2. *Planaria lactea*, showing the branched intestine: 3. Larva of one of the marine Turbellarians: 4. *Pilidium*, the larva of one of the *Nemertidæ*.

"nettle-cells" of the *Cœlenterata*. The intestine may be either straight or branched, but always terminates behind in blind pouches, and is never provided with an anus. The water-vascular system communicates with the exterior. The nervous system consists of two ganglia, placed in front of the mouth, and united by a cord. There are generally rudimentary eyes or pigment-spots, which vary in number from two to sixteen.

The remaining members of the *Turbellaria* are known as ribbon-worms (*Nemertidæ*), and are not uncommonly found on the sea-shore. They differ from the *Planarida* in being worm-like in shape, by the fact that the alimentary canal is furnished with a distinct anus, and by the absence of an external opening to the water-vascular system of the adult, in some cases at any rate. Their development sometimes shows phenomena very similar to what occurs in the *Echinodermata*, the larva (Fig. 41, 4) being a free-swimming, ciliated organism, of which only a portion is employed in producing the adult animal, the remainder being cast off as useless.

ORDER IV. ACANTHOCEPHALA (Gr. *akantha*, thorn; *kephale*, head).—The "*thorn-headed* worms" included in this order are all internal parasites. They are worm-like in shape, marked with transverse wrinkles, and destitute of any mouth or alimentary canal. The anterior extremity of the body forms a kind of proboscis or snout, which is armed with recurved hooks, and has placed at its base a single nervous ganglion. Beneath the skin is a net-work of canals, containing a clear fluid, and believed to represent the water-vascular system. The thorn-headed worms include some of the most formidable parasites with which we are as yet acquainted, the best known being the various forms of *Echinorhynchus*, which are found inhabiting the alimentary canal in many mammals, birds, and fishes, but not as yet in man.

ORDER V. GORDIACEA.—The *Gordiacea*, or "hair-worms," are thread-like parasites which in the earlier stages of their existence inhabit the bodies of various insects, chiefly beetles and grasshoppers. They possess a mouth and alimentary canal. The sexes are distinct, and they leave the bodies of the insects which they infest to breed, subsequently depositing their eggs in long chains either in water or in some moist situation. In form the *Gordiacea* are singularly like hairs, and they often attain a length very many times greater than that of the insect in which they live.

ORDER VI. NEMATODA (Gr. *nema*, a thread).—In this order are the "round-worms" and "thread-worms," both of which are parasitic, together with a number of worms which lead a permanently free existence. All the *Nematoda* (Fig. 42) are elongated and cylindrical or *thread-like* in shape. They possess a distinct mouth, and an alimentary canal which is freely suspended in an abdominal cavity, and which terminates in a distinct anus. They possess a system of canals which are believed to represent the water-vascular system;

and the nervous system is in the form of a gangliated cord surrounding the gullet, and sending filaments backward. Among the best known of the parasitic Nematodes are the common round-worm (*Ascaris lumbricoides*) and the thread-worm (*Oxyuris*) of the human subject, both of which inhabit the alimentary canal, and the guinea-worm (*Filaria*), which spends a portion of its existence in the cellular tissue of man, especially of the legs, and which attains a length of several feet. More dangerous than any of these is the *Trichina*, which spends its immature stages encysted in the muscles of some such animal as the pig, and only attains maturity and becomes capable of producing eggs, when introduced into the alimentary canal of some other warm-blooded vertebrate animal. When this takes place, a train of symptoms are originated which sometimes resemble rheumatic fever, and appear to be very generally fatal.

Of the free Nematode worms, which are never parasitic at any time of their lives, about two hundred species have been described, most of which inhabit fresh water or the shores of the sea. One of the most familiar is the so-called "vinegar-eel" (*Anguillula aceti*, Fig. 42, A).

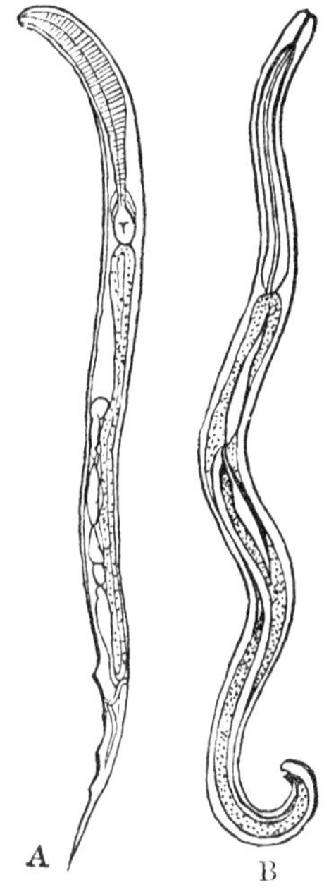

Fig. 42.—Nematoda. A. Vinegar-eel (*Anguillula aceti*). B. *Dorylaimus stagnalis*, a free Nematode, living in stagnant water.

Order VIII. Rotifera (Lat. *rota*, wheel; *fero*, I carry).—The *Rotifera*, or "wheel-animalcules," derive their popular name from the fact that the anterior end of the body is furnished with one or two circlets of cilia (Fig. 43) which, when in motion, vibrate so rapidly as to produce the illusory impression of a quickly-rotating toothed *wheel*. The *Rotifera* are almost all aquatic, and are mostly inhabitants of fresh water. They are all microscopic in size, none attaining a greater length than one-thirty-sixth of an inch. In the females there is a distinct mouth, intestinal canal, and anus. A nervous system is also present, consisting of ganglia placed near the anterior extremity of the body and sending filaments backward. There is, finally, a well-developed water-vascular system.

Most of the *Rotifera* are free-swimming, active little animals (Fig. 43, A), but some are permanently fixed, as in *Melicerta* (Fig. 43, B), or in the crown-animalcule *Stephanoceros*). They are usually simple, but they are sometimes composite, forming colonies. As a rule, the male and female *Rotifera* differ greatly from one another, the males being smaller than the females, devoid of any masticatory or diges-

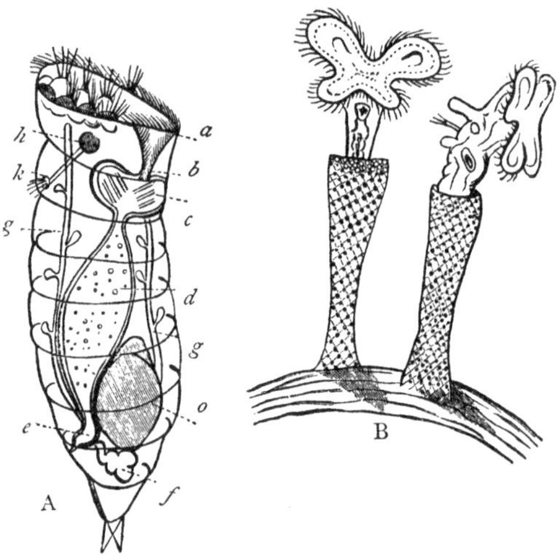

FIG. 43.—Rotifera. A. Diagrammatic representation of *Hydatina senta* (generalized from Pritchard): *a* Depression in the ciliated disk leading to the digestive canal; *b* Mouth; *c* Pharyngeal bulb with masticatory apparatus; *d* Stomach; *e* Cloaca; *f* Contractile bladder; *g g* Respiratory or water-vascular tubes; *h* Nerve-ganglion, giving filament to ciliated pit (*k*); *o* Ovary. B. *Melicerta ringens* (after Gosse).

tive apparatus, and more or less closely resembling the young forms of the species. The males, in fact, merely lead a transient existence, and die as soon as they have succeeded in fertilizing the females. The body in most cases is very distinctly ringed or annulated (Fig. 43, A), but is not composed of distinct rings separated by partitions. The integument is usually provided with bundles of muscular fibres taking a longitudinal and transverse direction. In the free forms the anterior ciliated disk acts somewhat like the propeller of a screw-steamer in driving the organism through the water—in all cases it has the action of producing currents in the water by which particles of food are brought to the mouth. The posterior end of the body is usually developed in the free forms into a kind of tail or foot (Fig. 43, A), which may take the shape of a kind of pincers or of a little suctorial disk.

As regards their internal anatomy, in the females of almost all the *Rotifera* there is a well-developed alimentary canal, which is completely shut off from the general cavity of the body. The mouth (Fig. 43, A *b*) opens into a dilated chamber (*c*), which contains a complicated apparatus of horny teeth. This in turn opens into a capacious stomach (*d*), continued into an intestine which terminates by a chamber known as the "cloaca" (*e*), which forms the common outlet for the water-vascular and generative systems. In both sexes there is a well-developed water-vascular system consisting of a contractile chamber or bladder (*f*), opening into the cloaca, and giving origin to two complicated tubes which are known as the "respiratory tubes" (*g g*), and which terminate near the anterior end of the body, apparently by blind extremities. The nervous system is in the form of a large double ganglion placed above the gullet, and having one or two eye-specks placed upon it. The ovaries (*o*) constitute conspicuous organs in the female *Rotifera*, but in summer the young Rotifers appear to be produced by the females without having access to the males.

The *Rotifera* were long confounded with the *Infusoria*, in consequence of their great similarity in external appearance. They are, however, of an obviously much higher grade of structure. One of the most remarkable phenomena presented by the *Rotifera* is found in the undoubted fact that, in spite of their complex organization and aquatic habits, they can be dried, and again brought to life by the addition of a little water, and that this desiccation and restoration to life can be apparently repeated many times in succession without injury.

SUB-KINGDOM IV.—ANNULOSA.

CHAPTER XII.

Anarthropoda.

Sub-kingdom Annulosa.—In this sub-kingdom are comprised an enormous number of animals which agree in the following characters (Fig. 44): The body is composed of a number of segments or rings arranged along a longitudinal axis. There is a distinct alimentary canal (*b*), placed centrally as compared with the other organic systems, and completely shut off from the general cavity of the body. The

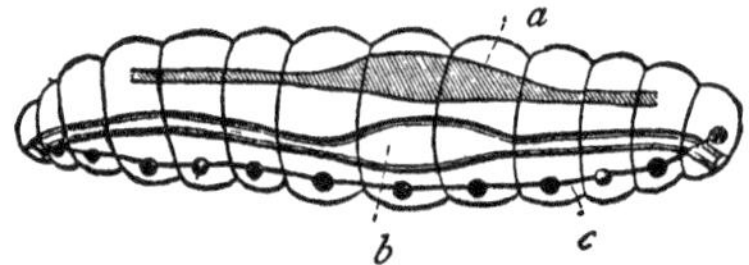

Fig. 44.—Diagram of an Annulose animal. *a* Blood-vascular or hæmal system; *b* Digestive system; *c* Neural system.

blood-vascular system may be absent, but, when present, it is always situated on the dorsal aspect of the body (*a*). The nervous system is always present, and is placed along the ventral surface of the body. In its typical form it consists of two nervous cords running along the whole length of the ventral surface, and having a pair of ganglia developed in each ring. The first pair of ganglia is always placed above the gullet, and the second below, so that the gullet is surrounded by the two cords uniting these ganglia (constituting the so-called *œsophageal collar*). The limbs (when present) are always turned toward the neural aspect—that is to say, toward that side of the body upon which the nervous system is situated. (See also the transverse section of an Annulose animal, Fig. 1.) The entire sub-kingdom of the *Annulosa* is divided into two great divisions termed *Arthropoda* and *Anarthropoda*, according as the body is provided with jointed appendages or not. In the *Arthropoda*, in which the body-rings are furnished (some or all) with jointed appendages, are included the Crustaceans (lobsters, crabs, etc.), the spiders and scorpions, the centipedes, and the insects. In the *Anarthropoda*, in which there are no true jointed appendages, are included the spoon-worms, leeches, earth-worms, tube-worms, and sand-worms.*

DIVISION I. ANARTHROPODA (Gr. *a*, without; *arthros*, joint; *podes*, feet).—In this division of the *Annulosa*, *the locomotive appendages are never distinctly jointed or articulated to the body*. In this division are included two principal classes—the *Gephyrea* and the *Annelida*.†

CLASS I. GEPHYREA.—This class is a very small one, and includes a number of worm-like animals, which in most respects are very similar to the following class of the *Annelida*, but are distinguished by having no locomotive appendages attached to the sides of the body. They were long placed among

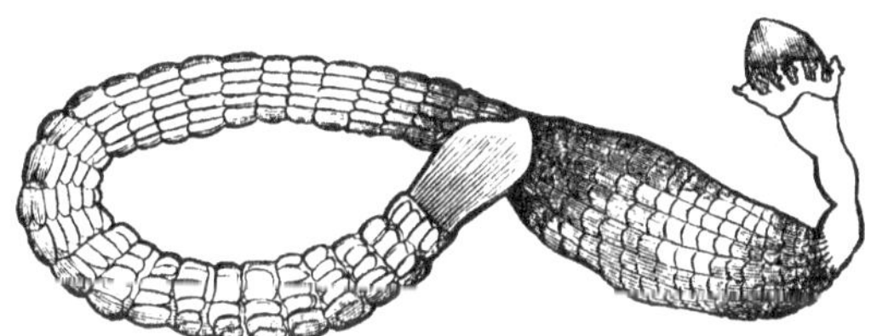

FIG. 45.—Gephyrea. *Syrinx nudus* (after Forbes).

the *Echinodermata*, having a decided relationship to the worm-like *Holothurians*. They are distinguished, however, by never secreting calcareous matter in the skin, and by having no water-vascular or ambulacral system. There can be no doubt,

* The *Anarthropoda* are often united with the *Scolecida* into a common sub-kingdom under the name of *Vermes;* in which case the *Echinodermata* are retained apart in a special sub-kingdom.

† A third class has been constituted under the name of *Chætognatha* for some singular marine animals, transparent and worm-like in form, with lateral fins at the hinder end of the body, and having the mouth armed with bristles. They form the genus *Sagitta*.

however, that the *Gephyrea* are, on the whole, very nearly related to the *Holothurians*, and it is chiefly from the total absence of any radiate arrangement of the nervous system and internal organs that they appear to be more properly classed with the worms. The *Sipunculus* or spoon-worm is found burrowing in the sand of many sea-coasts, or inhabiting the cast-away shells of univalve shell-fish. A considerable number of species of this class have been recorded as occurring in European seas, and one of the more characteristic forms is figured above (Fig. 45).

Class II. Annelida (Lat. *annulus*, a ring).—The *Annelida* or ringed-worms are distinguished from the preceding by the possession of definite segmentation, the body being composed of a number of *rings* which are all similar to each other except at the two ends of the body. All the *Annelida* are more or less worm-like in shape, and in all, except the leeches, the segments are (some or all) provided with lateral appendages which mostly subserve locomotion, but which are never jointed to the body. In the typical *Annelida* each segment (Fig. 46)

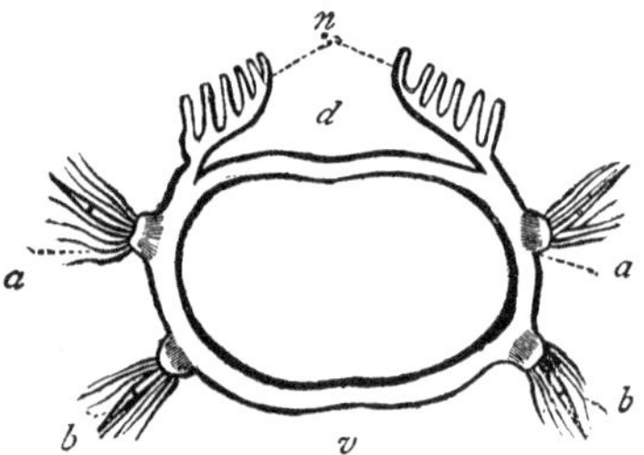

Fig. 46.—Diagrammatic transverse section of a typical Annelide. *d* Dorsal arc; *v* Ventral arc; *n* Branchiæ or gills; *a* Dorsal oar; *b* Ventral oar—both carrying bristles and a jointed filament.

consists of two arches, termed, from their position, respectively the "dorsal arc" (*d*), and the "ventral arc" (*v*). Each segment carries a lateral process on each side, which are known as the "foot-tubercles" (*parapodia*). Each foot-tubercle in turn may consist of an upper piece or "dorsal oar" (*a*), and a lower piece or "ventral oar" (*b*), both carrying a tuft of bristles and a soft jointed filament.

The nervous system consists essentially of a double gangliated chain placed along the ventral surface of the body, and traversed in front by the gullet, so that the first ganglion lies above the gullet (Fig. 44). The digestive system consists of a mouth, generally with a protrusible proboscis, and sometimes horny jaws, a gullet, stomach, intestine, and a distinct anus. As a rule, the alimentary canal runs straight from one end of the body to the other without describing any convolutions in its course. In almost all cases the alimentary tube is placed in a distinct cavity, which contains a fluid with solid particles in it, believed to correspond to the blood of the higher *Annulosa*. In most, if not in all, there is an additional system of vessels which carry a fluid containing solid particles, which are contractile, and which send branches to the respiratory organs, when these exist. This system is believed not to correspond to the blood-vascular system of the higher animals, and it has, therefore, been termed the "pseudo-hæmal" system (Gr. *pseudos*, falsity; and *haima*, blood). It is believed, on the other hand, to be truly homologous with the water-vascular system of the *Annuloida*. Respiration is effected by the general surface of the body, or by distinct gills or branchiæ. In most cases, also, there exists a series of peculiar involutions of the integument, which are known as the "segmental organs" or "respiratory pouches," and which are believed to be partially concerned in the respiratory process. The sexes in the *Annelida* are sometimes distinct, sometimes united in the same individual. The embryos are almost always ciliated, and many of them pass through a metamorphosis.

The *Annelida* may be divided as follows:

SECTION A. ABRANCHIATA.—Without gills or branchiæ.
1. *Hirudinea.*—(Leeches.)
2. *Oligochæta.*—(Earth-worms.)

SECTION B. BRANCHIATA.—With branchiæ.
3. *Tubicola.*—Tube-worms.)
4. *Errantia.*—(Sand-worms.)

ORDER I. HIRUDINEA (Lat. *hirudo*, a horse-leech).—This order comprises only the leeches, some of which are marine, while others inhabit fresh water. The leeches (Fig. 47) are all characterized by the fact that the body is destitute of lateral bristles or foot-tubercles, but is provided with a sucking-disk at one or both extremities. In the typical forms, as in the common medicinal leech, there are sucking-disks at both ends of the body, and in those in which only the hinder sucker is present, the head can be converted into a suctorial cavity. Locomotion is effected either by means of the alternate fixation and detachment of the suckers, or by a serpentine bending of the body.

The body is obviously ringed or annulated, but none of the rings carry lateral appendages of any kind. The mouth is sometimes destitute of teeth, but is occasionally armed with complex jaws. The alimentary canal is short, with lateral dilatations, and united to the skin by means of a spongy vascular tissue, so that the body-cavity is obliterated. The pseudo-hæmal system is well developed, and consists essentially of four great longitudinal vessels. Respiration appears to be effected, in part, at any rate, by means of the segmental organs, which have the form of little sacs opening externally by minute apertures. The nervous system has its usual form, and the ganglia in front of the gullet ("*præ-œsophageal*" ganglia) give off branches to a number of simple eyes which are placed on the head. The sexes are united in the same individual.

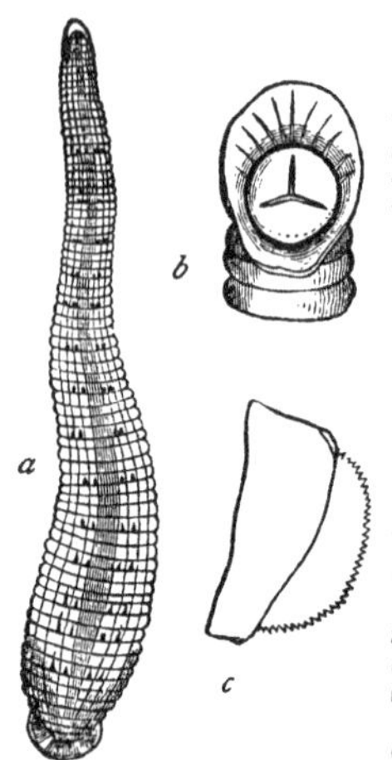

Fig. 47.—Hirudinea. *a* The medicinal leech (*Sanguisuga officinalis*), natural size; *b* Anterior extremity of the same magnified, showing the sucker and triradiate jaws; *c* One of the jaws detached, showing the semicircular toothed margin.

The most familiar of the leeches are the common horse-leech (*Hæmopsis*), and the medicinal leech (*Sanguisuga officinalis*, Fig. 47). The former has small and blunt teeth, but the latter is provided with three semicircular toothed jaws (Fig. 47, *b*, *c*), which meet in a point, and are sufficiently powerful to cut through the human skin. The medicinal leech is a native of fresh waters throughout the south and east of Europe, and it is imported in large numbers from Hungary, Bohemia, and Russia.

Order II. Oligochæta.—In this order are included the earth-worms (*Lumbricidæ*), and the water-worms (*Naïdidæ*). They are all distinguished from the preceding by the fact that the body is furnished with rows of bristles which take the place of the foot-tubercles of the higher *Annelida*, and which are the organs of locomotion. They are distinguished from the higher forms by the fact that the locomotive bristles are comparatively few in number, hence the modern name of the order (Gr. *oligos*, few; and *chaite*, a bristle). In the common earth-worm (*Lumbricus terrestris*) the body is cylindrical, attenuated at both ends, and furnished with eight rows of locomotive bristles. The mouth is destitute of teeth, and opens into a gullet which leads to a muscular crop, succeeded by a second muscular dilatation or gizzard. The intestine is continued straight to the anus, and is constricted in its course by numerous transverse partitions springing from the walls of the body-cavity. The pseudo-hæmal system is well developed; and there exists in even greater numbers than in the leeches the series of segmental organs, or lateral pouches, which open externally by pores. The *Naïdidæ* are chiefly noticeable on account of their power of producing fresh individuals by a process of budding before they attain sexual maturity. One of the commonest of them is a little worm which occurs abundantly in many pools and streams (*Tubifex rivulorum*), and which exhibits a fine red color, owing to the pseudo-hæmal system being visible through the transparent integument.

ORDER III. TUBICOLA.—The Annelides included in this group derive their name from the fact that they have the power of protecting themselves by means of tubes (Lat. *tuba*, a tube; and *colo*, I inhabit). In some cases (Fig. 48) the tube is composed of carbonate of lime, and is a genuine secretion from the body. In all the *Tubicola* the respiratory organs are in the form of branched filamentous external gills, in which the fluid of the pseudo-hæmal system is subjected to the action of the outer water. They are, therefore, "branchiate" Annelides. As they live in tubes, however, and do not voluntarily expose more than the anterior end of the body, the branchiæ are all placed on or near the head. The filaments of which the gills are composed (Fig. 48, *a*) are richly ciliated, and, as the pseudo-hæmal fluid is usually red, they have generally a beautiful scarlet color.

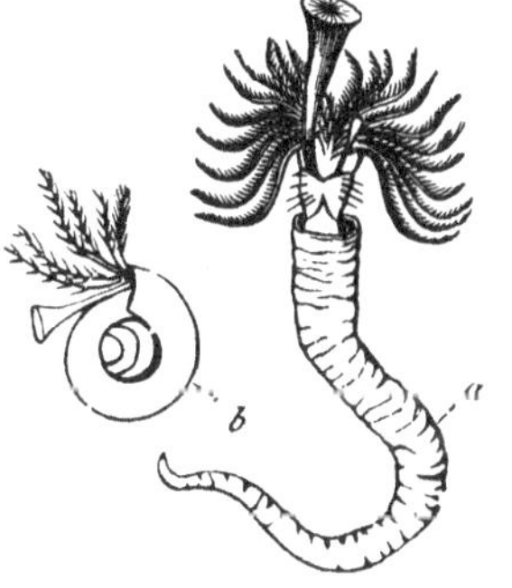

FIG. 48.—Tubicola. *a Serpula contortuplicata*, showing the branchiæ and operculum; *b Spirorbis communis.*

The most familiar of the *Tubicola* is the *Serpula* (Fig. 48, *a*), the contorted and winding tubes of which must be known to every one as occurring on shells or stones on the sea-shore. One of the cephalic filaments in *Serpula* is much developed, and its extremity forms a kind of conical plug which serves to close the mouth of the tube when the animal is retracted within it. In *Spirorbis* (Fig. 48, *b*) the shelly tube is coiled into a flat spiral, which is fixed to some solid object. It is of extremely common occurrence on the fronds of sea-weed, and on other submarine objects.

ORDER IV. ERRANTIA.—The Annelides comprised in this order are called "errant" (Lat. *erro*, I wander), or "roving," from the fact that they all lead a free existence, and are never confined in tubes. They have always lateral unjointed appendages, or foot-tubercles (Fig. 49), which carry tufts of

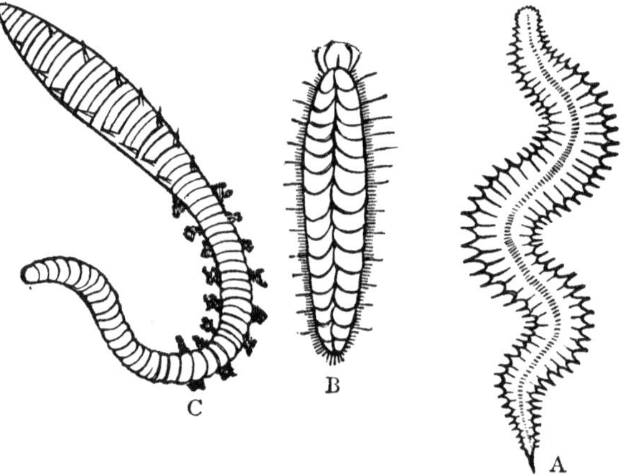

FIG. 49.—Errant Annelides. A. Hairy Bait (*Nephthys*); B. Sea-Mouse (*Aphrodite*); C. Lobworm (*Arenicola*). (After Gosse.)

bristles and a soft, jointed filament. The anterior rings of the body are usually so modified as to form a sort of head, which is provided with eyes and with two or more feelers, which differ from the antennæ of insects and Crustaceans in not being jointed. The mouth is placed on the inferior surface of the head, and is sometimes furnished with one or more pairs of horny jaws, which work from side to side. The upper part of the alimentary canal is muscular, and can be turned inside out, or protruded beyond the true opening of the mouth. The pseudo-hæmal system is well developed, and its contained fluid is mostly red. Respiration is effected by external processes, gills, or branchiæ, arranged in tufts placed along the sides or back of the body, and not confined to the immediate neighborhood of the head, as in the *Tubicola.* The sexes are in different individuals, and the young pass through a metamorphosis.

Among the best known and commonest of the Errant Annelides are the common lob-worm (*Arenicola piscatorum*) of our coasts, which is constantly used by fishermen for bait; and the sea-mice (*Aphrodite* and *Polynoe*), some of which attain a large size, and are conspicuous for their iridescent bristles. Other less abundant forms may be readily obtained by searching under stones at low water.

CHAPTER XIII.

ARTHROPODA.

DIVISION II. ARTHROPODA OR ARTICULATA.—The members of the sub-kingdom *Annulosa* comprised under this head are generally known as *Articulate* animals, or as *Arthropoda* (Gr. *arthros*, a joint; and *podes*, feet). They are all distinguished by the possession of *jointed appendages articulated to the body.* The body is composed of a series of distinct rings or segments (technically called "somites") arranged longitudinally one behind the other. The skin is more or less completely hardened by a horny deposit of "chitine," with or without lime, so as to form a resisting shell, to the inner surface of which the muscles are attached. There is consequently no necessity for any internal skeleton. The nervous system in the young of all Articulate animals has its typical form of a chain of ganglia placed along the ventral surface of the body, and traversed in front by the gullet. In the adult, however, this typical state of the nervous system is often lost or modified. The blood-circulatory system may be absent; but, when it is present, it is placed dorsally (Fig. 44), and consists of a true blood-system containing corpusculated blood, and furnished with a contractile cavity or heart. Respiration is sometimes effected simply by the general surface of the body, but there are generally special organs adapted for breathing air, either directly or through the medium of water. Jointed appendages are always present, and may be developed from any segment of the body.

The *Arthropoda* are divided into four great classes—viz., the *Crustacea* (crabs, lobsters, etc.), the *Arachnida* (mites, spiders, and scorpions), the *Myriapoda* (centipedes and gally-worms), and the *Insecta* (or true insects). These are roughly distinguishable from one another by the following characters

1. Crustacea.—Animal more or less truly aquatic; respiration by gills, or by the general surface of the body; two pairs of antennæ (feelers); locomotive appendages more than eight in number, borne by the segments of the thorax, and usually of the abdomen also.

2. Arachnida.—Respiration aërial, by pulmonary sacs, by air-tubes (tracheæ), or by the general surface of the body; head and thorax amalgamated; antennæ (as such), absent; legs eight; abdomen without jointed appendages.

3. Myriapoda.—Respiration by air-tubes (tracheæ); head distinct; remainder of the body composed of nearly similar segments; one pair of antennæ; legs numerous.

4. Insecta.—Respiration by air-tubes (tracheæ); head, thorax, and abdomen distinct; one pair of antennæ; three pairs of legs borne on the thorax; abdomen destitute of limbs; generally two pairs of wings on the thorax.

CHAPTER XIV.

CRUSTACEA.

CLASS I. CRUSTACEA (Lat. *crusta*, a crust, or external shell).—The members of this class are commonly known as crabs, lobsters, shrimps, prawns, king-crabs, barnacles, acorn-shells, wood-lice, etc. They are nearly allied to the succeeding class of the *Arachnida* (spiders and scorpions), but are distinguished by their adaptation to a more or less purely aquatic life, by having jointed appendages upon the hinder segments of the body (abdomen), and by the possession of two pairs of antennæ. As a class, the *Crustacea* are distinguished by being usually furnished with branchiæ or respiratory organs adapted for breathing air dissolved in water, by having more than four pairs of legs, by having a well-developed chitinous or partially calcareous "crust" or external skeleton, by the fact that some of the appendages are generally so modified as to act as organs of mastication, and by passing through a metamorphosis before attaining their adult condition.

The body in a typical Crustacean is composed of *twenty-one* (or, according to some writers, *twenty*) distinct segments or somites, placed one behind the other. These segments are distributed in three distinct divisions, known respectively as the "head," the "thorax" or chest, and the "abdomen" or tail, each of which is usually regarded as being composed of seven segments. In very many cases, however, the fourteen segments belonging to the head and chest are amalgamated together into a single mass, which is termed the "cephalo-thorax," thus leaving seven segments to the abdomen. It will be unnecessary, however, to dwell here longer upon the structure of the *Crustacea*, as the general morphology of the class will be given at somewhat greater length in speaking of the lobster. The classification, also, of the *Crustacea* is so complex that it will be as well to omit altogether the less important orders, merely giving the names and leading characters of these in an appendix. It has also been thought advisable to invert the usual order here adopted, and to commence with the consideration of the highest sections of the class first.

ORDER DECAPODA.—The *Crustacea* included in this order derive their name from the fact that they all possess five pairs of legs (Gr. *deka*, ten; *podes*, feet). They belong to a large section known as the "stalk-eyed" Crustaceans, from the fact that the eyes are supported by long, movable stalks. They include the lobsters, shrimps, cray-fish, crabs, hermit-crabs, and other forms, and are the most highly organized and most familiar of the whole class of the *Crustacea*. They are divided into three very well marked groups or tribes, all of which can be exemplified by the well-known British species.

A. Macrura.—The name of *Macrura* (Gr. *makros*, long; and *oura*, tail) is given to those ten-footed Crustaceans which have a long and well-developed tail. Among these are the lobster, shrimp, prawn, and cray-fish, of which the lobster may be selected as a good typical example.

In the lobster (Fig. 50) the body is at once seen to be composed of two parts, familiarly called the "head" and "tail." The so-called head is covered by a great shield termed the "carapace" (Fig. 50, *ca*), and it is in reality the cephalo-thorax, being composed of the amalgamated segments which belong to the true head and to the thorax. The so-called tail is really the abdomen, and it is composed of a number of segments which are not immovably united together, as in the cephalo-thorax, but are movably jointed together. The various appendages of the animal are arranged in pairs on the under surface of the body; and, where the segments are completely amalgamated (as in the cephalo-thorax), their existence may, nevertheless, be determined by the presence of a pair of appendages. The first segment of the head carries a pair of compound eyes, made up of a number of simple lenses aggregated together, and supported upon long and movable eye-stalks. Behind these come two pairs of jointed organs of touch, which are known as the "antennæ." The front pair is much smaller than the hinder pair, and they are known respectively as the "lesser antennæ," or "antennules," and the "great antennæ." Behind these, again, comes the mouth, which is placed on the under surface of the head, and is provided with a complicated series of masticatory organs. It is unnecessary to describe these minutely, but it should be noticed that they are all modified limbs, and therefore differ

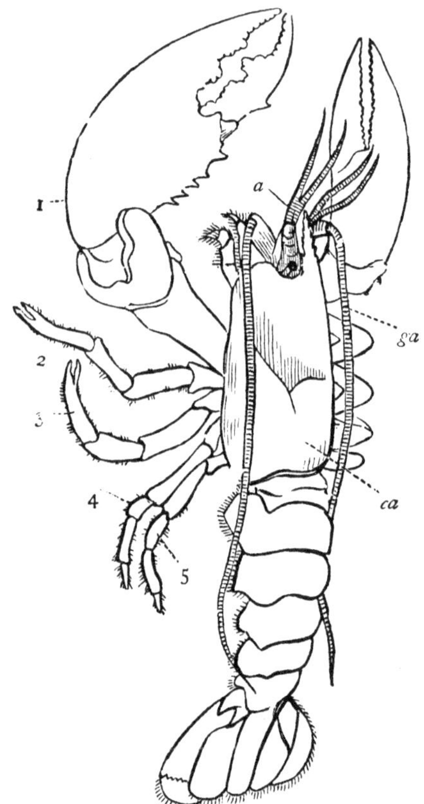

FIG. 50.—Common Lobster (*Homarus vulgaris*). 1. First pair of legs, constituting the great nipping-claws; 2 and 3. Second and third pairs of legs, also ending in nipping-claws; 4 and 5. Last two pairs of legs; *a* Smaller antennæ; *ga* Greater antennæ; *ca* Carapace.

altogether from the jaws of the Vertebrate animals. That this is their real nature is shown most obviously in the hindmost pairs of these jaws, which are so little altered from ordinary legs that they are known as "foot-jaws." The last five segments of the thorax carry five pairs of walking-legs, hence the name *Decapoda* applied to the order. Of these legs, the first three pairs have their extremities converted into nipping-claws or "chelæ," and the first pair is much larger than the others, and constitutes the well-known great claws of the lobster. The last two pairs of legs simply terminate in pointed extremities, and not in pincers. The segments of the abdomen, with the exception of the hindmost, carry each a pair of paddle-like appendages, which are used in swimming, and are called the "swimmerets." The last pair of swimmerets are attached to the last segment but one, and are very greatly expanded, so as to form a very powerful tail-fin. The last segment of all is known as the "telson," and is not provided with any lateral appendages.

The mouth in the lobster leads by a short gullet into a globular stomach, which is furnished with a calcareous apparatus for grinding down the food, commonly called the "lady in the lobster." The intestine is continued backward from the stomach without convolutions, and opens by a distinct anus placed in front of the telson. A well-developed liver is also present. The heart is placed dorsally, and is filled with aërated blood derived from the gills, which it propels through every part of the body. The gills, or branchiæ, are pyramidal bodies attached to the bases of the legs, and placed in a kind of chamber formed beneath the great shield, or carapace, on each side of the body. They consist each of a central stem supporting numerous lateral branches, and they are richly supplied with blood. The water which fills the gill-chambers is constantly renewed by the movements of the legs, and thus the gills are kept constantly supplied with fresh water. The nervous system is placed along the ventral surface of the body, and has its usual form. The organs of sense are the two pairs of feelers or antennæ, the compound eyes, and two organs of hearing.

B. Anomura.—The most familiar members of this tribe are the hermit-crabs (*Paguridæ*) which occur so commonly on every shore. They are distinguished by the fact that the abdomen is quite soft, and is not protected by a chitinous crust. The animal, therefore, is compelled to protect the defenceless part of the body in some artificial manner, and this it effects by appropriating the empty shell of some dead mollusk, such as the common periwinkle or whelk. The abdomen is provided with special appendages to enable the intruder to retain firm hold of his borrowed dwelling, at the same time that he can change it at will when too small or otherwise inconvenient. The first pair of legs are developed into pretty powerful nipping-claws or chelæ, and one of them is always much larger than the other, and acts as a kind of plug, blocking up the entrance of the shell when the animal is retracted within it.

C. Brachyura.—The decapod Crustaceans included in this tribe are familiarly known as crabs, and they derive their name of *Brachyura* (G. *brachus*, short; and *oura*, tail) from

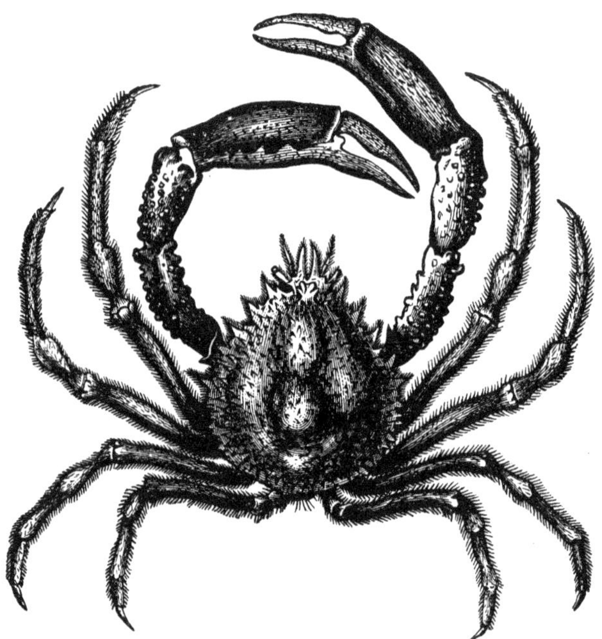

FIG. 51.—Brachyura. The Spiny Spider-crab (*Maia squinado*).

the rudimentary condition of the abdomen. The abdomen, in fact, is not only extremely short, but it is always tucked up beneath the greatly-developed cephalo-thorax, so that it is not visible at all, except when the animal is looked at from below (Fig. 51). The crabs are very various in their habits, but they are mostly denizens of the shore, hiding beneath stones or sea-weed, in cracks of rock, or in pools near the line of low water. Some of them, however, can swim with tolerable activity, and some of them (the land-crabs) even live habitually inland. One group, the "pea-crabs," is distinguished by the singular habit of living semi-parasitically within the shells of bivalve mollusks, such as the great horse-mussel.

The young or larval crab is exceedingly unlike the adult, and has a long and well-developed abdomen, thus approximating to the type of structure which is permanently retained in the *Macrura*.

ORDER ISOPODA (Gr. *isos*, equal; *podes*, feet).—In this order are a number of Crustaceans of which some inhabit the sea, others are parasitic in their habits, and others are terrestrial. The best known are the common wood-lice (*Oniscus*, Fig. 52), which are found so commonly under stones, or in the crevices of old walls. The Isopods all belong to a group of Crustaceans in which the eyes are not supported upon stalks, and they are therefore said to be "sessile-eyed." The head is distinct from the segment bearing the first pair of feet. The thoracic feet are all *similar* to one another, and the branchiæ are developed on the abdominal legs.

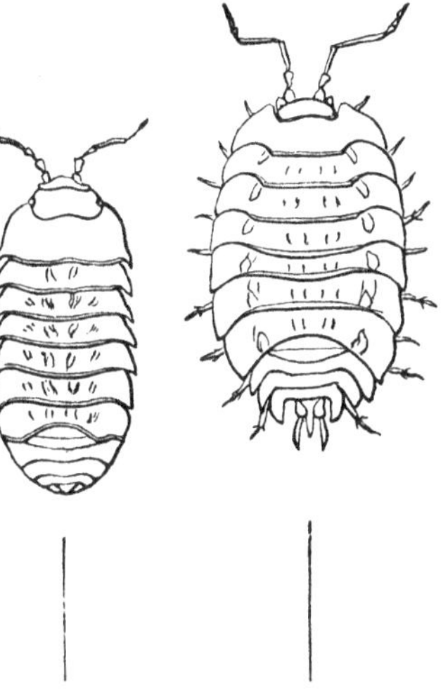

FIG. 52.—Isopoda. Wood-lice (*Oniscus*).

ORDER MEROSTOMATA.—In this order are only the living king-crabs (*Limulus*), and some large extinct forms nearly allied to them. They are all distinguished by the fact that the appendages which are placed round the mouth act by their bases as jaws, but have their extremities developed into swimming-paddles, walking-feet, or nipping-claws.

The King-crabs or Horseshoe crabs (Fig. 53) constitute a special group called *Xiphosura* (Gr. *xiphos*, a sword; and

7

oura, tail), from the fact that the end of the abdomen is furnished with a long sword-like spine (Fig. 53, *t*). The mouth is surrounded by six pairs of appendages, the bases of which are spinous and act as jaws, while their free extremities are developed into nipping-claws or chelæ. The whole of the upper surface of the body is protected by a kind of buckler, composed of an anterior semicircular shield, and a posterior somewhat hexagonal plate, the under surface of which carries

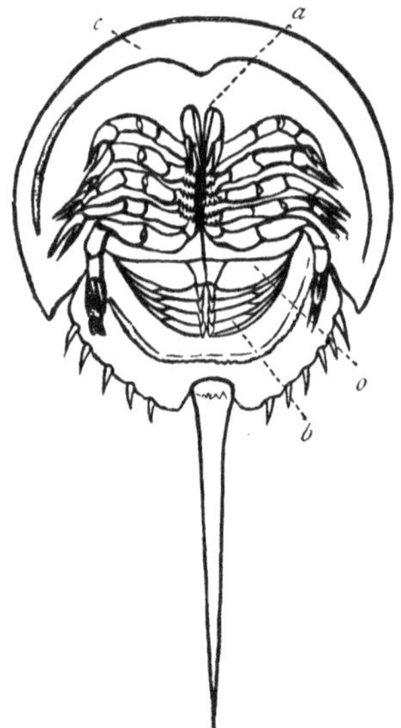

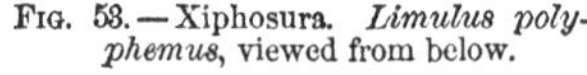

Fig. 53.—Xiphosura. *Limulus polyphemus*, viewed from below.

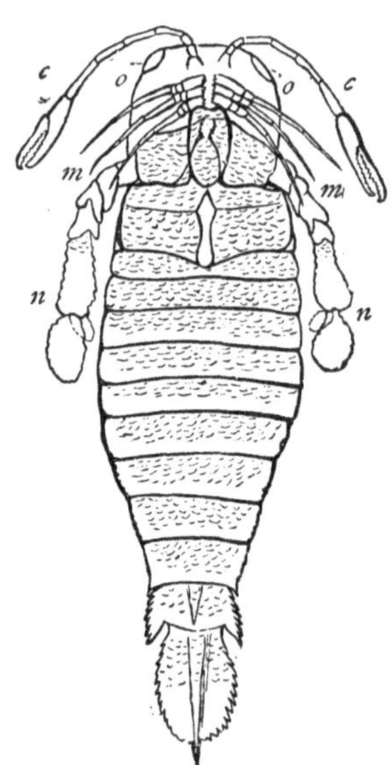

Fig. 54.—Eurypterida. *Pterygotus Anglicus*, restored (after H. Woodward).

branchial plates, while the sword-like telson is jointed to its hinder margin. The king-crabs attain a large size, and are often called "Molucca crabs" from their occurrence in the Moluccas. Both the eggs and the flesh are eaten by the Malays.

Closely allied to the king-crabs is the extinct family of the *Eurypterida*, an example of which is figured above (Fig. 54). This species is supposed to have attained a length of probably six feet, but other forms were very much smaller.

ORDER TRILOBITA. — The Trilobites constitute another wholly extinct order of the *Crustacea*, and deserve a short notice from their great geological importance. They derive their name from the fact that the body exhibits a more or less conspicuous division into a central and two lateral lobes (Fig. 55, 1). The entire shell or crust is composed of an anterior

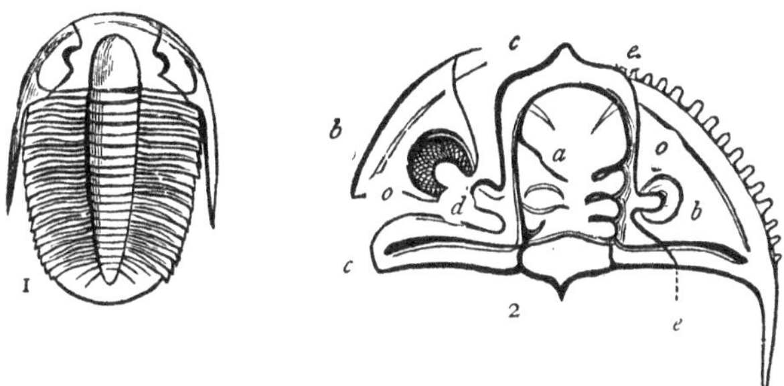

FIG. 55.—Trilobita. 1. *Angelina Sedgwickii;* 2. Diagram of the cephalic shield of a Trilobite (after Salter).

semicircular shield, covering the head (Fig. 55, 2), a series of movable rings, constituting the thorax, and a tail-piece composed of amalgamated segments, and representing the abdomen. On the under surface of the shell nothing had ever been discovered except the upper lip, but recently traces of limbs have been made out. The cephalic shield usually bears a pair of compound eyes (Fig. 55, 2 *o*), but these are sometimes wanting. It is probable that most of the Trilobites possessed the power of rolling themselves up into a ball, much as our modern wood-lice. The Trilobites are only known as occurring in the older rocks of the earth's crust, and they are chiefly characteristic of the period known to geologists as the "Silurian."

ORDERS CLADOCERA, COPEPODA, AND OSTRACODA.—These orders deserve mention more from the extreme abundance of their commoner forms than for any other reason. They include a number of minute Crustaceans, most of which are commonly called "water-fleas," and abound in fresh waters in most parts of the world. They are, however so small that, though visible to the naked eye, they can only be satisfactorily examined under the microscope. As an example of the

Cladocera may be taken the "branched-horned water-flea" (*Daphnia pulex*, Fig. 56, *b*), thousands of which may be captured in any pond in summer. In this pretty little species the whole body is enclosed in a bivalve shell, which is so transparent that the whole organization of the animal is clearly visible through it. The head is distinct, and carries a single

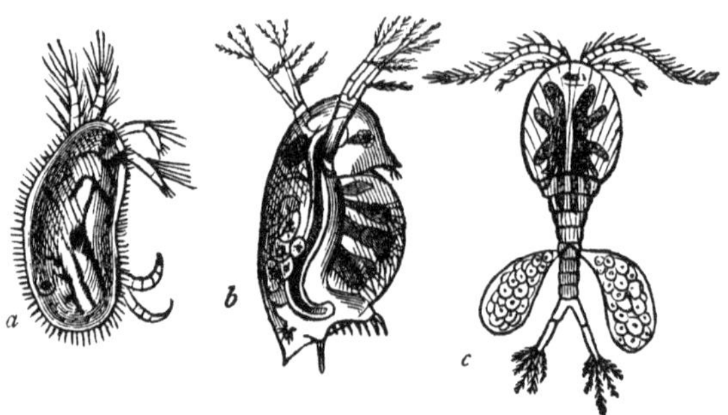

FIG. 56.—Fresh water Entomostraca. *a Cypris tris-striata; b Daphnia pulex; c Cyclops quadricornis.*

eye. The greater antennæ are branched. The males are smaller than the females, and much fewer in number; and it appears to be a well-established fact that the female, when once fertilized by the male, can not only lay eggs for the rest of her life, but can transmit the power of producing fertile ova to her young for several generations. Of the *Copepoda* one of the commonest is the *Cyclops* (Fig. 56, *c*), in which the cephalo-thorax is covered by a shield, and there is a well-developed abdomen. The female carries on either side a kind of pouch or ovisac, in which the eggs remain till they are hatched. The little *Ostracoda* (Fig. 56, *a*) are all minute Crustaceans, which occur in both fresh and salt water. They are distinguished by the fact that the body is entirely enclosed in a shell, which is made up of two lateral halves or valves. The valves of the shell are united by a membrane along the back, but can be opened below, so as to allow of the protrusion of the feet.

ORDER CIRRIPEDIA.—The last order of *Crustacea* which requires mention is that of the *Cirripedia* (Lat. *cirrus*, a curl; and *pes*, foot), comprising the so-called barnacles and acorn-shells, both extremely unlike Crustaceans to look at. All the Cirripedes are distinguished by the fact that, while they are quite free when young, and very similar to some of the little Crustaceans just described, when adult they are immovably fixed by their heads to some solid object. In this fixed condition the body and internal organs are, in most cases, protected by means of a calcareous shell, composed of many pieces, and the only part of the body which remains movable is the legs, which are constantly thrust out of the shell and again drawn in in quest of food. The *Cirripedia* were formerly described as "multivalve" shell-fish (*Mollusca*), owing to their possession of a regular calcareous shell. Two distinct types of structure are known among the *Cirripedia* (Fig. 57), constituting the two families of the barnacles (*Lepadidæ*), and the acorn-shells (*Balanidæ*).

In the barnacles (Fig. 57, *b*), the anterior end of the body

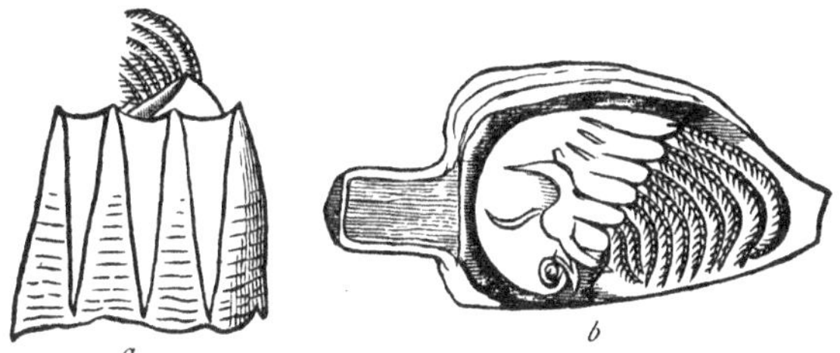

Fig. 57.—Cirripedia. *a* Sessile Cirripede (*Balanus*); *b* Stalked Cirripede (*Lepas*).

is much elongated, and is converted into a kind of stalk, by means of which the animal is attached to some solid object, such as a rock, a floating log of timber, or even some marine animal. In the acorn-shells (Fig. 57, *a*), which occur in myriads upon every solid object between tide-marks, there is no stalk, but the head is firmly cemented to the centre of a membranous or shelly plate. The body is enclosed in a limpet-shaped or conical shell, composed of several pieces, and having an aperture at its summit. This opening is closed by a movable lid, and from it the animal can protrude its delicate legs or "cirri," which look like a "glass hand," and are constantly employed in sweeping the water in search of food.

In accordance with the fixed condition of the adult, almost all the *Cirripedia* are hermaphrodite, possessing both male and female organs of reproduction. In some cases, however, males exist, but these are much smaller than the females, and quite different to them in appearance, and they spend their existence within the shell of the female.

Appendix, giving the remaining Orders of Crustacea.

Order Rhizocephala.—Minute Crustaceans, free when young, but when adult parasitically attached to the abdomen of various crabs. When adult they are completely deformed, destitute of limbs, and attached to their host by means of numerous branched tubes or roots which ramify deeply among the internal organs. *Ex. Peltogaster.*

Order Ichthyophthira.—Minute Crustaceans, free when young, but when adult parasitic upon various kinds of fishes; adult usually deformed and soft; young with eyes and swimming-feet. *Ex. Lerñæa.*

Order Phyllopoda.—Thoracic feet leaf-like and acting as branchiæ. *Ex. Apus.*

Order Læmodipoda.—Eyes sessile; abdomen rudimentary; respiration by means of little vesicles attached to the thoracic segments or legs. *Ex. Cyamus* (the whale-louse).

Order Amphipoda.—Eyes sessile; abdomen well developed; respiratory organs in the form of vesicles attached to the thoracic limbs. *Ex.* Sand-hopper (*Talitrus*); Fresh-water shrimp (*Gammárus*).

Order Stomapoda.—Eyes stalked; gills unprotected, usually suspended beneath the abdomen. *Ex.* Locust-shrimp (*Squilla*).

CHAPTER XV.

ARACHNIDA.

CLASS II. ARACHNIDA (Gr. *Arachne*, a spider).—This class includes the mites, ticks, scorpions, and spiders, and, as a whole, is very nearly related to the preceding. The *Arachnida*, however, are distinguished from the *Crustacea* by being adapted in most cases for a strictly terrestrial life, so that when any distinct breathing-organs are present these are never in the form of gills, but are always either pulmonary sacs or air-tubes (*tracheæ*). In none of the *Arachnida*, further, are there ever more than *four* pairs of legs, and *the segments of the abdomen never carry limbs of any sort.* The eyes are always sessile, and never supported upon stalks; if antennæ exist at all, they are much modified, and *the head is always amalgamated with the thorax, so as to form a cephalo-thorax.*

The integument usually produces chitine more or less abundantly, so as to constitute a resistent shell; but in some cases the skin remains permanently soft. The mouth is situated in the anterior portion of the body, and in the higher forms is furnished with a pair of prehensile jaws, called "mandibles," a pair of chewing-jaws, called "maxillæ, and a lower lip. In the scorpions an upper lip is present as well. In the true spiders each mandible terminates in a sharp movable hook, perforated by a canal which communicates with a poison-gland situated near its base. By means of this poisonous fluid the spiders kill such animals as they capture. In the scorpions the mandibles are short, and terminate in strong pincers. In them, too, the maxillæ are furnished with enormously-developed nipping-claws or chelæ.* In all the *Arachnida* the mandibles are believed to correspond to the antennæ of the

* These nipping-claws in the scorpions are produced not by the maxillæ themselves, but by two appendages to the maxillæ, which are known as "maxillary palpi."

Crustacea. In the lower *Arachnida*, such as the ticks, the organs of the mouth are modified to enable them to imbibe fluids.

The mouth in the *Arachnida* opens into a pharynx, which is of extraordinarily small diameter in the true spiders, which live simply on the juices of their prey. The intestinal canal is usually short and straight, and is continued without convolutions to the aperture of the anus. Salivary glands are also present, as well as ramified tubes which are believed to act as kidneys.

The circulation is maintained by means of a dorsal heart, which is situated above the alimentary canal. All the *Arachnida* breathe air directly, and the function of respiration is performed by the general surface of the body (as in the lowest members of the class), or by branched air-tubes termed "tracheæ," or by distinct pulmonary chambers or sacs, or, lastly, by a combination of tracheæ with pulmonary sacs. The tracheæ are essentially similar in structure and function to the breathing-tubes of the *Myriapoda* and *Insecta*, and consist of tubes, which open on the surface of the body by distinct apertures called "spiracles." The walls of the tube are prevented from collapsing by means of a spirally-coiled thread or filament of chitine, which is wound round their walls within their inner lining. The pulmonary sacs which occur in the *Arachnida* are simple chambers formed by an inversion of the skin, which constitutes a number of closely-set plates or folds. The whole of the interior of the pulmonary sacs is richly supplied with blood, and air is admitted by means of minute external openings.

The nervous system is of the regular articulate type, but the ganglia of the ventral chain are often massed together in particular situations. In no case are compound eyes present; and, when distinct organs of vision exist, these are in the form of from two to eight simple eyes.

Orders of the Arachnida.

Order I. Podosomata.—In this order are included the "Sea-spiders," which are wholly marine, and were long believed to be referable to the *Crustacea* on this account. As they have no respiratory organs of any kind, the question cannot be definitely settled, but they have no more than four pairs of legs, and would therefore seem to be properly referable to the *Arachnida.* In some forms the legs attain an extraordinary length, and contain prolongations from the stomach. They are all grotesque-looking animals, found at low water upon stones or marine plants, or parasitically attached to marine animals. One of the commonest forms is figured below (Fig. 57½, *a*).

ORDER II. ACARINA.—The most familiar members of this order are the Mites and Ticks (Fig. 57½, *b*, *c*). They are distinguished by the fact that the abdomen is amalgamated with the cephalo-thorax to form a single mass. Respiration is effected by the general surface of the body or by air-tubes (tracheæ).

The habits of the mites are extremely varied. Some are found upon different plants (Fig. 57½, *b*); others are parasitic upon water-insects when young, but swim about freely when adult (Fig. 57½, *c*); others are permanently parasitic upon

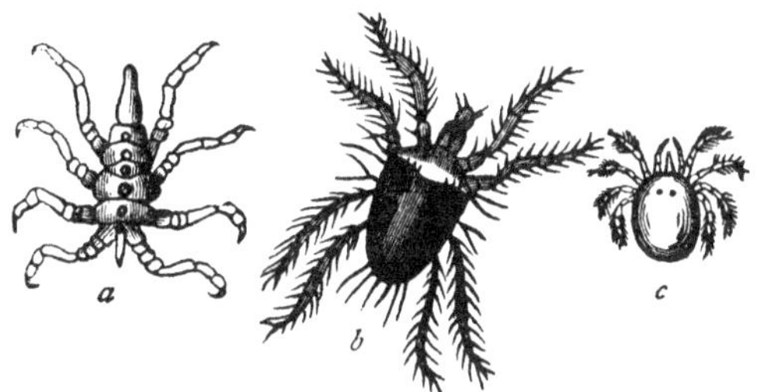

FIG. 57½.—Arachnida. *a Pycnogonum littorale; b Tetranychus telarius*, one of the "Sociable" mites; *c Hydrachna globulus*, one of the "Water-mites."

other animals, such as sheep, dogs, insects, etc.; and others inhabit decaying provisions, as is the case with the well-known "cheese-mite" (*Acarus domesticus*). Two species have a considerable medical interest as attacking man. One of these causes the skin-disease which is known as the "itch," and the other is found inhabiting certain glandular follicles of the skin, probably without an exception even in favor of the most cleanly people.

ORDER III. PEDIPALPI.—In this family are the most formidable of all the *Arachnida*—namely, the Scorpions. They are all distinguished by the fact that the abdomen is divided into distinct segments, and is continued into the cephalothorax without any well-marked boundary or constriction. In the true scorpions the end of the abdomen (Fig. 58) is composed of a hooked telson, which is perforated for the duct of a poison-gland, situated at its base. It is by means of this that the scorpions sting; and the poisonous fluid which they secrete

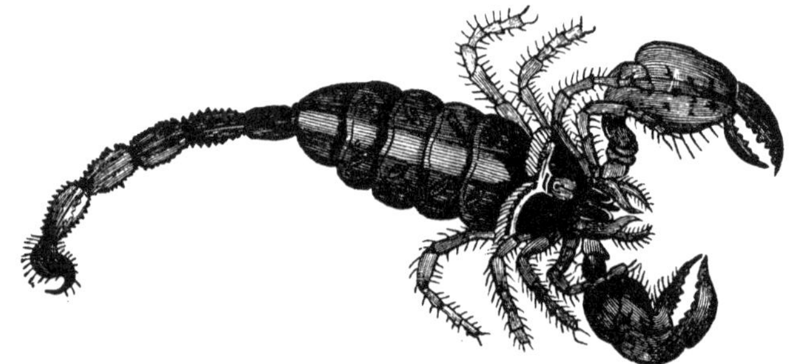

FIG. 58.—Scorpion (reduced).

is sufficiently powerful to render their wounds troublesome and painful, if not positively dangerous. The mandibles in the scorpions, as already said, are developed into pincers, and the so-called "maxillary palpi" constitute powerful nipping-claws. The respiratory organs are in the form of pulmonary sacs, four on each side, opening on the under surface of the abdomen by as many distinct apertures or spiracles.

The scorpions live in the warmer regions of the temperate zone and in tropical countries, and are generally found hiding under stones or in crevices of walls. Their sting, though much exaggerated, is certainly capable of producing very unpleasant symptoms.*

ORDER IV. ARANEIDA.—In this order are the true Spiders, readily distinguished from the insects, with which they are popularly confounded, by having four pairs of legs, as well as by other characters. In all the true spiders (Fig. 59) the segments of the thorax and head are united to form a single mass or cephalo-thorax, to which the soft and unsegmented abdomen is joined by a constricted stalk, or neck. Respiration is effected by means of pulmonary sacs, usually conjoined with tracheæ. The pulmonary sacs are two or four in number, and open on the surface of the abdomen by as many apertures.

* Nearly allied to the Scorpions are the so-called "Harvest-spiders" (*Phalangidæ*), and the diminutive "Book-scorpion" (*Chelifer*), which is commonly to be found among old books.

The head bears from six to eight simple eyes; the mandibles are hooked, and carry the duct of a poison-gland; and the maxillary palpi are not developed into nipping-claws. The spiders are all predaceous and rapacious animals, and many of

Fig. 59.—Araneida. *Theridion riparion* (female).

them possess the power of constructing webs, either for the capture of their prey, or simply for lining their habitations. For the production of the web, spiders are furnished with special glands, situated at the extremity of the abdomen. The secretion of these glands is a viscid fluid, which hardens rapidly on exposure to air, and which is cast into its proper thread-like shape by passing through what are called the "spinnerets." These are little conical or cylindrical organs placed at the end of the abdomen, and perforated by a number of extremely minute tubes, through which the secretion of the glands has to pass before reaching the air. Many spiders, however, do not construct any web, unless it be for their own habitations, but simply hunt their prey for themselves.

The spiders are oviparous, and their young pass through no metamorphosis, but they cast their skin, or "moult," repeatedly before they attain the size of the adult.

CHAPTER XVI.

MYRIAPODA.

Class III. Myriapoda (Gr. *murios*, countless; *podes*, feet).—This class is an extremely small one, and includes only the Centipedes and the Millipedes. In all the *Myriapoda the head is distinct*, and not amalgamated with the thorax. *There is no clear boundary-line between the thorax and the abdomen*, both being composed of nearly similar segments. *The body, with one exception, always consists of more than twenty rings*, and the hinder segments, which correspond to the abdomen, always carry locomotive appendages, whereas the abdominal rings in *Arachnida* and *Insecta* are always destitute of locomotive appendages. *One pair of antennæ is present, and the number of the legs is always more than eight pairs.* Respiration is carried on by branched air-tubes or tracheæ.

In most of their characters the *Myriapoda* closely resemble the true insects, with which, indeed, they are not uncommonly classed. The true insects, however, always have the head, thorax, and abdomen, distinct from each other, and have never more than three pairs of legs. In most of the *Myriapoda* the young, or "larvæ," are more like insects than the adult, since they have only three pairs of legs, or are altogether destitute of feet. In some cases, however, the young Myriapod, on escaping from the egg, possesses nearly all the characters of the parents, except that the number of body-rings, and consequently of legs, is smaller, and increases with every change of skin ("moult"). The class is divided into two leading families, represented by the common Centipedes and Millipedes.

The Centipedes (Fig. 60) are carnivorous in their habits, and the organs of the mouth are adapted for a life of rapine.

In addition to the parts of the mouth proper, they have two pairs of "foot-jaws," of which the second is hooked and perforated for the discharge of a poisonous fluid. The bite of the common European species is perfectly harmless to man, but some of the tropical forms attain a length of a foot or more, and are consequently able to inflict extremely severe and even

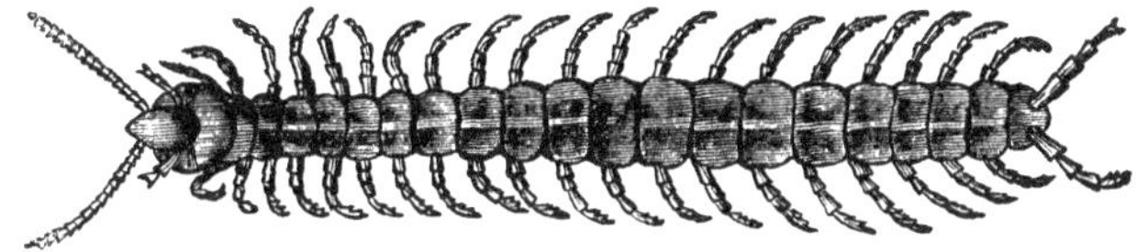

FIG. 60.—Centipede (*Scolopendra*).

dangerous bites. The true centipedes are further distinguished by the number of legs not being indefinitely great (usually from fifteen to twenty pairs), and by the fact that the antennæ are composed of not less than fourteen joints each.

The Millipedes (Fig. 61) are repulsive-looking but perfectly

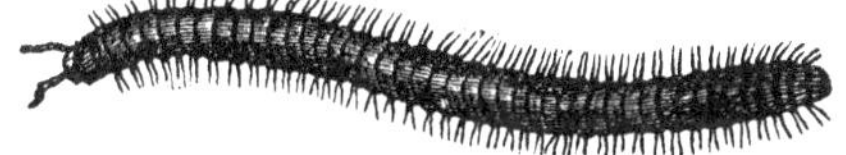

FIG. 61.—Millipede (*Iulus*).

innocent animals, which feed principally upon decaying vegetable matter. The body, in the ordinary millipedes, is rounded and worm-like, and the segments are so amalgamated that each apparent body-ring gives origin to *two* pairs of minute, thread-like feet. The mouth is destitute of the powerful jaws which are found in the centipedes, the legs are indefinitely numerous, and the antennæ are short, and are composed of no more than six or seven joints each.

The European millipedes are all of small size, but an American species is stated to attain a length of more than half a foot.

A third family has been established for a curious little creature called *Pauropus*. In this the body consists of only ten segments, and there are no more than nine pairs of legs. The antennæ are five-jointed, forked, and provided with jointed appendages. There are no tracheæ, and respiration is carried on by the skin. It is very small, and is found inhabiting decayed leaves and damp situations.

CHAPTER XVII.

INSECTA.

Class IV. Insecta.—The true Insects are distinguished from the preceding classes of articulate animals by the fact that *the three divisions of the body, namely, the head, thorax, and abdomen, are always distinct from one another; there are never more than three pairs of legs in the adult, and these are borne upon the thorax; the abdomen is destitute of locomotive appendages.* Respiration is effected by means of air-tubes or tracheæ, and, in most insects, two pairs of wings are developed from the back of the second and third segments of the thorax.

The integument in insects is more or less hardened by the deposition of chitine in it, and the body is deeply cut into segments—hence the name *Insect* (from the Latin *insectus*, cut into). The head in insects (Fig. 62, *a*) is composed of several segments amalgamated together, and carries a pair of jointed feelers or antennæ, a pair of eyes, usually compound, and the appendages of the mouth. The thorax in insects (*b*, *c*, *d*) is composed of three segments, which are amalgamated together, but are generally pretty easily recognized. Each of these segments of the thorax carries, in perfect insects, a single pair of jointed legs, so that there are three pairs in all. To the back of the two hinder segments of the thorax, in most insects, there are also attached two pairs of wings. In their typical form, the wings are membranous expansions, supported by more or less numerous hollow tubes, known as the "nervures." One or both pairs of wings may be wanting, and when all are present the anterior pair may be much modified by the deposition of chitine in it. These modifications will be treated of in speaking of the orders of insects. The abdomen in insects (*e*) is properly composed of nine segments,

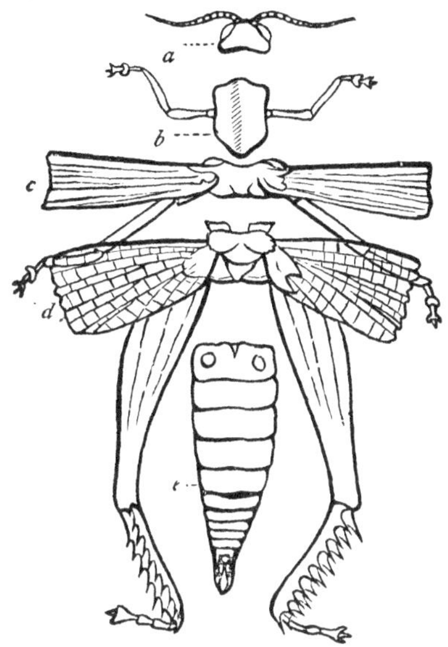

FIG. 62.—Diagram of the external anatomy of an insect: *a* Head, carrying the eyes and antennæ; *b* First segment of the thorax, with the first pair of legs; *c* Second segment of the thorax, with the second pair of legs and the first pair of wings; *d* Third segment of the thorax, with the third pair of legs and the second pair of wings; *e* Abdomen, without limbs, but carrying terminal appendages concerned in reproduction.

which are usually more or less freely movable upon one another, and which never carry locomotive limbs, as is so commonly the case in the *Crustacea*. The extremity of the abdomen is, however, often furnished with appendages which are primarily connected with reproduction, but which are often converted into weapons of offence and defence. Of this nature are the "ovipositors" of ichneumons, the stings of bees and wasps, and the forceps of the common earwig.

The organs of the mouth in insects require a brief consideration, as being in the closest possible relation with their habits and mode of life. Two chief types of mouth are recognizable in insects, termed respectively the "masticatory" and "suctorial," according as the mouth is fitted for biting and chewing, or simply for imbibing fluids. The masticatory mouth is seen in perfection in the beetles, in which the following organs are present: 1. An upper lip or "labrum" attached below the front of the lead. 2. A pair of biting-jaws or "mandibles." 3. A pair of chewing-jaws or "maxillæ" provided with jointed filaments, called the "maxillary palpi." 4. A lower lip or "labium" which also carries a pair of jointed filaments, known as the "labial palpi." In the typical suctorial mouth, as seen in the butterflies and moths, the following is the arrangement of parts: The upper lip and mandibles are quite rudimentary; the maxillæ are greatly lengthened, and form a spiral tube fitted for sucking up the juices of flowers; and the labial palpi are much developed, and form two hairy cushions between which the trunk can be coiled up when not in use. In many insects, the organs of the mouth are essentially adapted for suction, but are also fitted for piercing solid substances, such as the skin of animals or the stems of plants. In these the lower lip forms a kind of sucking-tube or sheath, within which are contained the maxillæ and mandibles, which are modified so as to form piercing organs or lancets. In the common bee, the masticatory and suctorial types of mouth are combined. The mandibles or biting-jaws are retained, to enable the honeycomb to be manufactured, and there is also a tubular trunk fitted for sucking up the juices of flowers. In the butterflies, too, in which the mouth of the adult is strictly adapted for suction, the caterpillar is furnished with a masticating mouth, so that it can feed upon leaves or other solid substances.

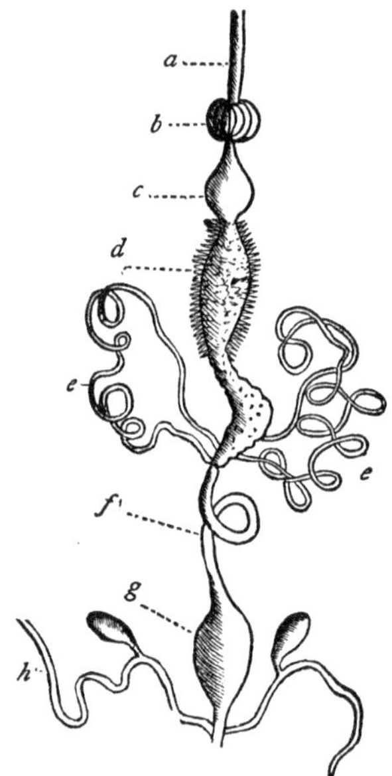

FIG. 63.—Digestive apparatus of a Beetle (*Carabus auratus*). *a* Gullet; *b* Crop; *c* Gizzard; *d* Chylific stomach; *e* Malpighian tubes; *f* Intestine; *g* Cloaca; *h* Renal vessels.

The mouth in the masticating Insects (Fig. 63, *a*) leads into a membranous and often folded cavity, termed the "crop" (*b*), from which the food passes to a second muscular cavity or "gizzard" (*c*). The gizzard is adapted for crushing the food, and often has plates or teeth of chitine developed in its walls. It is succeeded by the true digestive cavity (*d*), which is termed the "chylific stomach." From this there proceeds an intestine (*f*), of variable length, which usually terminates in a chamber (*g*) called the "cloaca" (Lat. *cloaca*, a sink), into which the ducts of the reproductive organs open. The commencement of the gullet is furnished with glandular appendages, which are believed to discharge the functions of salivary glands. Immediately behind the posterior aperture of the stomach are a variable number of cæcal convoluted tubes (*e*), which are known as the "Malpighian vessels," after their discoverer Malpighi, and which are generally looked upon as representing the liver. Close to the cloaca may be other tubes, which are believed, from their position, to exercise the functions of kidneys (*h*).

The circulation in insects is mainly carried on by a long, contractile tube, placed along the back, and termed the "dorsal vessel." The blood, collected from the various tissues and organs of the body, enters the dorsal vessel from behind, and is driven forward to the anterior extremity of the body. Respiration is effected by means of air-tubes or tracheæ, which commence at the surface by so many apertures or spiracles, and branch repeatedly as they proceed inward through the tissues. They have essentially the same structure as in the *Arachnida*, consisting of membranous tubes strengthened by means of a spirally-coiled filament of chitine.

The nervous system in insects, though sometimes somewhat modified, has essentially the regular annulose form of a ventral chain of ganglia, traversed in front by the gullet. The organs of sense are the eyes and antennæ. The eyes are usually "compound," and are composed of numerous six-sided lenses, united together, and each supplied by a separate nervous filament. As many as eight thousand of these lenses have been counted in one of the eyes of the common cockchafer, and this number is sometimes greatly exceeded. Besides these compound eyes there are sometimes "simple" eyes, identical in structure with the single lenses of the compound eyes; and in rare cases these are the only organs of vision. The feelers or antennæ, with which all insects are furnished, are jointed filaments attached close to the eyes, and assuming very different shapes in different insects. They appear to be certainly organs of touch, but they very probably minister to other senses as well, and there is some reason to suppose that they are connected with the sense of hearing in particular.

The sexes in insects are distinct, and most of them are

oviparous. Generally speaking, the young insect is extremely different in external character from the adult, and it requires, before reaching maturity, to pass through a series of changes which collectively constitute what is called the "metamorphosis. In some insects, however, there is no proper metamorphosis, and in some the changes which take place are not so complete and striking as in others. By the absence of metamorphosis, or by its completeness when present, insects are divided into three convenient, though perhaps not strictly natural, sections, as follows:

Section I. Ametabolic Insects.—The insects belonging to this section are said to be "Ametabolic" (Gr. *a*, without; *metabole*, change), because they pass through no metamorphosis. The young, on their escape from the egg, resemble the adult in every respect except in size, and they undergo no alteration in reaching maturity, except that they grow larger. All the insects of this section are destitute of wings in the adult state, and they are therefore often called "Aptera" (Gr. *a*, without; *pteron*, a wing).

Section II. Hemimetabolic Insects.—In the insects belonging to this section (Gr. *hemi*, half; and *metabole*, change), there is a metamorphosis consisting of three stages, but these stages do not differ much from one another in appearance. The young, on escaping from the egg, is known as the "larva," and it is not only much smaller than the adult, but is destitute of wings. After several changes of skin, the larva enters into the second stage, when it is termed the "pupa." The pupa is active and locomotive, and rarely differs much from the larva, except that it is bigger, and rudimentary wings have now appeared on the back of the thorax. After a certain period, and after some changes of skin, the wings burst from their sheaths, and the pupa is now converted into the third and final stage, when it is known as the "imago" or perfect insect. In all the insects belonging to this section—such as grasshoppers, dragon-flies, etc.—the second stage or pupa is active and locomotive; and for this reason the metamorphosis is said to be "incomplete.'

Section III. Holometabolic Insects (Gr. *holos*, entire; and *metabole*, change).—The insects belonging to this section—such as butterflies, moths, and beetles—pass through three stages, just as do the preceding, but these stages differ from one another very much in appearance, and the metamorphosis is therefore said to be "complete" (Fig. 64). In these insects the larva is worm-like, segmented, and usually furnished with locomotive feet, which do not correspond with the three pairs proper to the adult (see Fig. 71, *b*), though these are usually present as well. The larva is also provided with masticating organs,

FIG. 64.—Metamorphosis of the Magpie-moth (*Phalæna grossulariata*).

and eats voraciously. In this stage of the metamorphosis, the larvæ constitute what are popularly known as "caterpillars" or "grubs." Having remained in this condition for a longer or shorter time, and having undergone repeated changes of skin, necessitated by its rapid growth, the larva passes into the second stage, and becomes a pupa (Fig. 64—see also Fig. 71). In this stage the insect remains quiescent, unless irritated, and it is very often attached to some foreign object, so as to be quite incapable of changing its place. In the case of the butterflies and moths, the pupa constitutes what is so familiarly known as the "chrysalis." The body is protected by a chitinous pellicle, and in some cases this is still further protected by the dried skin of the larva; while in other cases the larva—immediately before entering the pupa stage—spins for itself a protective case of silken threads, which surrounds the chrysalis, and is known as the "cocoon." Having remained for a variable time in this inanimate, quiescent pupa-stage, during which rapid changes have been going on in the interior of the animal, the insect now frees itself from the envelope which obscured it, and appears as the perfect winged adult or imago.

CHAPTER XVIII.

ORDERS OF INSECTS.

THE known number of insects is so enormous, their forms are so various, and their habits and instincts are not only so remarkable but have been so fully described, that it were hopeless to attempt here to do more than give the briefest possible outline of the leading characters which distinguish the different orders. The student desirous of further information on this head must have recourse to treatises specially devoted to entomology.

SECTION I. AMETABOLIC INSECTS.—*Young not passing through a metamorphosis, and differing from the adult in size only. Perfect insect (imago) destitute of wings; eyes simple, sometimes wanting.*

ORDER I. ANOPLURA (Gr. *anoplos*, unarmed; *oura*, tail). —The insects comprised in this order are parasitic upon man and other animals, and they are commonly known as Lice. They are all very minute in size, destitute of wings in the adult state, having a mouth formed for suction, and having either two simple eyes or none.

ORDER II. MALLOPHAGA (Gr. *mallos*, a fleece; *phago*, I eat).—These are known as "Bird-lice," and are all minute parasites on different birds. They are distinguished from the true lice by not living upon the juices of their host, but upon the more delicate and tender appendages of the skin. The mouth is, consequently, not suctorial, but fitted for biting.

ORDER III. THYSANURA (Gr. *thusanoi*, fringe; *oura*, tail). —The most familiar members of this order are the "Springtails" (*Poduræ*), which are commonly found under stones or in cellars and such like situations. They are distinguished by having the extremity of the abdomen furnished with bristles, by the sudden straightening of which the insect can effect powerful leaps. In many cases the body is covered with delicate scales which form beautiful objects under high powers of the microscope.

SECTION II. HEMIMETABOLIC INSECTS. — *Metamorphosis incomplete; the larva differing from the perfect insect chiefly in the absence of wings and in size; pupa usually active, or, if quiescent, capable of movement.*

ORDER IV. HEMIPTERA (Gr. *hemi*, half; *pteron*, wing).— In this order the mouth is formed for suction; the eyes are compound, but simple eyes are often present in addition. Two pairs of wings are always present.

The *Hemiptera* live upon the juices of plants or animals, which they are enabled to obtain by means of their suctorial mouths. All the four wings are generally present, but the condition of these varies in different sections of the order. In one group all the four wings are membranous (Fig. 65); but in the other the posterior wings and the tips of the anterior

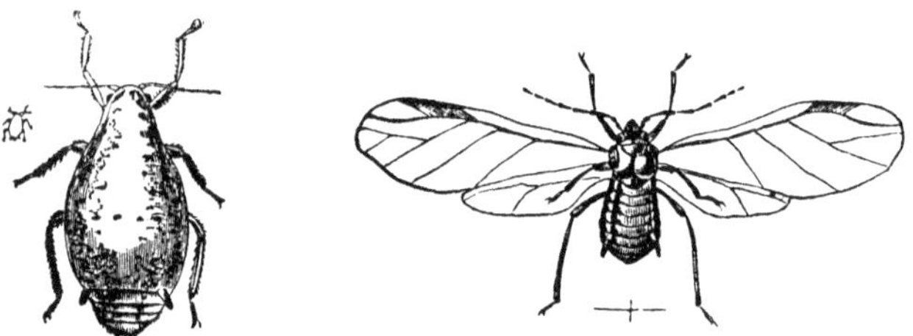

FIG. 65.—Hemiptera. Bean Aphis (*Aphis fabæ*); winged male and wingless female.

wings alone are membranous. The inner portions of the anterior wings are hardened by chitine, and they are known as "hemelytra" (Gr. *hemi*, half; and *elutron*, a sheath). Among the more familiar examples of the *Hemiptera* are the numerous species of Plant-lice (*Aphides*), the Field-bug (*Pentatoma*), the Boat-fly, the Cochineal insects, and the Cicadas.

The Cochineal insects are of considerable commercial importance, as the dried and powdered bodies of the female constitute the substance known as Cochineal, from which is obtained the brilliant pigment carmine. The male insect is winged, and is smaller than the female, which is wingless. They live upon different species of Cactus (*Opuntia*,) and are mainly imported from Mexico, Algeria, and the Canary Islands.

Numerous species of *Aphides* or Plant-lice (Fig. 65) are known, and they are among the greatest pests of the gardener and farmer, as they are extraordinarily prolific, and live upon the juices of plants. One of the most curious points about the Plant-lice is that they secrete a sweet and sticky fluid, which is expelled from the body by two little tubular filaments placed near the end of the abdomen. Ants are excessively fond of this fluid, and hunt after *Aphides* in all directions in order to obtain it; and it is a well-established fact that the Plant-lice are actually pleased with this, and voluntarily yield up the coveted fluid to the importunity of the ants.

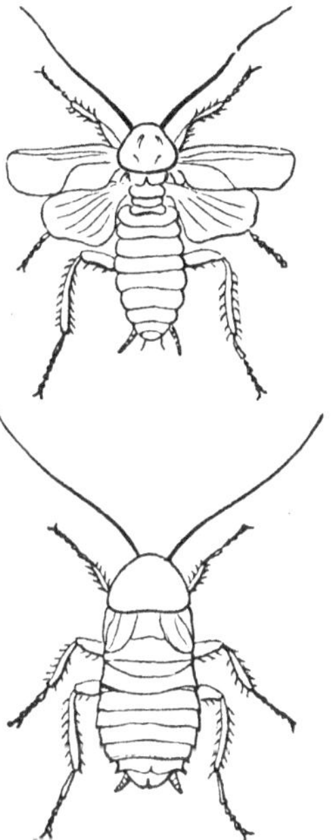

FIG. 66.—Orthoptera. The common Cockroach (*Blatta orientalis*), male and female.

ORDER V. ORTHOPTERA (Gr. *orthos*, straight; *pteron*, wing).—The mouth in this order is strictly masticatory; there are four wings present in most, but the anterior pair is smaller than the posterior, and of a different texture. The posterior wings are membranous, and are folded lengthwise, like a fan; the anterior wings are leathery, and constitute cases for the posterior wings (elytra). This order includes the Crickets (*Achetina*), Grasshoppers (*Gryllina*), Locusts (*Locustina*), Cockroaches (*Blattina*, Fig. 66), and others. Some of them are formed for running, all the legs being nearly equal in size; others have the first pair of legs greatly developed, and constituting powerful organs of prehension; while others, such as the Locusts and Grasshoppers, have the hindmost pair of legs much longer than the others, giving them a considerable power of leaping. All the *Orthoptera* are extremely voracious, and every one is acquainted with the terrible ravages occasionally caused in hot countries by swarms of locusts.

The most destructive species is the Migratory Locust (*Acrydium migratorium*, Fig. 67), which is very abundant in Africa, India, and throughout the whole of the East. Owing to the rapidity with which they devour every thing they can possibly eat, and owing to their enormous numbers, the Locusts are compelled to be constantly on the move, looking for "fresh fields and pastures new." It is from these migrations

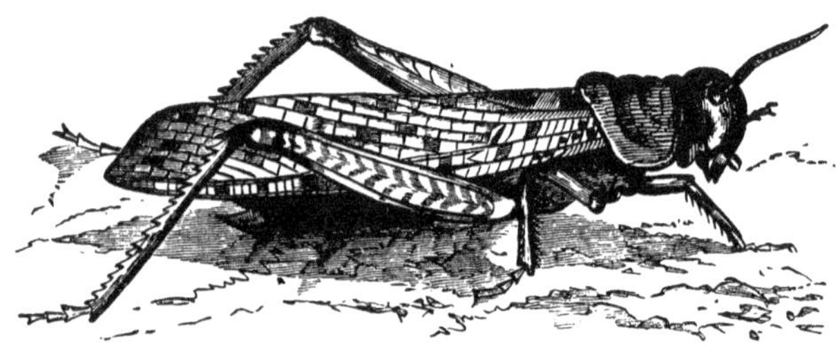

FIG. 67.—The Migratory Locust (*Acrydium migratorium*).

in vast bodies in search of food that the Migratory Locust takes its name. When one of these destructive hosts visits a district, it only needs a few hours to convert the most fertile country into a howling wilderness. In an incredibly short space of time, every green thing on their line of march is destroyed, every leaf is stripped from every tree, every blade of grass and corn is eaten down, and it is not until the ground is utterly bare and brown that the locusts take wing and seek out some fresh region to devastate.

ORDER VI. NEUROPTERA (Gr. *neuron*, nerve; *pteron*, wing).—The mouth in this order is fitted for mastication; the wings are four in number, generally nearly equal in size, all membranous, and traversed by numerous delicate nervures, which interlace so as to form a delicate net-work (Fig. 68). The metamorphosis is generally incomplete, but is sometimes complete.

This order includes the well-known and rapacious Dragonflies (*Libellulidæ*), the Caddis-flies (*Phryganeidæ*), the May-flies (*Ephemeridæ*), the Ant-lion (*Myrmeleo*), the Aphis-lion (Fig. 68), Termites, etc. The last of these—namely, the Termites or white ants—are social insects, living in organized communities, and exhibiting many remarkable phenomena. They are mostly inhabitants of hot countries, and cause immense mischief by destroying wood-work of all descriptions.

Though called "white ants," it is to be remembered that they are not related in any way to the true ants. They build mounds of different shapes and sizes, sometimes several feet in height, formed of "particles of earth worked into a material as hard as stone." Each family of Termites (Fig. 69) possesses a king and queen, which are always kept together closely guarded in a chamber placed in the centre of the nest. The king (Fig. 69, *a*) and the queen (*b*) both originally possessed wings, but they lose these as soon as they found a colony. Both are much larger than the bulk of the community, the queen immensely so, owing to the enormous distention of her abdomen with eggs. The ordinary Termites are all sterile females, incapable of laying eggs, and they are divided into two distinct sets or "castes," both destitute of wings, and differing in the armature of the head. The one caste includes the so-called "workers," who perform all the ordinary work of the colony, while the "soldiers" have greatly-developed jaws, and are simply occupied in defending the nest against all enemies.

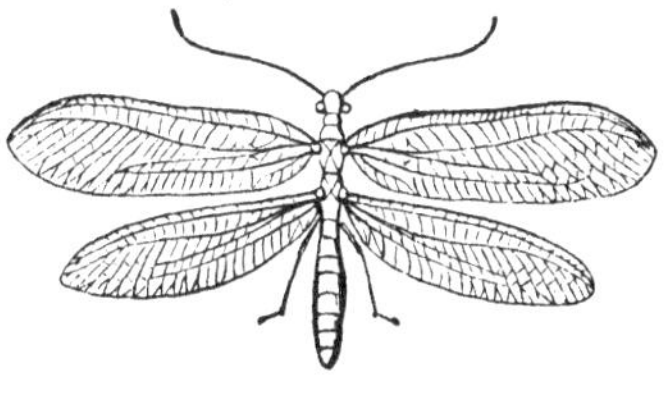

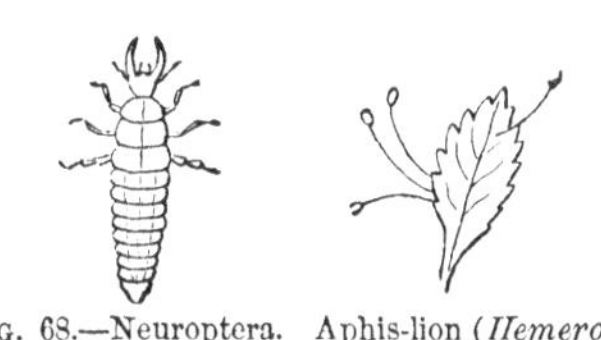

Fig. 68.—Neuroptera. Aphis-lion (*Hemerobiidæ*), imago, larva, and eggs.

Fig. 69.—Termites (*Termes bellicosus*); *a* King, before the wings are cast off; *b* Queen, with the abdomen distended with eggs; *c* Worker; *d* Soldier.

SECTION III. HOLOMETABOLA.—*Metamorphosis complete; the larva, pupa, and imago, differing greatly from one another in external appearance. The larva worm-like, and the pupa quiescent.*

ORDER VII. APHANIPTERA (Gr. *aphanos*, inconspicuous; *pteron*, wing).—In this order are only the Fleas (*Pulicidæ*), in which the mouth is suctorial, the metamorphosis is complete, and the wings are rudimentary, being represented by four minute scales placed on the last two segments of the thorax. The larva of the common flea is a footless grub, which in about twelve days spins a cocoon for itself, and becomes a quiescent pupa, from which the imago emerges in about a fortnight more.

ORDER VIII. DIPTERA (Gr. *dis*, twice; *pteron*, wing).—The insects of this order, as implied by its name, have only a single pair of wings—namely, the anterior pair. The posterior wings are rudimentary, and are represented by two

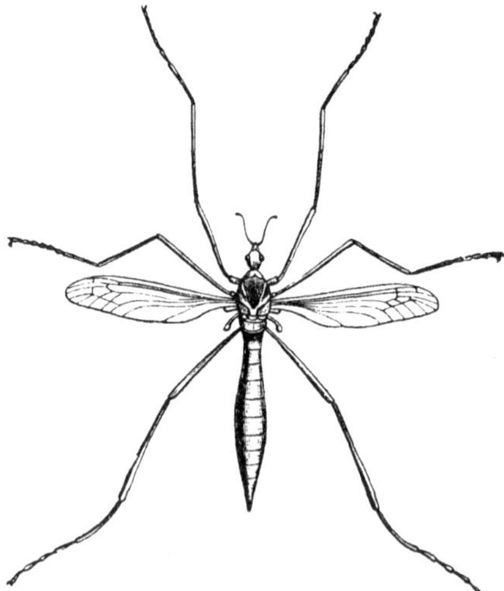

FIG. 70.—Diptera. Crane-fly (*Tipula oleracea*).

clubbed filaments called "balancers" or "poisers" (Fig. 70). The mouth in the *Diptera* is suctorial.

The *Diptera* constitute one of the largest orders of insects; the House-flies and Flesh-flies (*Musca*), the Gnats (*Culex*), the Crane-flies (*Tipula*), the Forest-flies (*Hippobosca*), and the Gad-flies (*Tabanidæ*), constituting good examples.

ORDER IX. LEPIDOPTERA (Gr. *lepis*, scale; *pteron*, wing).—This well-known and most beautiful of all the orders of insects comprises the Butterflies and Moths, the former being active by day (*diurnal*), and the latter mostly toward twilight (*crepuscular*), or at night (*nocturnal*). In all the *Lepidoptera* the mouth of the adult insect is purely suctorial, and is provided with a spiral trunk fitted for imbibing the juices of flowers. The wings are four in number, and are covered more or less completely with modified hairs or *scales*, which are pretty objects under the microscope, and from which the wings derive their beautiful colors. The larvæ of the *Lepidoptera* (Fig. 71) are generally known as caterpillars. They are worm-like, provided with masticatory organs fitted for dividing solid

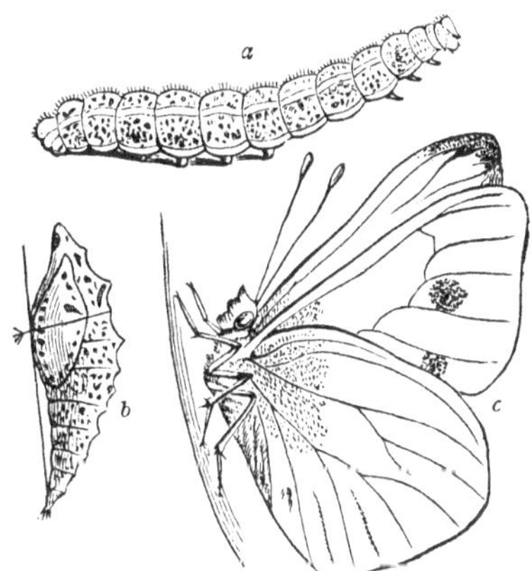

FIG. 71.—Large white Cabbage-butterfly (*Pontia brassicæ*). *a* Larva or caterpillar; *b* Pupa or chrysalis; *c* Imago or perfect insect.

substances, possessing false legs in addition to the three pairs proper to the adult, and having attached to the under lip a tubular organ or spinneret, by which silken threads can be manufactured.

The butterflies or diurnal *Lepidoptera* are characterized by being active during the daytime, by keeping their wings mostly erect when at rest (Fig. 71, *c*), by having club-shaped antennæ, and by having a chrysalis (*b*), which is almost always naked and angular, and is generally attached to some solid object by silken threads variously disposed.

The Moths are mostly active during the night-time, when they are said to be nocturnal. Many of them, however, are "crepuscular"—that is to say, they are active during the hours of twilight; and a few come out in broad daylight and in the brightest sunshine. The pupæ, or chrysalides, are never angular, as in the case of the butterflies.

Apart from the destruction committed by the Caterpillars of some of the *Lepidoptera*, the only members of the order which are of importance to man are the various species of *Bombyx*, from which *silk* is derived. Several species are cultivated for this purpose, but by far the most valuable is the common Silk-moth (*Bombyx Mori*), which owes its name to the fact that the larva feeds upon the leaves of the common Mulberry (*Morus nigra*). It is hardly necessary to say that raw silk is derived from the "cocoon," or silken case in which the caterpillar enwraps itself before becoming a chrysalis. Most of the raw silk is derived from France, Italy, China, and the East Indies.

ORDER X. HYMENOPTERA (Gr. *humen*, membrane; *pteron*, wing).—In this order all the four wings are present, as a rule, and they are all membranous in texture, with few nervures (Fig. 72). The mouth is always furnished with biting-jaws or mandibles, but often is adapted for suction as well. The females have the extremity of the abdomen furnished with an instrument connected with the process of laying eggs (ovipositor); and in very many cases this becomes the powerful defensive weapon known as the sting. The metamorphosis is complete.

The *Hymenoptera* form a very extensive order, comprising the Bees (*Apidæ*), the Wasps (*Vespidæ*), the Ants (*Formicidæ*), the Saw-flies (*Tenthredinidæ*, Fig. 72), and the Ichneumons. The Bees and Wasps are well known as forming social communities, though solitary members of both are not uncommon. In both groups these organized communities consist of a vast number of undeveloped females, or "neuters"—the so-called "workers"—presided over by a single fertile female, or "queen," or containing several such. The males are only produced at certain seasons, and they constitute the so-called "drones" of a hive of bees. The workers discharge all the duties necessary for the preservation of the colony, such as procuring food, building the nest, and feeding the young. As there is only one set, or "caste," of neuters, the duty of

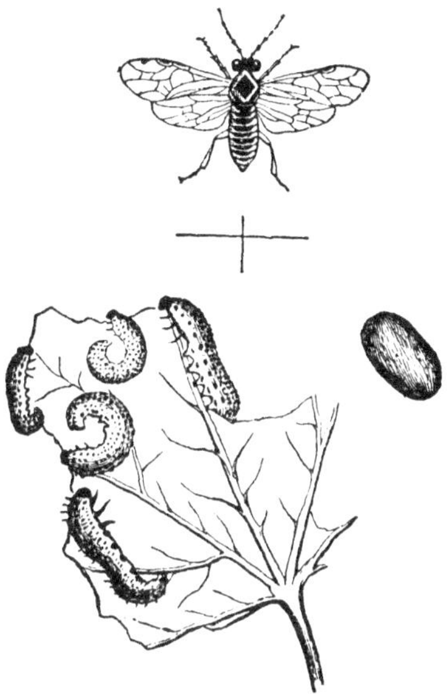

Fig. 72.—Gooseberry Saw-fly (*Tenthredo grossulariæ*), larva, pupa, and imago.

defending the nest falls to the lot of all the workers, and is not delegated to a special class of soldiers. The queen is the founder of the colony, and her sole function, after starting the community, is to lay eggs. The drones, or males, do no work, as a rule, and they either die, or are killed by the workers, as soon as the female is fertilized.

The Ants likewise form communities, consisting of males, females, and neuters. The males and females, like those of the very different "White Ants," or Termites, are winged (Fig. 73, *a*), and are produced in great numbers at particular times of the year. They then quit the nest and pair, after which the fecundated females lose their wings and form fresh societies. The workers (Fig. 73, *b*) are sometimes all of one kind, but they are often divided into two, or even three, distinct classes or "castes." The Ants exhibit many most extraordinary and interesting instincts and habits, of which

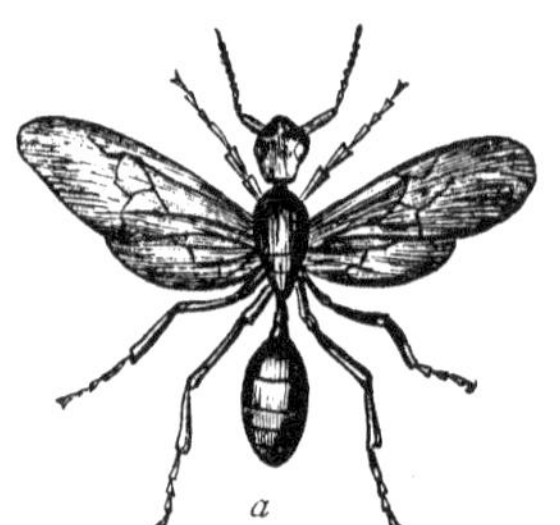

FIG. 73.—The Red Ant (*Myrmica rufa*). *a* Winged male; *b* Wingless female. Magnified.

their custom of "milking" the little Plant-lice has been already mentioned. Another very singular habit of some Ants may be just alluded to—their habit, namely, of capturing the pupæ of other species of Ants and bringing them up as slaves. The relations, however, between the masters and slaves vary a good deal in different cases. In the case, for instance, of the Russet Ant (*Formica rufescens*) the masters are so entirely dependent upon their slaves that they cannot even feed themselves, and the only work which they perform unassisted is the capturing of fresh slaves. In the Blood-red Ant (*Formica sanguinea*), on the other hand, the slaves are much fewer in number, and the masters are much less dependent upon their good offices. In all cases, the slaves exhibit the greatest devotion to their masters, and are invariably taken the greatest care of by their captors.

ORDER XI. STREPSIPTERA (Gr. *strepho*, I twist; *pteron*, wing).—This is an extremely small order of insects, which merely requires to be mentioned. It includes only certain minute parasites, which are found on bees and other *Hymenoptera*. The females are destitute of wings or feet, and are merely soft, worm-like grubs. The males are active, and possess a single pair of large membranous wings. Unlike the *Diptera*, it is the *posterior* pair of wings which is present, and the *anterior* pair is quite rudimentary, and is only represented by curious twisted filaments, from which the name of the order is derived.

ORDER XII. COLEOPTERA (Gr. *koleos*, a sheath; *pteron*, wing).—The twelfth and last order of insects is that of the *Coleoptera*, including the well-known insects familiar to every one under the name of "beetles." The leading peculiarity of the *Coleoptera* is to be found in the fact, that though all the four wings are present, only the posterior pair are membranous, and perform the function of wings. The anterior pair of wings are no longer capable of being used in flight, but are hardened by the deposition of chitine, and constitute pro-

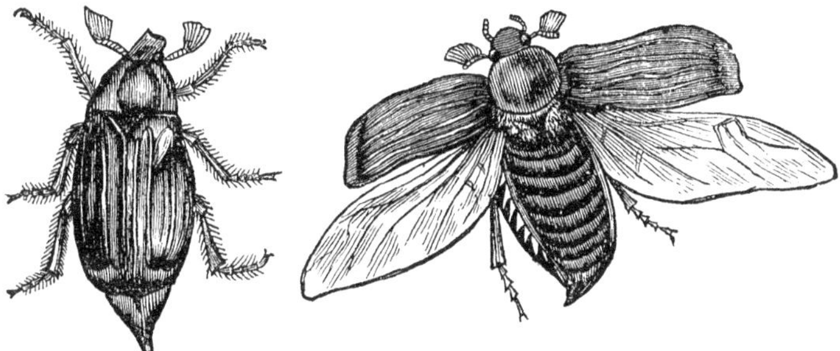

FIG. 74.—Coleoptera. The common Cockchafer (*Melolontha vulgaris*), with the elytra closed, and in flight.

tective cases, which cover the hind-wings, and are known as "elytra" (Gr. *elutron*, a sheath). The mouth in all the beetles is masticatory, and is furnished with biting and chewing jaws.

The larvæ of the beetles are all worm-like grubs, with masticatory mouths, and they all pass through a complete metamorphosis, generally requiring a protracted period for its completion. The known number of different kinds of beetles cannot be estimated with any certainty, but it is probably little short of 50,000 species, and this estimate has been doubled by some writers. They are, as a general rule, remarkable for their hard, chitinous skin, their glittering, often metallic, colors, and their voracious habits, though many of them feed upon vegetable matters.

Of the enormous number of known Beetles, the only one which can be said to be of any decided use to man is the so-called "Blister-beetle," or "Spanish Fly" (*Cantharis vesicatoria*). This handsome insect is a native of Southern Europe, especially of Italy, Spain, and France, and lives upon the leaves of the ash, lilac, elder, and poplar. It is largely collected and exported for medicinal purposes, as it yields one of the most generally used and efficient of blisters.

SUB-KINGDOM V.—MOLLUSCA.

CHAPTER XIX.

SUB-KINGDOM MOLLUSCA. — The *Mollusca* (Lat. *mollis*, soft), as implied by their scientific name, are mostly soft-bodied animals, but their popular name of "shell-fish" expresses the fact that their soft body is usually protected by an external skeleton or "shell." All the *Mollusca* are furnished with a distinct alimentary canal, which is completely shut off from the general cavity of the body (Fig. 75, *a*). There is sometimes no distinct blood-circulatory apparatus; but, when there is, its central portion (i. e., the heart) is placed upon the dorsal aspect of the body. The chief peculiarity, however,

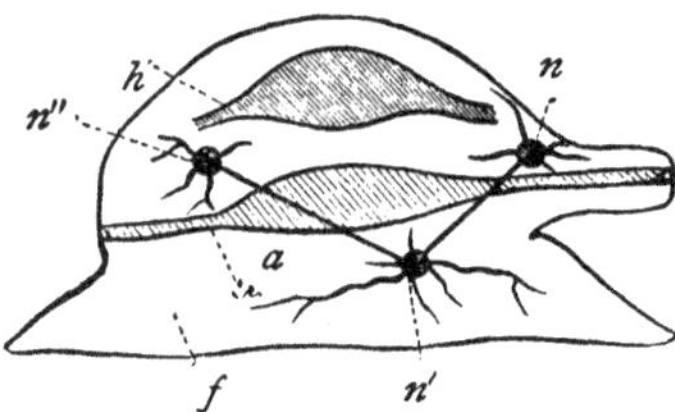

FIG. 75.—Diagram of a Mollusk. *a* Alimentary canal; *h* Heart; *f* Foot; *n* Cerebral ganglion; *n'* Pedal ganglion; *n''* Parieto-splanchnic ganglion.

of the *Mollusca* is found in the nature of the nervous system. In the lower forms (Fig. 76, 2 *d*), the nervous system consists essentially of a single ganglionic mass, giving off filaments in various directions. In the higher *Mollusca* (Fig. 75, *n*), the nervous system consists of three scattered ganglia, united to one another by nervous cords. One of these ganglia is placed above the gullet or œsophagus, and is known as the "supra-œsophageal" or "cerebral" ganglion. A second supplies nerves to the great locomotive organ of most Mollusks, the "foot," and is therefore called the "pedal" ganglion. The third is known by the cumbrous name of the "parieto-splanchnic" ganglion, because it supplies nervous filaments to the walls (*parietes*) of the body, and also to the internal organs (*splanchna*). In all the higher Mollusks it is this *scattered* condition of the nervous masses which distinguishes them so sharply from all other animals. Distinct respiratory organs may or may not be present, and they may be adapted for breathing air directly or through the medium of water. All the higher *Mollusca* are simple animals, and perpetuate their kind by means of the sexes, but many of the lower forms have the power of producing colonies by continuous gemmation, much as we have formerly seen in the Hydroid Zoophytes.

The digestive system in all the *Mollusca* consists of a mouth, gullet, stomach, intestine, and anus, with the exception of a few forms in which the intestine ends blindly. In some the mouth is surrounded by ciliated tentacles (*Polyzoa*, Fig. 77); in others, it is furnished with two long ciliated arms (*Brachiopoda*); in the bivalves (*Lamellibranchiata*), it is mostly furnished with four membranous processes or "palpi" (Fig. 80, *p*); in others, it is furnished with a complicated toothed organ or "odontophore" (*Gasteropoda*, Fig. 83, and *Pteropoda*); and lastly, the *Cephalopoda*, in addition to an odontophore, possess horny mandibles, forming a kind of beak, very like that of a parrot.

The blood is colorless, or nearly so. In the lowest class of the *Mollusca* (*Polyzoa*), the circulation is carried on by means of cilia, and there is no distinct heart, nor any definite course of the circulating fluid. In the Sea-squirts (*Tunicata*), there is a distinct heart, but the structure of this is very simple, consisting of a mere tube, open at both ends, so that the course of the circulation is periodically reversed. In the higher *Mollusca*, there is a distinct heart, consisting of two chambers, of which one (the *auricle*) receives the aërated blood from the gills, while the other (the *ventricle*) drives it through the body.

Respiration is very variously effected among the *Mollusca*. In the *Polyzoa* (Fig. 77) respiration is discharged mainly by the crown of ciliated tentacles surrounding the mouth. In the sea-squirts (Fig. 78), respiration is effected by a greatly-developed pharynx, which is perforated by numerous ciliated apertures. In the lamp-shells and their allies (*Brachiopoda*), the long, ciliated arms, which spring from the sides of the mouth, seem to be the main agents in respiration. In the bivalve-shell-fish, the cuttle-fishes, and most of the univalves, the breathing-organs are in the form of gills or branchiæ, adapted for breathing air dissolved in water. In the remainder of the univalves (e. g., snails and slugs), the breathing-organs are adapted for breathing air directly, and have the form of an air-chamber or pulmonary sac, produced by the folding of a portion of the mantle. The air is admitted to the chamber by a round opening, situated on the side of the neck, and capable of being closed at will. The lining membrane of the chamber is richly supplied with blood-vessels, and thus the necessary purification of the blood is carried out.

In accordance with the scattered or rudimentary condition of the nervous system, the *Mollusca* are not characterized by acuteness of senses, nor by any great power of locomotion. Organs of sight exist in some of the lower and many of the higher *Mollusca*, attaining in the cuttle-fishes (Fig. 89) an extremely high type of organization. The common bivalve shell-fish, such as the scallop, possess numerous simple eyes placed along the margins of the mantle, but, in many cases, even these are absent. Locomotion is very variously effected, but seldom with much vigor or activity. The lowest classes of the *Mollusca* are, in the great majority of instances, fixed when adult. The common univalve shell-fish, such as whelks, snails, slugs, etc., creep about slowly by means of a flattened disk, developed on the under surface of the body, and known as the "foot." Other Univalves and many Bivalves can effect short leaps by means of the foot, but many of the latter are permanently fixed to solid objects, or buried in the sand. The minute *Mollusca*, known as the *Pteropoda* (Fig. 88), swim freely at the surface of the ocean by means of two fins, formed by a modification of the foot, and attached to the sides of the head. The only Mollusks which enjoy really active powers of locomotion are the predacious cuttle-fishes, which swim rapidly by means of fins, or by ejecting a jet of water from the cavity of the mantle, and which can also creep about by means of the "arms" placed around the mouth (Fig. 89).

The last feature in the *Mollusca* which requires to be mentioned is the "shell." The shell is not invariably and universally present in the *Mollusca*, many being either destitute of a shell altogether, or having one so small that it would not commonly be recognized as such. In these cases, as in the common slugs, the animal is said to be "naked." In all the *Mol-*

lusca which possess a shell, this is secreted by the integument, or by what is technically called the mantle, and, in all cases, it is composed of carbonate of lime. The methods in which the lime is arranged differ in different cases, but all living shells have an outer covering of animal matter, which is known as the "epidermis." In a great many of the higher *Mollusca*, such as the whelks, periwinkles, snails, and others, the shell consists of only a single piece, when it is said to be "univalve." In many others, such as oysters, mussels, scallops, etc., the shell is composed of two pieces, and is then said to be "bivalve." In a few forms, the shell consists of several pieces, and it is then said to be "multivalve." The more important variations in the shells of the *Mollusca* will be noticed in speaking of the different classes of the sub-kingdom.

In accordance with the nature of the nervous system, the *Mollusca* are divided into two great divisions, known respectively as the *Molluscoida* and *Mollusca proper*. In the *Molluscoida*, the nervous system consists of a single ganglion, or principal pair of ganglia, and there is either no circulatory organ or an imperfect heart. In this division are included the three classes of the Sea-mosses (*Polyzoa*), the Sea-squirts (*Tunicata*), and the Lamp-shells and their allies (*Brachiopoda*). In the *Mollusca proper*, the nervous system consists of three principal pairs of ganglia, and there is a well-developed heart, consisting of at least two chambers. Under this head come all the ordinary forms of shell-fish.

CHAPTER XX.

MOLLUSCOIDA.

Class I. Polyzoa (Gr. *polus*, many; *zoön*, animal.) — The members of this class are the lowest of all *Mollusca*, and they are generally known by the popular names of "Sea-mosses" and "Sea-mats." They are invariably compound, forming associated growths or colonies, each consisting of a number of distinct but similar zoöids, produced by gemmation from a single primordial individual. The colonies thus produced are very generally protected by a horny or chitinous integument, and they are so like the Hydroid Zoophytes that they were long described as such. The only absolute distinction between the two classes is to be found in the internal structure of the zoöids of each; but they may be generally separated by the fact that the separate cells in a compound Hydroid are all united to one another by means of a common flesh or cœnosarc; whereas in the *Polyzoa* the separate cells composing the colony are merely connected externally, but very rarely have any direct communication with each other. The separate beings or zoöids which collectively constitute the colony of any *Polyzoön* are spoken of as "polypides"—the term *polypite* being only used in connection with the *Hydrozoa*, and the term *polype* being similarly restricted to the *Actinozoa*.

Each polypide in a typical *Polyzoön* has the following structure (Fig. 76, 2): The body of the animal is enclosed in a double-walled sac, of which the outer layer is usually chitinous or calcareous, and constitutes a "cell" in which the zoöid is contained. This outer layer is known as the "ectocyst," to distinguish it from the ectoderm of the *Cœlenterata*. The cell, thus formed, is lined by a much more delicate membranous layer, which is known as the "endocyst." This membranous sac, formed by the endocyst, is pierced by two openings. One of these is the mouth, and it is always surrounded by a circle or crescent of hollow ciliated processes or tentacles (Fig. 76, 2, *a*). These ciliated tentacles serve partly as respiratory organs, and partly to set up a current of water by which floating particles of food are brought to the mouth. The mouth and tentacular crown can be partially or completely pulled into the sac by means of a muscle which is fixed to the gullet (2, *g*). The mouth leads into a gullet, and that

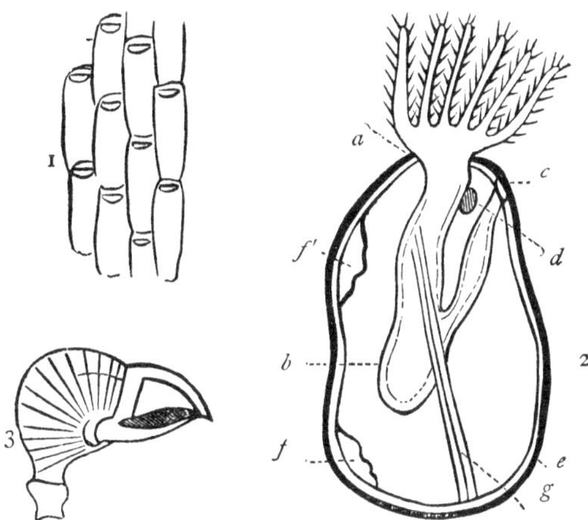

Fig. 76.—Morphology of Polyzoa. 1. Fragment of one of the Sea-mats (*Flustra truncata*), magnified to show the cells. 2. Diagram of a single polypide of a Polyzoön (after Allman): *a* Mouth surrounded by the ciliated tentacles; *b* Alimentary canal; *c* Anus; *d* Nervous ganglion; *e* Investing sac or "ectocyst;" *ff* Reproductive organs; *g* Muscle. 3. Bird's-head process.

again into a stomach, sometimes with a muscular gizzard between. From the stomach proceeds an intestine of variable length, which terminates by a distinct anus at the upper part of the sac (2, *c*). On one side of the gullet, between it and the anus, is placed a single nervous ganglion (*d*). Distinct reproductive organs (*ff*) are also present, and the whole cavity of the sac is filled with fluid. From the above description it will be evident that the typical polypide of a *Polyzoön* differs from the polypite of a *Hydrozoön* in having a distinct alimentary canal suspended freely in a body-cavity, and having both a mouth and vent, in having a distinct nervous system, and in having the reproductive organs contained within the body. On the other hand, in the *Hydrozoa*, there is no alimentary canal distinct from the body-cavity, there is no nervous system, and the reproductive organs are in the form of external processes of the body-wall.

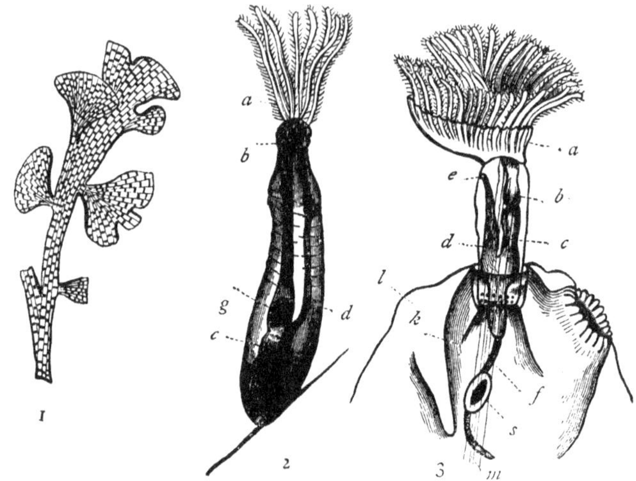

FIG. 77.—1. Fragment of *Flustra truncata*, one of the Sea-mats, natural size. 2. A single polypide of *Valkeria*, magnified, showing the circular crown of tentacles. 3. A polypide of *Lophopus crystallinus*, a fresh-water Polyzoön, highly magnified, showing the horse-shoe shaped crown of tentacles: *a* Tentacular crown; *b* Gullet; *c* Stomach; *d* Intestine; *e* Anus; *g* Gizzard; *k* Endocyst; *l* Ectocyst.

The foregoing gives the essential structure of the polypide of any *Polyzoön*, but in nature this simplicity is lost. In all cases in nature the primitive polypide possesses the power of producing fresh zoöids by a process of budding; and these zoöids remain attached to one another, so that ultimately there is produced a compound growth or colony. Further, in almost all the *Polyzoa*, the outer layer of the polypide is more or less hardened by the deposition in it of chitine or of carbonate of lime. The skeletons thus formed are the parts of the colony which are most familiarly known, and in the case of the common Sea-mats (Fig. 77, 1) they are very well known to sea-side visitors, and are generally regarded as sea-weeds. Examined in its dead state, such a skeleton only shows a number of little horny chambers or cells (Fig. 76, 1), each with a little aperture. When alive, however, each of these cells was tenanted by a single zoöid or polypide, capable of protruding its ciliated head from the aperture, and of again retiring within it, if alarmed. The skeleton is, in some cases, furnished with curious organs, which are known as "bird's-head processes" (Fig. 76, 3), from their resemblance to the beak of a bird. The parts of this beak keep constantly snapping together, very much like the little pincer-like organs called "pedicellariæ" in the sea-urchins and star-fishes; but it is difficult to see what service they perform. They continue their movements long after the death of the polypides, and this appears, in some cases, at any rate, to be due to a peculiar system of nerves known as the "colonial" nervous system. In addition, namely, to the single ganglion with which each polypide is furnished, it has been shown that in many forms the zoöids composing the colony are united together by a well-developed nervous system, and are thus brought into organic connection with one another.

The vast majority of the *Polyzoa* are fixed, and thus assume a very plant-like appearance. There is one fresh-water species, however (viz., *Cristatella*), in which the colony can creep about upon a flattened base very like the foot of a slug. In this same form, also, alone of all the *Polyzoa*, there is not any outer covering or ectocyst to the polypides.

The *Polyzoa* are partly inhabitants of the sea and partly of fresh water, and they are thus divided into two groups which differ from one another very much in anatomical structure. In most of the fresh-water *Polyzoa* the tentacles are borne upon a crescentic disk or stage (Fig. 77, 3), so that the crown of tentacles assumes the shape of a horseshoe. In almost all the marine forms, on the other hand, the tentacles (Fig. 77, 2) are simply arranged in a circle.

All the *Polyzoa* are hermaphrodite, each polypide being furnished with the reproductive organs proper to the two sexes. The eggs are simply liberated into the body-cavity, where they are fertilized; but it is uncertain how the fertilized ova escape into the external medium. Besides true sexual reproduction, and besides the power of producing colonies by continuous budding, fresh individuals can be produced in many cases by a process of discontinuous gemmation.

Class II. Tunicata (Lat. *tunica*, a cloak).—The members of this class are not uncommonly called *Ascidian Mollusks* (Gr. *askos*, a wine-skin) from the resemblance which many of them exhibit in shape to a two-necked leather bottle (Fig. 78, 2). They are popularly known as "Sea-squirts," from their power of forcibly ejecting water from the orifices of the bottle. Their scientific name, again, of *Tunicata*, is derived from the fact that the body is enveloped in a leathery elastic integument, which consists of different layers, and which takes the place of a shell. The outer covering of the animal is of a gristly or leathery consistence, and is known as the "test." It is remarkable for containing a considerable proportion of a substance apparently identical with *cellulose*, which is one of the most characteristic of all vegetable products. The test is lined by a second coat, which is highly muscular, and confers upon the animal its power of contracting itself and squirting out water. Of the two necks which are placed at the anterior end of a simple Ascidian (Fig. 78), one is per-

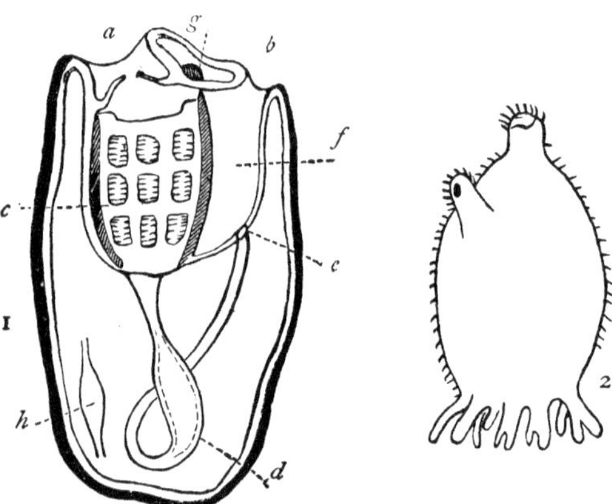

FIG. 78.—Morphology of Tunicata. 1. Diagram of a Tunicary (after Allman): *a* Oral aperture; *b* Atrial aperture; *c* Respiratory sac, with its rows of ciliated apertures; *d* Alimentary canal; *e* Anus; *f* Cloaca or atrium; *g* Nervous ganglion. 2. A simple Ascidian (*Cynthia papillosa*).

forated by the aperture of the mouth, while the other serves as an excretory aperture. These two apertures are known respectively as the "oral" and "atrial" apertures.

The oral aperture (*a*) is usually furnished with a circle of small non-retractile tentacles, and opens into a great chamber known by various names, but best as the "respiratory sac." This sac occupies the greater part of the cavity of the body (Fig. 78, 1, *c*), and has its walls perforated by numerous apertures, the sides of which are ciliated. At the bottom of the respiratory sac is a second opening (the *mouth* of some writers) which leads by a short gullet into a capacious stomach (*d*). From the stomach an intestine is continued to terminate by a distinct anus, which does not communicate directly with the exterior, but opens into a second great chamber, known as the "cloaca" or "atrium" (*e*). The cloaca, in turn, opens on the exterior by the second, or atrial aperture, in the test (*b*). These two great chambers—namely, the respiratory sac and the cloaca—occupy the greater part of the body-cavity, and, where their walls come into contact, a free communication is established between the two by means of the ciliated apertures already spoken of as perforating the respiratory sac. The cilia which fringe these apertures all work toward the cloaca, and thus a constant current of water is caused to set in by the oral aperture, through the respiratory sac, into the cloaca, and out again by the atrial aperture. In this way respiration is effected, the walls of the respiratory sac being almost made up of blood-vessels. A distinct heart is present in all the *Tunicata*, but it has a very simple structure (1, *h*). It consists of a simple tube, open at both ends, and not provided with valves. In consequence of this, the circulation in the majority of Tunicaries is periodically reversed, the blood being driven for a certain number of contractions in one direction, and then propelled for a like period in an opposite direction; so that "the two ends of the heart are alternately arterial and venous."

The nervous system in the *Tunicata* consists of a single ganglion placed on one side of the oral aperture.

With one or two exceptions all the *Tunicata* are hermaphrodite, the organs of reproduction being situated in a fold of the intestine, and opening into the cloaca. The embryo is at first free, and in most swims about by means of a long tail, so that it presents considerable resemblance to the tadpole of a frog.

The *Tunicata* are all marine, but differ a good deal from one another in form. In the so-called "simple" Ascidians the animal has the shape figured above, and is fixed to some solid object by one end of the test. In the "social" Ascidians the organism consists of a number of zoöids, produced by continuous budding, and connected together by a common tube, through which the blood circulates. In the so-called "compound" Ascidians the tests are fused together into a common gelatinous mass, in which the individuals are imbedded in groups. Some of the *Tunicata* are oceanic — that is to say, are found floating, or swimming, at the surface of the open ocean—and some exhibit the phenomenon of phosphorescence.

In the foregoing description it has been found impossible to convey even the most elementary outline of the anatomy of a Tunicate without having recourse to technical terms. There still remain a few points of homology which should be mentioned. In the foregoing, the so-called "oral aperture" of the animal has been regarded as truly the *mouth*, this being the simplest view, and the one held by Prof. Huxley. Upon this view the "respiratory sac," into which the mouth opens, must be regarded as a greatly-developed pharynx (i. e., the upper portion of the alimentary tube). Similarly, on this view, the *lower* aperture of the respiratory sac will have to be regarded as the opening of the *gullet*. By Prof. Allman, again, the respiratory sac is looked upon as formed by a great modification of organs corresponding to the ciliated tentacles of the *Polyzoa*, so that the *lower* aperture of the respiratory sac is the true *mouth*. Lastly, by Prof. Rolleston the respiratory sac is looked upon as corresponding to the gills of the Bivalve shell-fish (*Lamellibranchiata*), and the oral and atrial apertures are regarded as corresponding to the "respiratory siphons" of these same animals. On this view, the *lower* aperture of the respiratory sac is again looked upon as the true *mouth*. The question cannot be regarded as settled, and Huxley's view has been here adopted merely as being the most readily intelligible to learners.

CLASS III. BRACHIOPODA.—The members of this class are little known to the general public, being all marine, often inhabiting considerable depths in the sea, and being much more abundantly represented by fossil forms than by living examples. They are often placed with the ordinary Bivalve shell-fish (*Lamellibranchiata*), in consequence of their universally possessing a shell composed of two pieces or valves (Fig. 79), but they are really of a much lower organization. In their essential structure they show many points of affinity to the *Polyzoa*, but they are always simple animals, never forming colonies, and they always have a bivalve shell. The two pieces of which the shell is composed are always placed one in front and one behind, so that they are "ventral" and "dorsal," and not "right" and "left" as in the true Bivalves. The two valves of the shell are also always slightly, and sometimes greatly, different to one another in size, so that the shell is said to be "inequivalve." The ventral valve is usually the largest, and often possesses a prominent curved beak, which is generally perforated by an aperture through which there passes a muscular stalk by means of which the shell is attached to some solid object. In some cases, however, as in *Lingula* (Fig. 79), the stalk of attachment simply passes between the valves, and is not transmitted through a distinct aperture. In other cases the shell is simply attached by the substance of the ventral valve.

The inner surface of the valves of the shell is lined by expansions of the integument, which are called the "mantle-lobes," and which secrete the shell. The digestive organs and muscles occupy a small space near the apex or "beak" of the shell, which is partitioned off by a membranous partition, perforated by the aperture of the mouth. The remainder of the cavity of the shell is almost filled by two long processes, derived from the sides of the mouth, fringed with lateral branches, and termed the "arms." These arms are usually closely coiled up, and serve to obtain food for the animal. It is from these organs that the name of the class is derived (Gr. *brachion*, arm; and *podes*, feet). The arms also serve as respiratory organs, and in many forms they are supported on an internal calcareous framework or skeleton, sometimes called the "carriage-spring apparatus."

FIG. 79. — *Lingula anatina*, showing the muscular stalk by which the shell is attached.

The mouth is placed between the bases of the arms, and is not furnished with any apparatus of teeth. It conducts by a gullet into a distinct stomach, surrounded by a well-developed granular liver. The intestine may or may not be furnished with a distinct anus, but in no case does it open into the body-cavity. Within the lobes of the mantle, there is a remarkable system of branched tubes, which commence by blind extremities, and finally communicate with the mantle-cavity by means of certain organs, which were formerly believed to be hearts, and are now known as "pseudo-hearts." This system of tubes appears to be mainly, if not entirely, connected with reproduction. A true heart, however, is present in most, if not in all, of the *Brachiopoda*.

The nervous system consists of a single principal ganglion, connected in some cases with others so as to form a collar round the commencement of the gullet. In some cases, however, the nervous system appears to be very rudimentary.

The sexes appear to be sometimes distinct and sometimes united in the same individual. The embryo, in some cases, at any rate, is locomotive, moving from place to place by means of the ciliated arms or by ventral spines.

CHAPTER XXI.

MOLLUSCA PROPER.

THE higher *Mollusca* or *Mollusca proper* comprise those members of the sub-kingdom in which *the nervous system consists of three principal pairs of ganglia; and there is always a well-developed heart, consisting of at least two chambers.*

In this division are included the following classes:

1. *Lamellibranchiata*, without a distinct head.
2. *Gasteropoda*, } with a distinct head and a masticatory
3. *Pteropoda*, } apparatus or "odontophore."
4. *Cephalopoda*, }

CLASS I. LAMELLIBRANCHIATA.—These are well known as bivalve shell-fish, such as mussels, oysters, scallops, etc., and they are all either marine or inhabitants of fresh water. They are distinguished from the other Mollusks by having no distinct head, and by having the body more or less completely protected by a bivalve-shell composed of two pieces. They are called *Lamellibranchiata* (Lat. *lamella*, a plate; Gr. *bragchia*, gill), from the fact that the organs of respiration are in the form of leaf-like gills or branchiæ, two of which are placed at each side of the body, constituting what is known in the oyster as the "beard." The body of the *Lamellibranchiata* is more or less completely enclosed in an expansion of the integument which constitutes the "mantle," and which is divided into two halves or "lobes," which are placed on the *sides* of the animal, and secrete the shell. The shell, therefore, of the true bivalves is composed of two valves, which are "right" and "left," and not "dorsal" and "ventral," as in the *Brachiopoda*. Moreover, the valves of the shell are usually of the same size, so that the shell is "equivalve," and, lastly, the shell is more developed on one side than the other, so as to become "inequilateral" (Fig. 81, 2). The lobes of the mantle are sometimes quite free; but, at other times, they are more or less united to each other, and leave only two openings. Through one of these openings (the anterior) the "foot" is protruded (Fig. 80, *f*); and through the other pass the respiratory tubes or "siphons" (*s*). The foot in the bivalves is a muscular organ developed upon the lower surface of the body, but not forming a creeping flattened disk, as in the ordinary univalves. In many cases, it is quite rudimentary; and even when it is employed in locomotion it is usually small. Most generally, it is hatchet-shaped or pointed (Fig. 80, *f*), and serves to enable the animal to make short leaps. In many cases, as in the common mussels, the foot is subsidiary to a special gland, which secretes a viscous fluid, which hardens rapidly on exposure to the air. This fluid is moulded by the foot into silky threads (the so-called "byssus"), by means of which the shell is firmly fixed to some solid object. Besides the muscular foot, other muscles are present as well in the *Lamellibranchiata*. Of these, the most important are the muscles which close the shell, and are called the "adductor" muscles. In one group of the bivalves (Fig. 81, 3), there

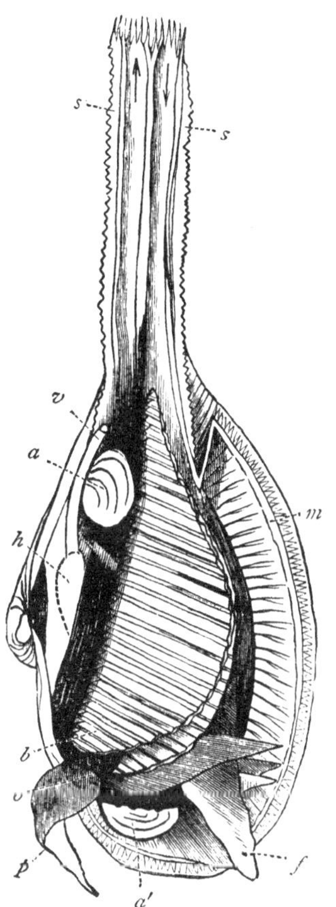

FIG. 80.—Anatomy of a Bivalve Mollusk. *Mya arenaria* (after Woodward). The left valve and mantle-lobe, and half the siphons are removed. *s s* Respiratory siphons, the arrows indicating the direction of the currents; *a a′* Adductor muscles; *b* Gills; *h* Heart; *δ* Mouth, surrounded by (*p*) labial palpi; *f* Foot: *v* Anus; *m* Cut edge of the mantle.

is only one adductor muscle, but ordinarily there are two (Fig. 81, 2). These muscles leave distinct scars or "muscular impressions" in the dead shell, so that it is easy to determine how many were present in any given shell. The margin of the mantle, too, is muscular, and leaves upon the shell a distinct line where it was attached, this being known as the "pallial line" (Lat. *pallium*, a mantle), as shown in Fig. 81.

As regards the shell of the bivalves, the following are the chief points to be noticed. Each valve of the shell (Fig. 81) is to be regarded as essentially a hollow cone, the apex of which is turned more or less to one side. The apex of the valve is known as the "umbo" or "beak," and is turned toward the mouth of the animal. Consequently, the side of the

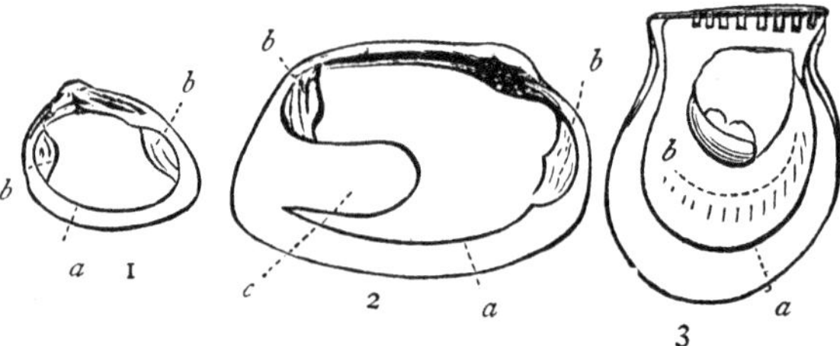

Fig. 81.—Shells of Lamellibranchiata. 1. *Cyclas amnica*, a shell with two adductor muscles, and an "entire" mantle-margin. 2. *Tapes pullastra*, a shell with two adductors, and an indented pallial line. 3. *Perna ephippium*, a shell with one adductor muscle: *a* Pallial line; *b* Scars left by the adductors; *c* Siphonal impression.

shell toward which the beaks are turned is known as the "anterior" side, and it is usually much shorter than the opposite or "posterior" side. The side of the shell at which the beaks are situated is known as the "dorsal" margin; and here the valves are united to one another for a longer or shorter distance along a line which is known as the "hinge-line." The union between the valves is usually effected by means of interlocking parts or "teeth," and there is often a band of horny fibres passing between the two valves just behind the beaks. In many cases, there is also a series of horny fibres placed perpendicularly between the beaks, so as to be compressed when the shell is shut. By the elasticity of these, and of the external ligament, when present, the valves of the shell are opened, without any effort of the animal, simply by relaxing the adductor muscles. The valves are shut again by the contraction of the adductor muscle or muscles.

As already said, the margin of the mantle leaves on the shell a distinct impression—the "pallial line"—and, by inspection of this, important conclusions can be drawn in any given case as to the mode and life of the animal. In certain shells, namely, the pallial line (Fig. 81, 1) is unbroken or "entire," and in these the mantle-lobes were either quite free, or if attached to one another and drawn out into respiratory tubes, these were not furnished with special muscles by which the tubes could be retracted within the shell. In other bivalves, on the other hand (Fig. 81, 2), the pallial line is indented to a greater or less extent, showing that the mantle-lobes were more or less united to one another, and were drawn out into long respiratory tubes or siphons, which were furnished with special muscles by which they could be withdrawn within the shell. This difference expresses a real distinction among the bivalves, due to their mode of life. In all alike, the respiratory organs are in the form of membranous leaf-like gills, of which there are generally two on each side of the body. The gills are composed generally of tubular rods (Fig. 80, *b*) richly supplied with blood-vessels, and covered with vibrating cilia. For the proper maintenance of respiration, however, it is necessary that the gills should be constantly supplied with fresh water. In those bivalves in which the animal is free and the mantle-lobes not attached to one another, this is effected without any special mechanism. In those forms, however, in which the animal lives buried in the mud and sand, and the mantle-lobes are more or less completely united, there are two orifices, one of which admits fresh water, while the effete water is got rid of through the other. These orifices, in the shells just spoken of, are extended into two long tubes which are known as the "respiratory siphons." The water passes in by one siphon, is swept over the surface of the gills, and then reaches the mouth (Fig. 80, *s s*), when it is returned in the opposite direction to escape by the other siphon. The same current of water, therefore, both carries oxygen to the gills, and serves to convey food to the mouth. The two siphons may be quite distinct from one another, but they are very often united together so as to look like a single tube (Fig. 80). They are often very small, and then they leave no traces of their existence in the dead shell; but, when they are very long, they are furnished with muscles to retract them within the shell, and it is the scar left by these muscles which causes the pallial line to be indented. This indentation, therefore, as seen in the dead shell, is an indication that the animal possessed long retractile respiratory siphons, and lived, therefore, most probably imbedded in sand or mud.

There is always a distinct heart, composed of two or three chambers, and in all cases acting as a mere *arterial* heart. That is to say, the heart propels the aërated blood derived from the gills through the body, and has nothing to do with the propulsion of the non-aërated or venous blood through the gills. There is never any distinct head in any of the Bivalves, and for this reason they are sometimes called the "headless" (*acephalous*) Mollusks. The mouth is simply placed at the anterior end of the body, and is never furnished with teeth, though usually provided with membranous processes or "palpi" (Fig. 80, *p*). The mouth opens into a gullet which conducts to a stomach. The intestine is convoluted, and usually perforates the ventricle of the heart, ultimately terminating in a distinct anus, which is always placed near the respiratory aperture. A large and well-developed liver is also present.

The nervous system has its normal form of three principal masses—the cerebral, the pedal, and the parieto-splanchnic ganglia.

The majority of the bivalve Mollusks have the sexes distinct, but they are sometimes united in the same individual. The young are hatched before they leave the parent, and, when first liberated, are ciliated and free-swimming.

The habits of the *Lamellibranchiata* are very various. Some, such as the Scallops (*Pecten*), habitually lie on one side, the lower valve being the deepest, and the foot rudimentary or wanting. Others are fixed to the bottom of the sea by the substance of one of the valves. Others, such as the common Mussel, are moored to some foreign object by a tuft of silky fibres, constituting a "byssus." Many, such as the Gapers (*Mya*) and Razor-shells (*Solen*), spend their existence sunk in the sand of the sea-shore or the mud of estuaries. Others, such as the *Pholades*, bore holes in rock or wood, in which they live. Finally, many are permanently free and locomotive.

CLASS II. GASTEROPODA (Gr. *gaster*, belly; *podes*, feet).—This class includes an enormous number of Mollusks, such as the land-snails, sea-snails, whelks, limpets, slugs, sea-lemons, etc., which agree in many fundamental characters, but nevertheless present many striking differences. From the very common occurrence of a shell composed of a single piece, the *Gasteropoda* are often spoken of in a general way as the "univalve"

Mollusks. In many, however, there is either no shell at all, or one so small that it would not generally be recognized as such; and in a few the shell is composed of several pieces ("multi-valve"). In none, however, is the shell composed of two pieces or "bivalve." The great majority of the *Gasteropoda* are further distinguished by the great development of the foot, which constitutes a broad, flattened disk upon which they creep about, as may readily be observed in the common slugs. Some, however, have the foot much modified and adapted for swimming. In many cases, also, the foot carries behind a horny or shelly plate which is known as the "operculum" (Fig. 82, *o*), and which serves to close the shell when the animal is withdrawn within it.

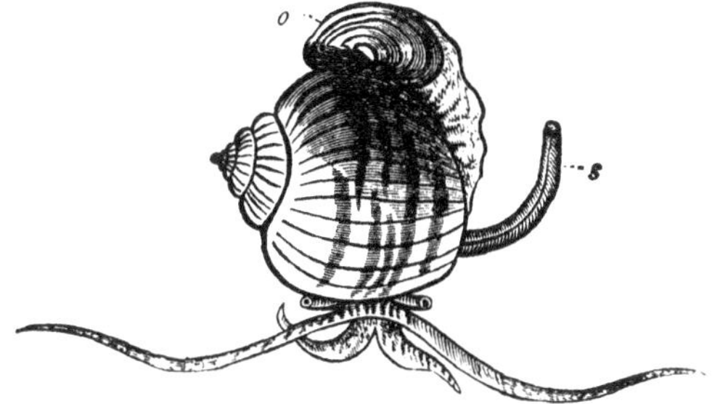

FIG. 82.—Gasteropoda. *Ampullaria canaliculata*, one of the Apple-shells: *o* Operculum; *s* Respiratory siphon.

The head in most of the *Gasteropoda*, unlike the Bivalves, is very distinctly marked out, and carries two long feelers, and two eyes, often placed upon stalks (Fig. 85). The mouth, also, differs from that of the Bivalves in being furnished with a singular apparatus of teeth, constituting what is known as the "odontophore" (Fig. 82), or "lingual ribbon." This consists essentially of a number of siliceous teeth, of different shapes in different species, supported upon a kind of strap which can be made to work backward and forward over a cartilaginous cushion, thus acting like a chain-saw. In addition to the odontophore there are sometimes horny jaws as well. The mouth leads by a gullet into a distinct stomach, which sometimes is provided with calcareous plates for grinding down the food. The intestine is long, and always terminates in a distinct anal aperture. Distinct salivary glands are usually present, and the liver is well developed.

A distinct heart is almost always present, and consists of two chambers, an auricle and a ventricle. Respiration is very variously effected—one great division being constructed to breathe air by means of water, while in another section the respiration is aërial. In the former of these—often spoken of as the "branchiate" Gasteropods—respiration may be carried on in three ways: Firstly, there may be no special breathing-organ, the blood being simply exposed to the action of the water, as it circulates through the thin walls of the mantle-cavity. Secondly, the breathing-organs may be in the form of outward processes of the skin, exposed to view on the back or sides of the animal (Fig. 85). Thirdly, the breathing-organs are in the form of plume-like gills, contained in a more or less complete chamber, formed by a folding of the mantle. In many members of this group the water obtains access to the gill-chamber by means of a tubular prolongation or folding of the mantle, forming a siphon (Fig. 82, *s*), and often the effete water is expelled by another tube which is similarly constructed. In the second great section—often called the "pulmonate" Gasteropods—respiration is effected by a pulmonary chamber or lung, formed by a folding of a mantle, and having air admitted to it by a distinct aperture.

Fig. 83.—Portion of the lingual ribbon of the common whelk, magnified (after Woodward).

The sexes in the *Gasteropoda* are mostly distinct, but they are sometimes united in the same individual. The young, when first hatched, are always provided with an embryonic shell, which may be entirely lost in the adult, or may simply become concealed by a fold of the mantle. In the water-breathing forms the young is protected by a small nautilus-shaped shell, within which it can entirely withdraw. It is enabled to swim about freely by means of two ciliated lobes springing from the sides of the head, and in this stage it is very like the permanent adult condition of the *Pteropoda* (Fig. 88).

As regards the shell of the *Gasteropoda*, the following points may be noticed: The shell is composed either of a single piece (univalve), or of a number of plates placed one behind the other (multivalve).

The univalve shell is to be looked upon as essentially a hollow cone, the apex of which is placed a little on one side. In the simplest forms, as in the Limpets, the conical shell is retained throughout life without any alteration. In the great majority of cases, however, the cone is considerably elongated so as to form a tube, which may retain this shape (as in the "tooth-shell"), but which is usually coiled up into a spiral. The "spiral univalve" may, in fact, be regarded as the typical

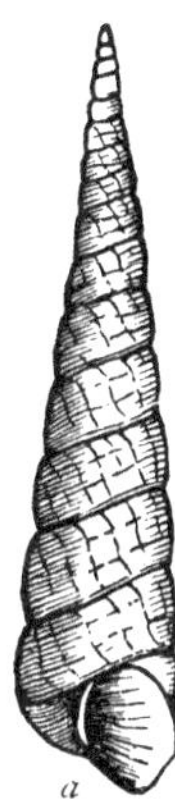

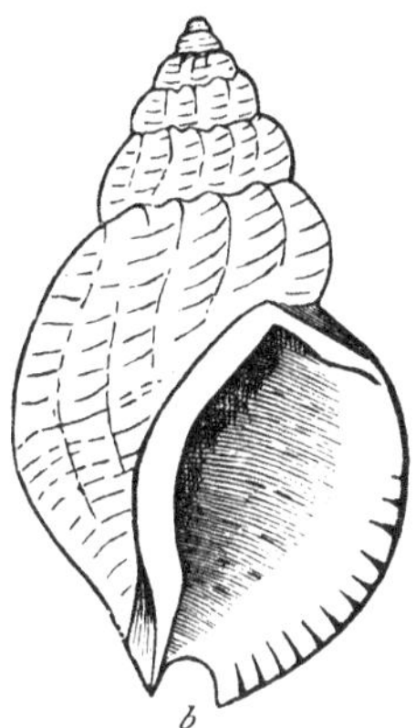

Fig. 84.—Gasteropoda. *a* Shell of the *Turritella communis*, showing a round mouth; *b* Shell of the common whelk (*Buccinum undatum*), showing the mouth notched for a respiratory siphon.

form of the shell in the *Gasteropoda* (Fig. 84). The coils of the spiral are termed the "whorls," and are usually more or less amalgamated on one side. In most cases, too, the whorls are wound obliquely round a central axis or pillar, increasing gradually in size to the mouth. The last whorl is the largest, and is termed the "body-whorl." The mouth of the shell in many forms is unbrokenly round or "entire" (Fig. 84, *a*), and it is found that most of these shells subsist upon vegetable food, as, for instance, the common periwinkles. In others, again (Fig. 84, *b*), the mouth of the shell is notched or is produced into a canal, as in the common whelk, and it is found that these live upon animal food, or are "carnivorous." There may be more than one of these canals or tubes, but they do not necessarily indicate the nature of the food, as their function is to protect the respiratory siphons.

The *Gasteropoda* are divided into a good many groups, of which the more important may be briefly noticed, the foregoing applying chiefly to the ordinary forms, which, therefore, need no further description. The remaining members of the water-breathing Gasteropods are divided into two sections, differing a good deal from the typical forms of the class in many respects.

As examples of the first of these may be taken the sea-slugs and sea-lemons (*Nudibranchiata*), specimens of which may at any time be found creeping about on sea-weeds, or attached to the under surface of stones at low water. These slug-like animals (Fig. 85) are wholly destitute of a shell when fully grown, but possess an embryonic shell when young. When there are any distinct respiratory organs, these are in the form of gills, placed, without any protection, upon the back or sides of the body. The head is furnished with tentacles, which do not appear to be used as organs of touch, but are more probably connected with the sense of smell; and behind the tentacles are generally two eyes. The nervous system is extremely well developed, and would lead to the belief that the Sea-slugs are among the highest of the *Gasteropoda*. Locomotion is effected, as in the true Slugs, by creeping about on the flattened foot.

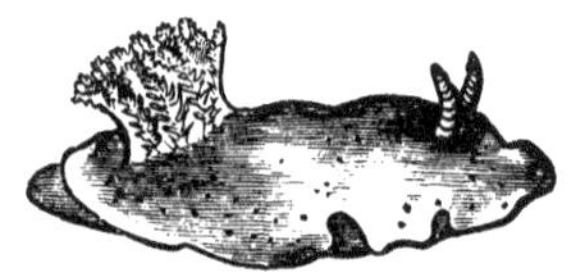

FIG. 85.—Nudibranchiata. *Doris Johnstoni*, one of the Sea-lemons.

The last remaining group of the "branchiate" Gasteropods is that of the *Heteropoda* (Fig. 86), comprising a number of curious forms which are found swimming at the surface of the open sea, instead of creeping about at the bottom of the sea. In order to adapt them for this mode of life, the foot, instead of forming a creeping disk, is modified to form a compressed fin (*f*). The *Heteropoda* are to be regarded as the most

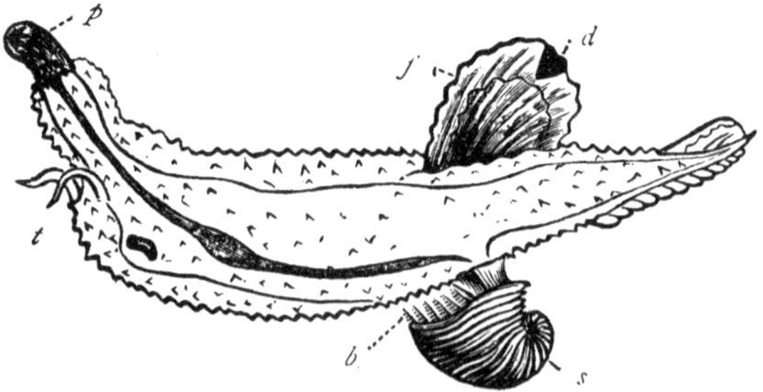

FIG. 86.—*Heteropoda*. *Carinaria cymbium*; *p* Proboscis and mouth; *t* Tentacles; *g* Gills; *s* Shell; *f* Foot; *d* Disk (after Woodward).

highly organized of all the *Gasteropoda*, at the same time that they are not the most typical members of the class. Some of them can retire completely within their shells, but others have large bodies, and the shell is either small or entirely absent. In *Carinaria*, which may be taken as a good example of the group, there is a little limpet-shaped shell protecting the gills (*b*) and heart. The animal swims, back downward, by means of a vertically-flattened ventral fin (*f*), on one side of which is a little sucking-disk (*d*), by which the animal can adhere at pleasure to floating sea-weed. *Carinaria* is found in the Mediterranean and other warm seas, and is so transparent that the course of the intestine can be seen along its whole length.

The last group of the class is that of the "air-breathing" Gasteropods, so well known as Land-snails, Pond-snails, and Slugs (Fig. 87). All the members of this group are formed to breathe air directly, instead of through the medium of water, and they, therefore, never possess gills or branchiæ.

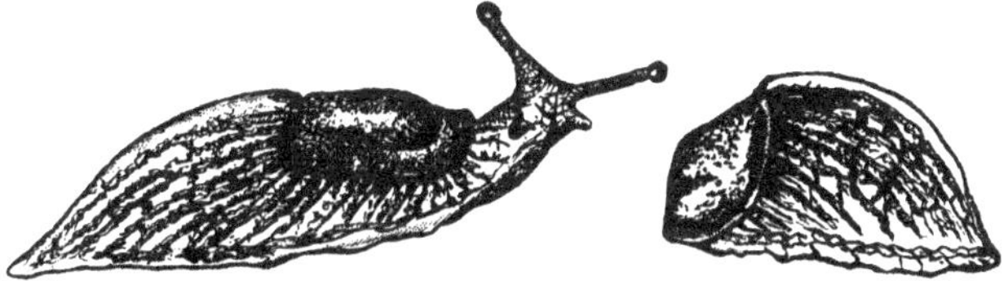

Fig. 87.—*Limax Sowerbyi*, one of the slugs (after Woodward).

In place of these they have a pulmonary chamber or lung, formed by a folding of the mantle, and having air admitted to it by a round hole on the right side of the neck, which can be opened and closed at will. Though thus adapted for breathing air directly, many of the members of this group can only live in damp or moist places, while others habitually live in fresh water. The common Pond-snails are examples of these last. The condition of the shell varies much. Some, such as the common Land-snails, have a well-developed shell within which the animal can completely withdraw itself for protection. Others, such as the common Slugs (Fig. 87), have a rudimentary shell which is completely concealed within the mantle. Others are entirely destitute of a shell. They all agree with the typical Gasteropods in creeping about on a broad, flattened foot.

Class III. Pteropoda (Gr. *pteron*, wing; *podes*, feet).—

This class is a very small one, and includes a number of minute *oceanic* Mollusks, which are found swimming near the surface in the open ocean, far from land, and often in enormous numbers. The organs of locomotion are two *wing-like* fins (Fig. 88) attached to the sides of the head, and formed by a

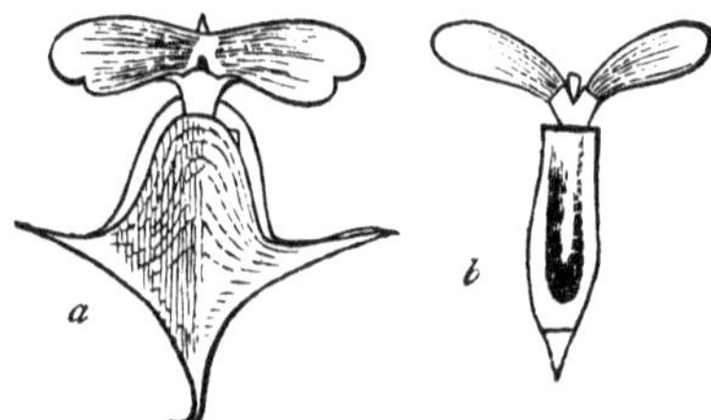

FIG. 88.—Pteropoda. *a Cleodora pyramidata; b Cuvieria columnella.* (After Woodward.)

modification of a portion of the foot. The body is usually protected by a symmetrical glassy shell (Fig. 88), consisting of two plates united along their edges, or in other cases forming a spiral. In some, however, there is no shell, and the body is quite naked. The head is rudimentary, and bears the mouth, which is furnished with an odontophore. The heart consists of an auricle and ventricle, and the respiratory organs are extremely rudimentary. The sexes are united in the same individual in all the *Pteropoda.*

The Pteropoda occur, as already said, in the open ocean, and they are found in all seas from the tropics to within the arctic circle, sometimes in such numbers as to discolor the water for many miles. Minute as they are, they constitute in high latitudes one of the staple articles of diet of the whale, and they themselves in turn are probably carnivorous, feeding upon small Crustaceans and other diminutive creatures. Though all the living forms are small, geology leads us to believe that formerly there existed comparatively gigantic forms, which appear to be truly referable to this class.

CHAPTER XXII.

CEPHALOPODA.

Class IV. Cephalopoda.—The last and highest class of the *Mollusca* is that of the *Cephalopoda*, comprising the Cuttle-fishes, Calamaries, Squids, and the Pearly Nautilus. They are all inhabitants of the sea, and are all carnivorous; and they are possessed of considerable powers of locomotion. At the bottom of the sea they can walk about, head downward, by means of the arms (Fig. 89), which surround the mouth, which are usually provided with numerous suckers, and which are really produced by a splitting up of the margin of the foot. It is from the presence of these arms that the class derives its name (Gr. *kephale*, head; and *podes*, feet). The Cuttle-fishes can also swim rapidly, either by means of expansions of the skin constituting fins, or by the forcible expulsion of water from the cavity of the mantle, the reaction of which causes the animal to move in the opposite direction. The majority of the living Cephalopods are naked, possessing only an internal skeleton, and this often a rudimentary one; but the Argonaut (Paper Nautilus) and the Pearly Nautilus are protected by an external shell, though the nature of this is extremely different in the two forms.

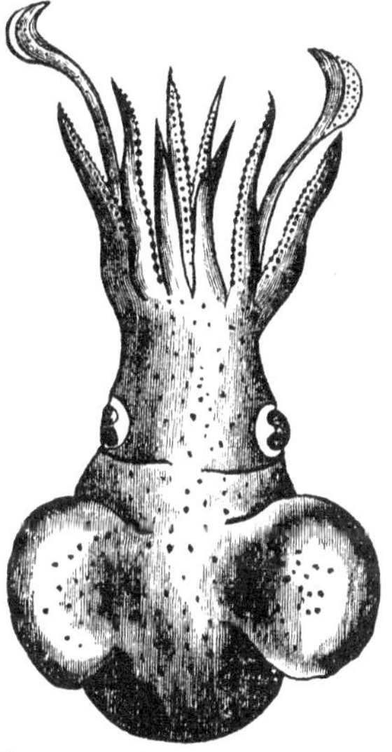

Fig. 89.—*Sepiola Atlantica*, one of the Cuttle-fishes (after Woodward).

The body in the *Cephalopoda* is symmetrical, and is enclosed in an integument which may be regarded as a modification of the mantle of the other *Mollusca*. Ordinarily there is a tolerably distinct division of the body into an anterior portion, carrying the head, and a posterior portion, in which the internal organs are enclosed. The head (Fig. 89) is very distinct, bearing a pair of large globular eyes, and having the mouth in its centre. The mouth is surrounded by a circle of eight, ten, or more, long muscular processes, or arms, which are generally provided with rows of suckers. Each sucker consists of a cup-shaped cavity, the muscular fibres of which converge to the centre, where there is a little muscular eminence. When the sucker is applied to any surface, the contraction of the radiating muscular fibres depresses the central eminence so as to produce a vacuum below it, and in this way each sucker acts most efficiently as an adhesive organ. The whole of this complex mechanism of suckers is completely under the control of the animal, and the irritability of the suckers is retained even for days after death. In most of the Cuttle-fishes (*Octopoda*) there are only eight arms, and these are nearly similar to one another. In others, however (Fig. 89), there are ten processes round the mouth, of which eight are like each other, and constitute the true arms, while two—called tentacles—are much longer than the others, and bear suckers only toward their extremities, which are enlarged and club-shaped. The Paper Nautilus (Fig. 90) has two of the arms webbed at their extremities and secreting a shell; and the Pearly Nautilus, alone of all living *Cephalopoda*, has numerous arms, more than ten in number, and destitute of suckers.

The mouth leads into a cavity containing two powerful horny or partially calcareous jaws working vertically, very like the beak of a bird, together with an "odontophore" or "tongue," the hinder part of which is furnished with recurved spines. This cavity leads by a gullet, furnished with salivary glands, into a stomach, from which an intestine is continued to terminate by a distinct anus, which opens on the ventral surface at the base of the so-called "funnel." The funnel is a muscular tube placed on the under surface of the head, and communicating on the one hand with the external medium, and on the other with the cavity of the mantle. In the *Nautilus* alone it is simply formed of two muscular lobes, which are in apposition, but are not united together so as to form a tube. In many cases there is also a special gland, known as the "ink-bag," for the secretion of an inky fluid, which the animal discharges into the water, so as to enable it to escape when menaced or pursued. The duct of the ink-bag opens at the base of the funnel near the anus, but the Pearly Nautilus and the allied fossil forms are without this means of defence, which the presence of an external shell renders unnecessary.

The respiratory organs are in the form of plume-like gills, placed on the sides of the body in a branchial chamber, which opens in front on the under surface of the body. In almost all the living *Cephalopoda* there are only two gills, one on each side, and hence this section is known as that of the "*Dibranchiata.*" In the Pearly Nautilus alone there are four gills, two on each side, hence the name of "*Tetrabranchiata*" applied to the order of which this is the only living representative. In the Cuttle-fishes, at the base of each gill is a special contractile cavity, called a "branchial heart," by which the venous blood, returned from the body, is driven through the gills. In addition to these branchial hearts there is a true arterial heart, by which the aërated blood received from the gills is driven through the body. The admission of water to the branchiæ is effected by the expansion of the mantle, which allows the entrance of the outer water into the mantle-cavity. The mantle then contracts, and the water is forcibly expelled through the funnel, which is often furnished with a valve, allowing the passage of water outward, but preventing its entrance inward. By a repetition of this process both respiration and locomotion are simultaneously effected, for the jets of water expelled from the funnel by their reaction drive the animal in the opposite direction. In this case, therefore, as in many others, the more active the animal is, the more perfectly is the respiratory process carried on.

The nervous system is formed upon essentially the same plan as in the other *Mollusca*, but the cerebral ganglia are protected by a cartilage, which is to be regarded as a rudimentary skull. This structure, therefore, is a decided approach to the Vertebrate type of organization.

The sexes in all the *Cephalopoda* are in different individuals, and the reproductive process in the Cuttle-fishes is attended with some singular phenomena. The most remarkable point in this connection is the modification of one of the arms of the male Cuttle-fishes, for the purpose of conveying the male element to the female. The details of the modification vary in different species of Cuttle-fish.

In some species one arm is simply so modified as to be

able to transmit the sperm-cells to the female, but it remains permanently attached to the animal. In the Paper Nautilus (*Argonaut*) the process goes still further. The female of this species (Fig. 90) attains a considerable size, and is protected by an external shell. The male is not more than an inch in length, is devoid of a shell, and has its third left arm metamorphosed. This arm is developed in a cyst, and is ultimately detached from the body, and deposited by the male within the mantle-cavity of the female. When first discovered in this position, it was described as a worm living parasitically on the Argonaut, under the name of "Hectocotylus" (Gr. *hekaton*, a hundred; and *kotulos*, a cup), from the suckers, or cups, with which it was furnished. Subsequently it was described as the entire male Argonaut; and it is only recently that it has been proved to be nothing more than one of the arms of the male, detached for the purpose of conveying the sperm-cells to the female.

The *shell* of the *Cephalopoda* is sometimes external, sometimes internal. The internal skeleton is seen in the various Cuttle-fishes, in which it is known as the "cuttle-bone" or "pen." It may be either horny or calcareous, and it is sometimes complicated by the addition of a chambered portion. The only living Cephalopods which are provided with an external shell are the Paper Nautilus (*Argonauta*) and the Pearly Nautilus (*Nautilus pompilius*); but not only is the structure of the animal different in each of these, but the nature of the shell itself is entirely different. The shell of the Argonaut (Fig. 90) is coiled into a spiral, but it is not divided into chambers, and it is secreted by the webbed extremities of two of the dorsal arms of the female. These arms are bent backward, so as to allow the animal to live in the shell; but there is no organic connection between the shell and the body of the animal. The shell of the Pearly Nautilus, on the other hand, is secreted by the mantle, and is organically connected to the animal. It is coiled into a spiral (Fig. 91), but it differs from the shell of the Argonaut in being divided into a series of chambers by means of shelly partitions, which are connected together by a tube or "siphuncle," the animal itself living in the last and largest chamber only of the shell.

The *Cephalopoda* are divided into two extremely distinct and natural orders, termed respectively *Dibranchiata* and *Tetrabranchiata*, according as they have *two* or *four* gills or branchiæ.

The *Dibranchiata* comprise the Cuttle-fishes, Squids, Calamaries, and Paper Nautilus, and they are characterized by being almost invariably destitute of any external shell; by never having more than eight or ten arms, which are always furnished with suckers; by having only two gills, which are provided with "branchial hearts;" by the possession of an "ink-bag;" and by the fact that the "funnel" forms a complete tube. They are divided into two sections—*Octopoda* and *Decapoda*—according as they have only eight arms, or eight arms with two additional longer processes or "tentacles" (Fig. 89). Among the *Octopoda* are the Paper Nautilus and the Poulpes (*Octopus*). The Paper Nautilus is found in the warmer seas of various parts of the world, generally floating at the surface. The two sexes differ, as already said, greatly in external appearance. The female (Fig. 90) inhabits a beau-

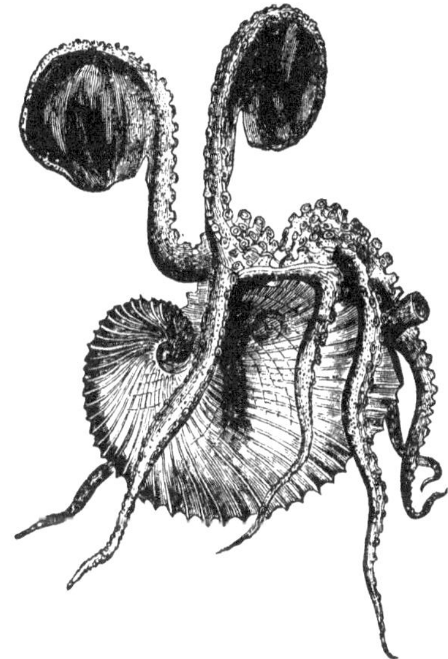

FIG. 90.—*Argonauta argo*, the Paper Nautilus, female. The animal is represented in its shell, but the webbed dorsal arms are separated from the shell which they secrete, and which they ordinarily embrace.

tiful one-chambered shell, which is secreted by the webbed extremities of two of the dorsal arms. The shell is not in any way attached to the body of the animal, but the webbed arms are turned backward, and the animal sits in the shell with the funnel turned toward the keel. It swims by the jets of water emitted from the funnel, and crawls upon the sea-bottom, head downward, carrying its shell on its back. The male Argonaut is only about an inch in length, has no shell, and has all its arms alike, except the one which is metamorphosed into the "hectocotylus." The Poulpes (*Octopi*) are universally distributed in the seas of both temperate and tropical regions. They are the "polypi" of Homer and Aristotle, and are voracious animals inhabiting rocky shores.

The *Decapoda* are chiefly found in the open sea, often in enormous numbers, and the best known are the Calamaries and Squids. The body is elongated, and is always furnished with lateral fins, with which they swim actively. The shell is internal, and differs considerably in different members of the group. To this section of the *Dibranchiata* belong the singular fossil forms which are known to the geologist as Belemnites. These singular forms are known almost solely by their complicated internal skeleton, and they appear to have abounded in the seas of the Secondary period.

The second order of the *Cephalopoda*—that of the *Tetrabranchiata*—comprises forms characterized by being creeping animals, protected by an *external*, *many-chambered shell*, the partitions between the chambers being perforated for the passage of a membranous or calcareous tube, termed the "siphuncle." The arms are more than ten in number, and are devoid of suckers; the gills are four in number, two on each side of the body; the funnel does not form a complete tube; and there is no ink-bag.

Though abundantly represented by many and varied fossil forms, the only living member of the *Tetrabranchiata* with which we are acquainted is the Pearly Nautilus, which has long been known by its beautiful chambered shell. The shell of the Pearly Nautilus (Fig. 91) is coiled into a spiral, and is many-chambered, the chambers being walled off from one another by curved shelly partitions or septa, perforated centrally by a foramen which transmits a membranous tube or siphuncle. The animal inhabits only the last and largest chamber of the shell, from which it can protrude its head at will. The function of the chambers of the shell is not very clearly understood; but it appears to be that of reducing the specific gravity of the shell to near that of the surrounding water; since they appear to be filled with some gas apparently secreted by the animal. The siphuncle does not communicate in any way with the chambers of the shell, and its functions are also unknown, except that it must certainly serve to maintain the vitality of the shell.

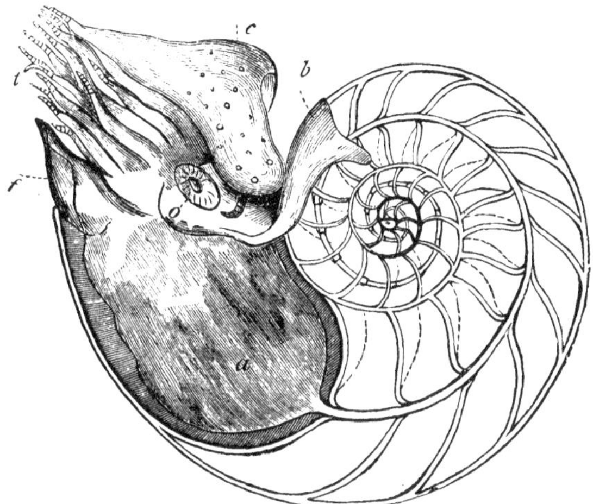

Fig. 91.—The Pearly Nautilus (*Nautilus pompilius*). *a* Mantle; *b* Its dorsal fold; *c* Hood; *o* Eye; *t* Tentacles; *f* Funnel.

Of the fossil *Tetrabranchiata* the most important are the *Orthocerata* and the *Ammonites*. The *Orthocerata* (Fig. 92) played a very important part in the seas of the Palæozoic or Ancient-life period of the earth's history, in which they apparently filled the place now taken by the predacious cuttle-

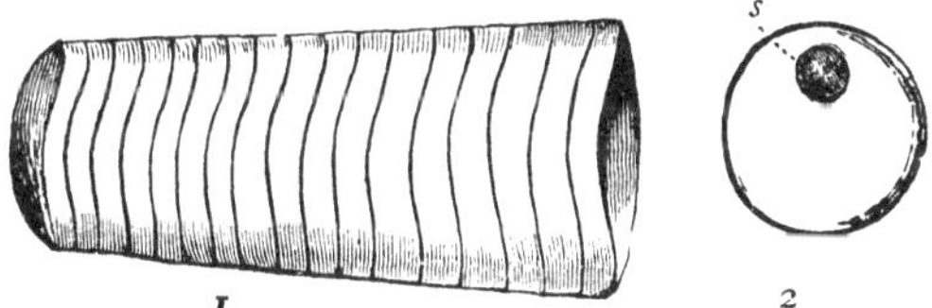

Fig. 92.—*Orthoceras explorator*. 1. Side view of a fragment, showing the edges of the septa. 2. Transverse section of the same, showing the siphuncle (*s*). (Billings.)

fishes. They agreed with the *Nautilus* in having a many-chambered shell, divided by curved partitions, perforated by a tube or siphuncle. The shell, however, differed from that of the *Nautilus* in not being curved or coiled up, but in being straight. In other nearly-allied forms the shell was bent or even partially coiled up, but never so completely as in the true *Nautilus*. Many of the *Orthocerata* were of small size, but some of them were colossal, shells having been found of six or seven feet in length, and as thick as the body of a man.

The *Ammonites*, with a number of allied forms of varied shapes and beautiful structure, appear to have taken the place of the *Nautilidæ*, to a great extent, in the seas of the Secondary period; at which time, too, Dibranchiate Cephalopods first made their appearance. The true Ammonites resembled the *Nautilus* in having a many-chambered shell, which was coiled up into a spiral, but the position of the siphuncle was different, and the partitions or septa between the various chambers of the shell were wonderfully folded and lobed instead of being simply curved. The numerous beautiful shells allied to the Ammonites cannot be even mentioned here; but it is to be remembered that they are almost all characteristic of the Secondary period in geology, and that they are hardly known as occurring in the older period (Palæozoic epoch).

VERTEBRATE ANIMALS.

CHAPTER XXIII.

GENERAL CHARACTERS OF THE VERTEBRATA.

THE five sub-kingdoms which we have previously considered, namely, the *Protozoa*, *Cœlenterata*, *Annuloida*, *Annulosa*, and *Mollusca*, were grouped together by Lamarck into one great division, which he termed the *Invertebrata*. The remaining sub-kingdom, that of the *Vertebrata*, is so well marked and compact a division, and its distinctive characters are so numerous and so important, that this mode of viewing the animal kingdom is, at any rate, a very convenient one.

The sub-kingdom *Vertebrata* includes the five great classes of the Fishes (*Pisces*), Amphibians, Reptiles, Birds (*Aves*), and Mammals; and the name of the sub-kingdom is derived from the very general, though not universal, presence of the bony axis known as the "vertebral column" or backbone. One of the most fundamental of the distinctive characters of Vertebrate animals is to be found in the fact that the main masses of the nervous system (that is to say, the brain and spinal cord) are completely shut off from the general cavity of the body. In all Invertebrate animals (Fig. 93, A), the body may be regarded as a single tube, enclosing all the viscera; and, consequently, when a distinct nervous system and alimentary canal are present, these are in no way shut off from one another. The transverse section, however, of any Vertebrate animal (Fig. 93, B) shows *two* tubes, one of which contains the great nervous axis (n') or brain and spinal cord, while the other contains the alimentary canal, the chief circulatory organs, and certain portions of the nervous system (n), which are known to anatomists as the "sympathetic" system. Leaving the brain and spinal cord out of sight for a moment, we see that the lower or visceral tube of a Vertebrate animal contains the digestive canal (*b*), the blood-vascular system (*c*), and a system of nervous ganglia (*n*). Now, this is

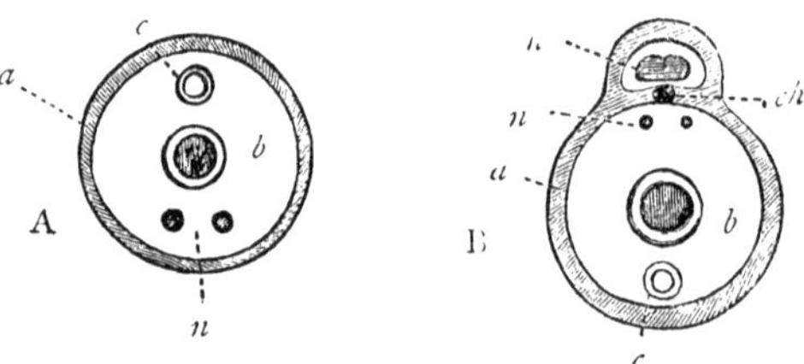

FIG. 93.—Diagrams representing transverse sections of one of the higher Invertebrata, A, and one of the Vertebrata, B. *a* Wall of the body; *b* Alimentary canal; *c* Hæmal or blood-vascular system; *n* Nervous system; *n'* Cerebro-spinal axis, or brain and spinal cord, enclosed in a separate tube; *ch* Notochord, or chorda dorsalis.

exactly what is contained within the visceral cavity of any Invertebrate animal; and it follows from this that it is the "sympathetic" system of Vertebrate animals which is truly comparable with the nervous system of the Invertebrata. The brain and spinal cord, or "cerebro-spinal axis," are to be looked upon as something not represented at all in the Invertebrata.

Another peculiarity which is present in all the *Vertebrata* is, that at an early period of life there is developed, in the lower wall of the tube which contains the cerebro-spinal axis, a singular structure known as the "notochord" (Gr. *notos*, back; *chordé*, string) (Fig. 93, B, *ch*). This is a semi-gelatinous rod, tapering at both ends, and extending along the floor of the cerebro-spinal tube. In some cases, the notochord remains permanently in this condition, but, in most cases, it is replaced at maturity by the bony column or backbone, from which the *Vertebrata* derive their name. The general structure of the vertebral column will be described shortly, and it is sufficient to state here that it consists of a series of more or less completely bony segments or "vertebræ," arranged so as to form a longitudinal axis upon which the spinal cord is supported. It is to be remembered, however, that all Vertebrate animals do not possess a vertebral column. They all possess a notochord, but this may remain persistent throughout life, and, in many cases, the development of the spinal column is very imperfect.

The skeleton of all Vertebrate animals is *internal*, and the muscles are attached to its several parts. The value of this character is in no way affected by the fact that many Vertebrates, such as the Tortoises, Crocodiles, and others, possess an *external* skeleton as well. The *limbs* of Vertebrate animals are always articulated or jointed to the body, and they are always turned away from that side of the body (the "neural" side) upon which the great masses of the nervous system are placed. The limbs may be altogether wanting, or partially undeveloped, but there are never *more than two pairs*, and they always have an internal skeleton for the attachment of the muscles of the limb.

A distinct blood-vascular system is present in all Vertebrates, and in all except one—the Lancelet—there is a single contractile cavity or heart, furnished with valvular openings.

Lastly, the masticatory organs of all Vertebrates are modifications of parts of the walls of the head, and are never modified limbs or hard structures developed in the mucous membrane of the digestive tube, as they are in the Invertebrates.

The above are the leading characters which distinguish the *Vertebrata* as a whole, and, before going on to consider the different classes, it may be as well to give a short and general sketch of the anatomy of the Vertebrates, commencing with their bony framework or skeleton.

The skeleton of the *Vertebrata* may be regarded as consisting of the bones which go to form the trunk and head on the one hand, and of those which form the supports for the limbs on the other hand. The bones of the trunk and head may be regarded as essentially composed of a series of bony rings or segments, arranged longitudinally. Anteriorly, these segments are much expanded and also much modified to form the bony case which encloses the brain and which is termed the *cranium* or skull. Behind the head, the segments enclose a much smaller cavity in which is contained the spinal cord, and they are arranged one behind the other, forming the "vertebral column." The segments which form the vertebral column are called "vertebræ," and they have the following general structure: Each vertebra (Fig. 94, A) consists of a central portion known as the "body," or "centrum" (*c*), placed immediately below the spinal cord, and giving origin to certain "processes." The ends of the bodies of the vertebræ are all united together in different ways, so as to give the col-

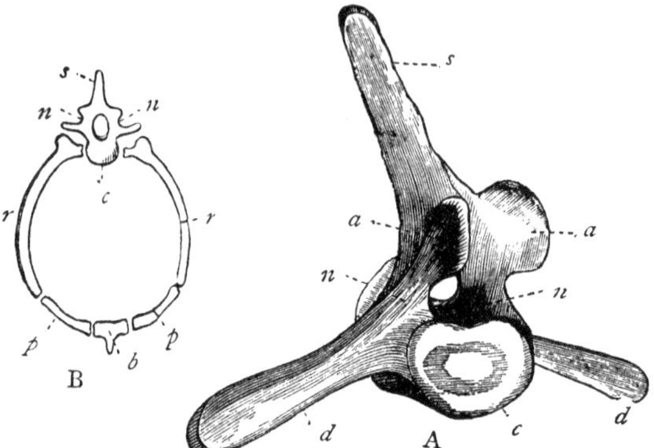

Fig. 94.—A. Vertebra (lumbar) of the whale. *c* Centrum or body; *n* Neural arches; *s* Spinous process; *a* Articular process; *d* Transverse processes. B. Thoracic segment or vertebra. *c* Centrum of vertebra; *n* Neural arches, enclosing the canal for the spinal cord; *e* Spinous process; *r* Ribs; *p* Costal cartilages; *b* Breastbone or sternum. (After Owen.)

umn great flexibility. From the back of the body of the vertebra proceed two bony arches which unite behind and thus form with the centrum a bony canal in which the spinal cord is contained. For this reason, these arches (*n*) are called the "neural" arches. From the point where the neural arches unite—that is to say, from the back of the neural canal—proceeds a long process, sometimes cleft at its extremity, termed the "spinous process" (*s*). Springing also from each neural arch is a second shorter process (*a*) termed the "articular process," since by means of these, as well as by the bodies, the vertebræ are jointed or "articulated" together. Also arising from the neural arches at their junction with the body of the vertebra, there may be two lateral processes (*d*) which are called "transverse processes." This is the ordinary structure of the vertebra of a Mammal, and the names here used are those applied to the parts of the vertebra in human anatomy. In philosophical anatomy, however, these parts have proper technical names which can be employed for them in all animals alike. The nature of this work, however, will not allow of the introduction of these here.

In the typical vertebra the segment is completed by a second arch, which is placed in front of or beneath the body of the vertebra, and which is known as the "hæmal" arch, as it includes and protects the principal organs of the blood circulation (Fig. 94, B). This second arch is often only recognizable with great difficulty, as its parts are generally much modified; but a good example may be obtained in the human chest. Here, attached to the front of the vertebræ, we find a series of bony arches, known as the ribs (*r*), followed by a series of cartilaginous pieces of a similar shape, termed the "costal cartilages" (*p*), the whole united in front by a central bone, known as the breastbone or "sternum" (*b*).

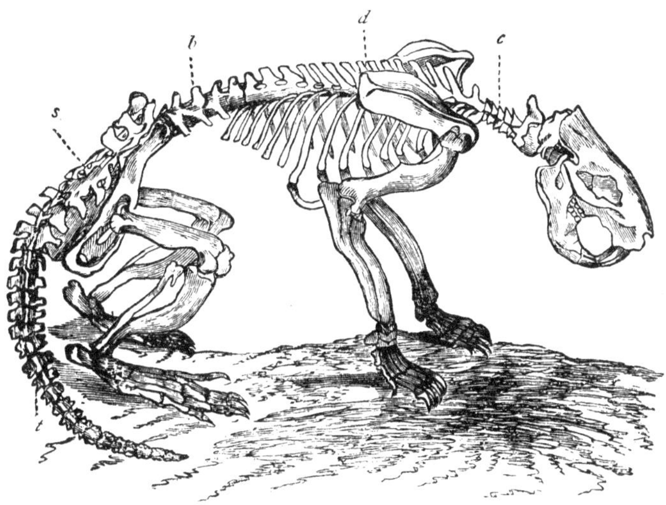

FIG. 95.—Skeleton of the Beaver (*Castor fiber*), showing the regions of the vertebra column. *c Cervical region*, or region of the neck; *d Dorsal region*, or region of the back; *b Lumbar region*, or region of the loins; *s* Sacrum; *t Caudal region*, or region of the tail.

As a general rule, among the higher Vertebrates, the following *regions* may be recognized in the vertebral column: Firstly, the *cervical region* (Fig. 95, *c*), comprising a variable number of vertebræ, which constitute the neck, and immediately follow the head. Secondly, the cervical region is succeeded by a variable number of vertebræ which usually carry ribs, and are known as the *dorsal* vertebræ (*d*), or vertebræ of the back. Thirdly, come certain vertebræ which constitute the *lumbar* region (*b*), or the region of the loins. Fourthly, there usually follows a series of vertebræ which are immovably united together to form a single bone, which is termed the *sacrum* (*s*). Lastly, there comes a variable series of vertebræ which are usually free and movable upon one another, and which constitute the *caudal* region, or the region of the tail (*t*).

The nature of the bones which enter into the composition of the limbs varies somewhat in different Vertebrates in accordance with their mode of life; but in all the higher members of the sub-kingdom the limbs are built upon a general

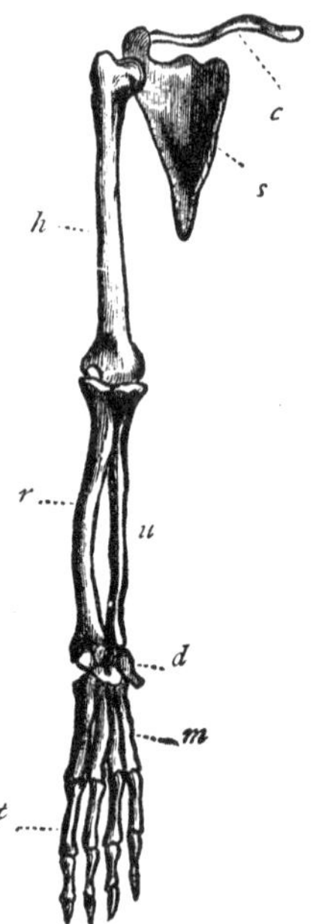

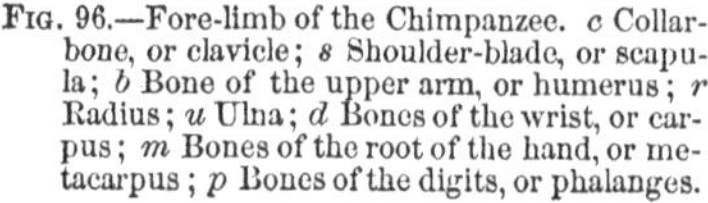

Fig. 96.—Fore-limb of the Chimpanzee. *c* Collar-bone, or clavicle; *s* Shoulder-blade, or scapula; *b* Bone of the upper arm, or humerus; *r* Radius; *u* Ulna; *d* Bones of the wrist, or carpus; *m* Bones of the root of the hand, or metacarpus; *p* Bones of the digits, or phalanges.

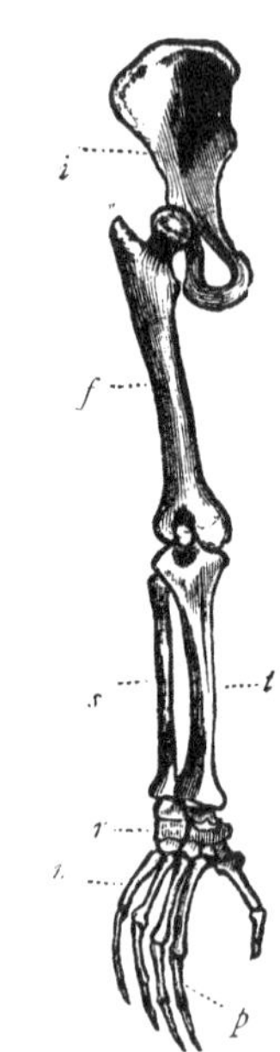

Fig. 97.—Hind-limb of the Chimpanzee. *i* Innominate bone; *f* Thigh-bone, or femur; *t* Tibia; *s* Fibula; *r* Bones of the ankle, or tarsus; *m* Metatarsus; *p* Phalanges.

and easily-recognizable type. The fore-limb consists generally of the following parts: 1. A series of bones uniting the limb to the trunk, the two most important being the shoulder-blade (*scapula*) and the collar-bone (*clavicle*) (Fig. 96, *s* and *c*). 2. The bone which forms the upper portion of the limb proper, and which is known as the *humerus* (*b*). 3. Two bones which form the lower portion of the limb (e. g., the forearm in man), and which are known as the *radius* and *ulna* (*r* and *u*), of which the former is the bone mainly concerned in carrying the hand or fore-foot. 4. A number of small bones, which form the wrist, and are termed the *carpus* (*d*). 5. The cylindrical bones (usually five in number) which form the root of the hand, and are known as the *metacarpus* (*m*). 6. The bones which form the fingers proper, and which are known as the *phalanges* (*p*).

Essentially the same parts can be traced in the hind-limb of a typical Vertebrate animal, but they are known by different names. The bones which unite the limb to the trunk are usually more or less completely united together, constituting a single mass, known as the *innominate bone* (Fig. 97, *i*). This is followed by a long, cylindrical bone, which forms the upper portion of the hind-limb, and is known as the "thigh-bone," or *femur* (*f*). Following this are the two bones of the shank, corresponding to the *radius* and *ulna* of the fore-limb, and known as the *tibia* and *fibula* (*t* and *s*). Of these, the *tibia* (*t*) corresponds to the *radius*, and is mainly concerned in carrying the foot. Next comes a series of small bones, which form the ankle, and are known as the *tarsus* (*r*). This is succeeded by a series of cylindrical bones (usually five in number), which form the root of the foot, and which are termed the *metatarsus* (*m*). Finally, the metatarsus is succeeded by the bones of the toes, which in this case are again termed the *phalanges* (*p*). In both limbs the usual number of phalanges to each toe or "digit" is three.

The digestive system of the *Vertebrata* does not require a lengthened notice. The mouth is usually furnished with *teeth*, which have for their chief function the reduction of the food to a condition in which it can be digested. In some animals, however, such as the snakes, the teeth are only used to hold the prey, and not for mastication; and in others, such as the turtles and birds, the jaws are not furnished with any teeth at all. The food is also usually subjected in the mouth to the action of a special fluid—the saliva—which acts chemically as well as mechanically upon the food, and which is secreted by special glands, known as the "salivary glands." From the mouth the food passes through a muscular tube—the gullet, or *œsophagus* (Fig. 98, *g*)—to the proper digestive cavity, or stomach (*s*). Here it is subjected to the action of a special digestive fluid—the "gastric juice"—and is converted into a thick, pasty fluid, which is called *chyme*. From the stomach the chyme passes into a long, convoluted, muscular tube, which is called the "small intestine" (*sm*). Here it is subjected to the action of two other digestive fluids, called the "bile" and "pancreatic juice," as well as to the fluids secreted by the intestine itself. The bile is secreted by a large gland, which is known as the "liver," while the pancreatic juice is produced by another, termed the "pancreas," both pouring their secretion into the upper part of the small intestine. By the combined action of these digestive fluids the chyme is ultimately converted into a milky fluid, which is called *chyle*, when it is fit to be taken up into the blood-vessels. The small intestine finally opens into a tube of larger diameter, which is called the "large intestine" (*lm*), and this opens on the surface of the body by an anal aperture. In the large intestine the last remaining portions of the food which can be rendered useful are absorbed into the blood, the indigestible portions being ultimately got rid of as useless. The fluid products of digestion (*chyle*) are chiefly absorbed from the intestinal canal by a set of special vessels, which are present in all Vertebrates, and which are called the *lacteals* (Lat. *lac*, milk) from the milky fluid they contain. These lacteals combine to form a large trunk, by which their contents are ultimately added to the circulating blood. Part of the products of digestion are absorbed by the veins which ramify on the intestinal canal, and which ultimately unite to form a great vessel, called the "vena portæ," which goes to the liver. The materials, however, which are taken up in this way also ultimately reach the circulating blood. In this way, therefore, fresh matter is being constantly added to the blood to replace the waste caused by the performance of the vital functions.

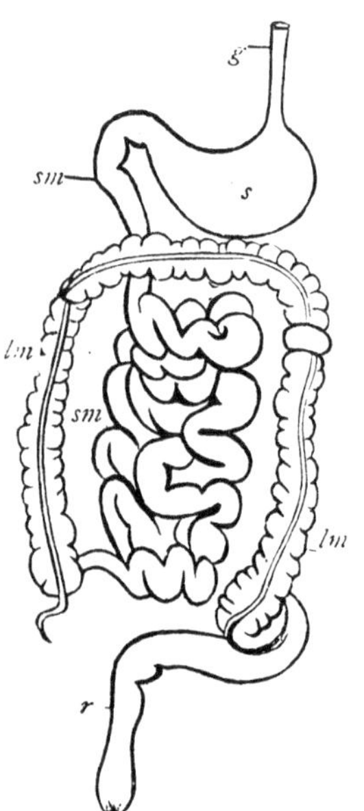

Fig. 98.—Digestive system of a Mammal. *g* Gullet, or œsophagus; *s* Stomach; *sm* Small intestine; *lm* Large intestine; *r* Large intestine terminating in its final portion, called the "rectum."

The blood is thus formed out of the materials which are taken into the alimentary canal as food; and in all the *Vertebrata* (with one exception) it is of a red color, when viewed in mass. This is due to the presence in it of numerous microscopical particles, which are known as the "blood-corpuscles," the fluid itself being colorless. In Fig. 99 are represented

FIG. 99.—Blood-corpuscles, magnified. *a* Man; *b* Goose; *c* Crocodile; *d* Frog; *e* Skate.

some of the forms of blood-corpuscles which are found in different divisions of the *Vertebrata.*

The blood is always distributed through the body by means of a system of closed tubes, which constitute the "blood-vessels," and, with the single exception of the Lancelet, it is always propelled by means of a contractile muscular cavity or "heart." The heart and other circulatory arrangements differ considerably in different classes of the *Vertebrata*, but these differences will be best considered at a later period. *Respiration* in all the *Vertebrata* is effected by means of distinct breathing-organs, assisted in many cases by the skin. In the water-breathing Vertebrates, such as fishes, the respiratory organs are in the form of *gills* or branchiæ, which are richly supplied with blood, and are exposed to the influence of water holding oxygen in solution. In the air-breathing Vertebrates, the breathing organs are in the form of *lungs.* These essentially consist of cellular or spongy organs, placed in the cavity of the chest, richly furnished with blood-vessels, and receiving constant supplies of fresh air by means of a tube which opens in the throat and is known as the "windpipe," or *trachea.* In the higher Vertebrates the heart becomes a double organ, one side being concerned wholly with driving the impure (*venous*) blood to the lungs, while the other side propels the pure oxygenated (*arterial*) blood to all parts of the body.

The waste substances of the body—of which the most important are *water*, *carbonic acid*, and the peculiar substance called *urea*—are got rid of by the skin and lungs, but principally by two glands which are called the *kidneys.* The excretion of *urea* from the body, as a general rule, is wholly effected by means of the kidneys alone; and this is their most important function, as the retention of this substance within the body rapidly causes death. The secretion of the kidneys is sometimes got rid of by means of special canals appropriated to this alone; but in the lower *Vertebrata* it is discharged into the hinder extremity of the alimentary canal, and is evacuated along with the undigested portions of the food.

The *nervous system* varies greatly in its development in the *Vertebrata*. In the little fish called the Lancelet, the main mass of the nervous system consists of a cord of nervous matter, representing the spinal marrow, but not having in front any enlargement which represents the brain. In all the other *Vertebrata* the central masses of the nervous system (termed the *cerebro-spinal axis*) consist of a nervous cord (the spinal cord) contained in the canal formed by the *neural* arches of the vertebræ, and of an anterior mass of nervous matter, which is protected by the skull, and is termed the *encephalon* or *brain*. The size and development, however, of the brain vary enormously in different Vertebrates; and in the lower forms the brain is little more than an aggregation or collection of nervous masses or "ganglia," which are connected with the special senses, sight, hearing, taste, and smell, special organs for which are present in almost all the *Vertebrata*.

Reproduction in the *Vertebrata* is always truly sexual, the sexes are always in different individuals, and in no case are compound organisms produced by a process of budding or fission. Most are *oviparous*, producing eggs from which the young are developed. Some retain the eggs within the body till the young are ready to be hatched, and these are sometimes said to be *ovo-viviparous*. The higher Vertebrates, however, bring forth their young alive, and are said to be *viviparous* (Latin, *vivus*, living; and *pario*, I bring forth).

PRIMARY DIVISIONS OF THE VERTEBRATA. — The *Vertebrata* are variously divided into great primary sections by different writers, and all of these divisions have more or less merit. Here, however, the classification proposed by Prof. Huxley will be followed, and it is not necessary to enter into any consideration of the others. It has also been thought advisable to give in this place a brief account of the leading characters which separate these divisions from one another, though it is not to be expected that the learner will be able to appreciate the full value of these characters till he has completed his study of the *Vertebrata* as a whole.

The *Vertebrata* are divided by Prof. Huxley into the following great divisions:

I. ICHTHYOPSIDA (Gr. *ichthus*, a fish; and *opsis*, appearance).—In this section are included the fishes (Class *Pisces*), and the frogs, newts, and their allies (Class *Amphibia*). They are all characterized by the fact that they possess gills or branchiæ, either throughout life or during the earlier stages of their existence; that they possess nucleated red blood-corpuscles (i. e., blood-corpuscles with a central particle or *nucleus*, Fig. 99, *d*, *e*), and by certain embryonic characters as well. From the temporary or permanent possession of gills, they are often spoken of as the *Branchiate* Vertebrates.

II. Sauropsida (Gr. *saura*, a lizard; and *opsis*, appearance).—In this division are the birds (Class *Aves*), and the true reptiles (Class *Reptilia*). They are characterized by the fact that at no time of their life are they ever provided with gills; that the skull is jointed to the vertebral column by a single articulating surface (or *condyle*); that the lower jaw is composed of several pieces, and is united to the skull by means of a special bone (called the *os quadratum*); that they possess nucleated red blood-corpuscles (Fig. 99, *b*, *c*), and by certain embryonic characters as well.*

III. Mammalia (Lat. *mamma*, the breast).—In this division are all the ordinary quadrupeds; characterized by the constant absence of gills; by the skull being jointed to the vertebral column by two articulating surfaces (or condyles); by the fact that the lower jaw is composed of only two pieces, and is not united to the skull by means of a special bone (the *quadrate* bone); by having non-nucleated red blood-corpuscles (Fig. 99, *a*); and by having special glands—the mammary glands—which secrete a special fluid—the milk—by which the young are nourished for a longer or shorter period after birth.

These three primary divisions comprise the five great classes into which the *Vertebrata* are divided:

1. Fishes (*Pisces*).
2. *Amphibia* (Frogs, Newts, etc.).
3. *Reptilia* (True Reptiles).
4. *Aves* (Birds).
5. *Mammalia*.

* Recent researches have led to the belief that the appearance of nuclei in the red blood-corpuscles of the Oviparous Vertebrates is due to changes taking place after death, and that these structures are not present during life.

ICHTHYOPSIDA.

CHAPTER XXIV.

CLASS I. PISCES.

THE fishes form the lowest class of the *Vertebrata*, and they may be broadly defined as being Vertebrate animals *provided with gills*, whereby they are enabled to breathe air dissolved in water; *the heart*, when present, *consists of a single auricle and ventricle* (with the exception of the mud-fishes); and the *limbs*, when present, are *in the form of fins, or expansions of the integument.*

In their external form, fishes are in most cases adapted for rapid locomotion in water, the shape of the body being such as to cause the least possible friction in swimming. To this end, as well as for purposes of defence, the body is generally enveloped in a species of chain-mail formed by overlapping scales, to which bony plates, tubercles, and spines, are sometimes added. Valuable characters can sometimes be drawn from the nature of the scales, and with a view to this the integumentary appendages of fishes have been divided by Agassiz as follows (Fig. 100):

1. *Cycloid* scales (*a*), consisting of thin, flexible, horny scales, which are circular or elliptical in shape, and have a smooth outline. These scales occur in most of our common fishes (e. g., the pike).

2. *Ctenoid* scales (*b*). These resemble the cycloid scales in being thin, flexible, horny scales, but they are distinguished by having their hinder margins cut into comb-like projections, or fringed with spines. The common perch supplies a good example of these scales.

3. *Placoid* scales (*c*). These are detached bony grains,

tubercles, or plates, scattered through the skin, and sometimes armed with projecting spines.

4. *Ganoid* scales (*d*,) composed of a layer of true bone, covered by a layer of hard polished enamel. These scales are usually much thicker and larger than the ordinary scales; they are often oblong or rhomboidal in shape; they are often connected together by little processes; and they generally are in contact by their edges, but rarely overlap one another. In most fishes there is also to be observed a line of peculiar scales, forming what is called the "lateral line." Each of the scales of this line is perforated by a minute tube, which leads into a longitudinal canal, believed to secrete the mucus with which the general surface is lubricated, or to have some sensory function.

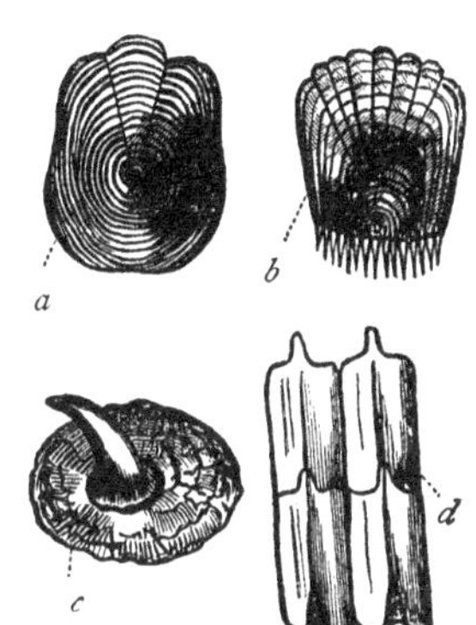

FIG. 100.—Scales of different Fishes. *a* Cycloid scale (Pike); *b* Ctenoid scale (Perch); *c* Placoid scale (Thornback); *d* Ganoid scale (*Palæoniscus*).

As regards the true internal skeleton, fishes differ very widely from one another, but the skeleton is so complicated that only a few of the most important points can be mentioned here. In one fish—the Lancelet—there can hardly be said to be any true skeleton, the vertebral column being represented permanently by the semi-gelatinous notochord (Fig. 105). In others, such as the Lampreys, Sturgeons, and Rays, the skeleton remains permanently in the condition of gristle (cartilage); in others it is partially cartilaginous and partially ossified; and, lastly, in most modern fishes it is completely converted into bone. The *vertebral column* in a bony fish consists of a number of vertebræ which are hollow or cup-shaped at both ends (bi-concave or "amphicœlous"), the cup-like margins being united together by ligaments. The cavities formed by the apposition of the vertebræ are filled with the gelatinous remains of the notochord. This gelatinous elastic substance acts as a ball-and-socket joint between the vertebræ, thus giving the whole spine the extreme flexibility which is essential to animals living in a watery medium. The entire spinal column is divisible into no more than two distinct regions, an *abdominal* and a *caudal*. The *ribs* are attached to the transverse processes or to the bodies of the abdominal vertebræ (Fig. 101, *r*); and they do not enclose any thoracic cavity, or protect the organs which are usually contained in the chest—namely, the heart and breathing-organs. The anterior or lower ends of the ribs of fishes are free, or are rarely united to hard productions of the integument; but there is never any breastbone or *sternum* properly so called.

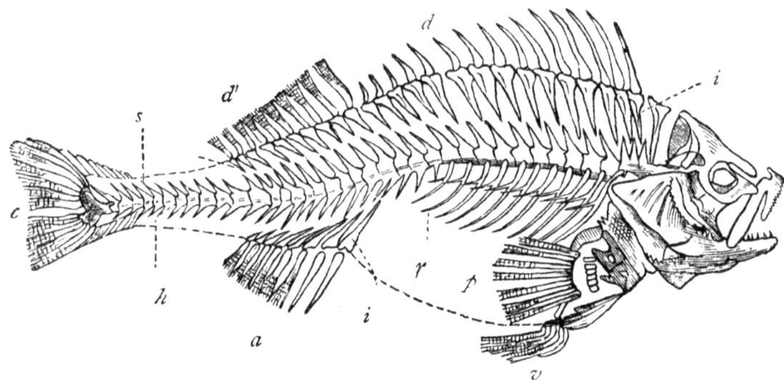

FIG. 101.—Skeleton of the common Perch (*Perca fluviatilis*). *p* Pectoral fin; *v* One of the ventral fins; *a* Anal fin, supported upon interspinous bones (*i*); *c* Caudal fin; *d* First dorsal fin; *d'* Second dorsal fin, both supported upon interspinous bones; *i i* Interspinous bones; *r* Ribs; *s* Spinous processes of vertebræ; *h* Hæmal processes of vertebræ.

The only remaining bones of the trunk proper are the so-called "interspinous bones" (Fig. 101, *i i*). These are a series of pointed, dagger-like bones, imbedded in the middle line of the body, between the great lateral muscles which form the greater part of the body of a fish. The inner ends or points of the interspinous bones are attached by ligament to the spinous processes of the vertebræ, and at their outer ends they support the framework (rays) of the so-called "median" fins. As a rule there is only one interspinous bone to each vertebra, but in the flat-fishes (Sole, Turbot, etc.) there are two. The limbs of fishes may be wholly wanting, or one pair may be absent, but in no case is the number greater than the regular vertebrate type—namely, two pairs. When developed, however, the limbs of fishes are very different from those of other Vertebrates, consisting of expansions of the integument, furnished with bony or gristly supports or rays, and thus constituting what are called "fins" (Fig. 102). The pair of limbs which correspond to the arms of man and to the fore-limbs of other Vertebrates are termed the *pectoral* fins, and they are attached to a bony arch which is attached either to the back of the skull or to the spinal column (Fig. 101, *p*, and 102, *p*). The hind-limbs in fishes are known as the *ventral* fins (Figs. 101, 102, *v*), and are not only often wanting altogether, but when present are less developed than the pectorals and less fixed in their position. They are united to an imperfect bony

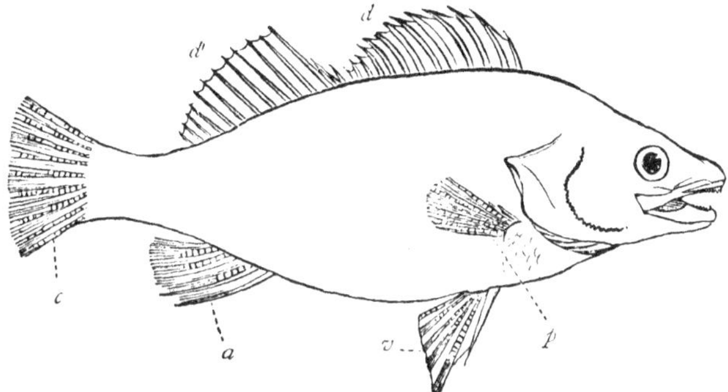

FIG. 102.—Outline of a Fish (*Perca granulata*), showing the "paired" and "median" fins. *p* Pectoral fin; *v* Ventral fin; *d* First dorsal fin; *d'* Second dorsal fin; *c* Caudal fin; *a* Anal fin.

arch, which represents the innominate bones, or pelvic arch, of the higher Vertebrates, but which is never joined to the spinal column. In some fishes the ventral fins are placed far back, and in these the bony arch which supports them is freely suspended in the muscles. In others the ventral fins are altogether out of position, and are placed beneath, or even in front of the pectoral fins; and in these cases the pelvic arch is attached to part of the pectoral arch. The pectoral and ventral fins represent, as just said, the fore and hind limbs, and consequently there are always *two* of each, when they are present at all. They are, therefore, spoken of as the "paired" fins. Besides these, however, or in the absence of one or other of these, there is also a series of what are called "median" fins; that is to say, fins which are placed in the middle line of the body, and which are *unpaired*, having no fellows. These median fins agree with the paired fins in being expansions of the integument, supported by bony or gristly supports or "rays," and they are carried by the heads of the "interspinous" bones, already described (Fig. 101, *i i*). They are variable in number, and in some cases there is only a single fringe running round the hinder extremity of the body. Commonly, however, the median fins consist of one or two expansions of the dorsal integument, called the "dorsal" fins (Fig. 101, *d d'*); one or two on the ventral or lower surface near the vent, called the "anal" fins (*a*); and a broad fin at the extremity of the vertebral column, constituting the "caudal" fin or tail (*c*).

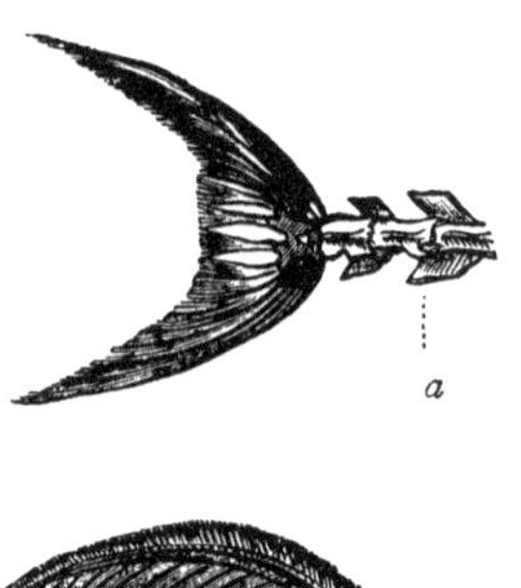

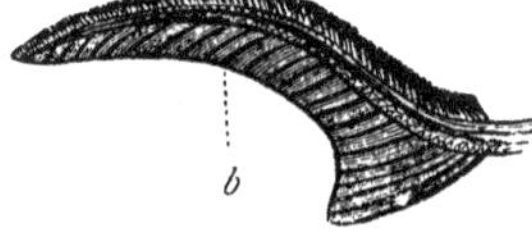

FIG. 103.—Tails of different Fishes. *a* Homocercal tail (Sword-fish); *b* Heterocercal tail (Sturgeon).

The tail in all fishes is placed vertically—that is to say, it strikes the water laterally, or from side to side, and it is the chief organ of progression in the fish. Two very distinct types of tail are found among the fishes. In one of these, found in most living forms, the tail is composed of two nearly equal lobes which spring from the end of the spine (Fig. 103, *a*). This form of tail is said to be "homocercal." In the other type of tail, found in the dog-fishes, sharks, and other living fishes, as well as in many extinct forms, the tail is unequally lobed, and is said to be "heterocercal" (Fig. 103, *b*). In these forms the vertebral column is prolonged into the upper lobe of the tail, and the greater portion of the tail is found below the spine.

In both the paired and the median fins the integument is supported by a series of spine-like bones, which are called "rays." These rays are sometimes simple undivided rays or spines, when they are called "spinous rays" (Fig. 101, *d*); but in other cases they are both divided by transverse joints, and split up into numerous longitudinal branches toward their extremities, when they are spoken of as "soft rays" (Fig. 101, *d'*). The soft rays occur in many fishes in different fins, but they are invariably present in the caudal fin or tail.

As regards the *digestive system* in fishes, the mouth is usually furnished with a complicated system of teeth, developed not only upon the jaws, but upon any or every bone which enters into the composition of the oral cavity. The gullet opens into a stomach, usually of large size, and its hinder aperture (the *pylorus*) is usually furnished with a valve.

Immediately behind the pyloric opening of the stomach there is usually a variable number of blind tubes (called the "pyloric cæca") which open into the intestine, and which are believed to represent the pancreas. In some fishes, however, there is a well-developed pancreas, and in others even these tubes are wanting. The intestinal canal is a longer or shorter, more or less convoluted tube, and its absorbing surface is sometimes largely increased by a spiral folding of the mucous membrane, which winds like a screw in close turns from the pylorus to the anus. The liver is usually of large size, and saturated with oil, but in the Lancelet it is doubtfully represented by a hollow, sac-like organ. The kidneys in fishes are of great comparative size, forming two elongated organs, situated beneath the spine, and extending along the whole length of the abdomen.

Respiration in all fishes is aquatic, and is effected by means of gills or *branchiæ*, in all except the Lancelet, in which respiration is effected by branchial filaments placed round the pharynx, and also by a greatly-developed pharynx perforated by ciliated apertures (Fig. 105). The arrangement and structure of the gills in fishes vary a good deal in different orders, and the leading modifications will be noticed hereafter. In the mean while it will be sufficient to give a short description of the branchial apparatus in one of the bony fishes. In such a fish the gills consist of a single or double series of flat cartilaginous leaflets, covered by mucous membrane, richly supplied with blood, and arranged on bony or cartilaginous arches which are connected with the tongue-bone (*hyoid* bone) below and with the under surface of the head above. The branchial arches and branchiæ are suspended in cavities placed on the side of the neck, and in the ordinary bony fishes there is only one such cavity on each side. The water is taken in at the mouth by a process analogous to swallowing, and it gains admission to the branchial chamber by means of a series of clefts or slits which perforate the sides of the pharynx. Having passed over the gills and lost its oxygen, the effete water makes its escape behind by an aperture called the "gill-slit," which is placed on the side of the neck. The opening of the gill-slit is closed in front by a chain of flat bones which constitute the "gill-cover," and by a membrane which is supported upon a variable number of slender bony spines. This is the general mechanism of respiration in one of the bony fishes, but different arrangements are found in other cases, which will be subsequently noticed.

The *heart* in fishes may be regarded as essentially a *branchial* or respiratory heart, being concerned chiefly with driving the venous and impure blood to the gills. It consists in almost all cases of two cavities, an auricle and a ventricle (Fig. 104). The auricle (*a*) receives the venous blood which has passed through all the various parts of the body, and propels it into the ventricle (*v*). From the ventricle proceeds a single great vessel (the "branchial artery"), the base of which is usually developed into a muscular cavity, the "bulbus arteriosus" (*m*), which acts as a kind of additional ventricle. By the ventricle and *bulbus arteriosus* the venous blood is driven to the gills, where it is subjected to the action of the water. The aërated blood is not returned to the heart, but is driven from the gills through all parts of the body, the propulsive force necessary for this being derived partly from the heart, and partly from the contractions of the muscles between which the blood-vessels pass. The essential peculiarity of the circulation of fishes consists in this, that the arterialized blood returned from the gills is propelled through the general vessels of the body (systemic vessels) without being sent back to the heart. In the Lancelet, alone of all fishes, there is no single heart, and the circulation is effected by means of contractile dilatations situated upon several of the vessels. In the Mudfish (*Lepidosiren*) the heart consists of two auricles and a ventricle. In all cases the blood is cold, or, in other words, has a temperature very little, or not at all, higher than that of the surrounding medium. The blood-corpuscles (Fig. 99, *e*) are always nucleated, and, except in the Lancelet, are most of them red.

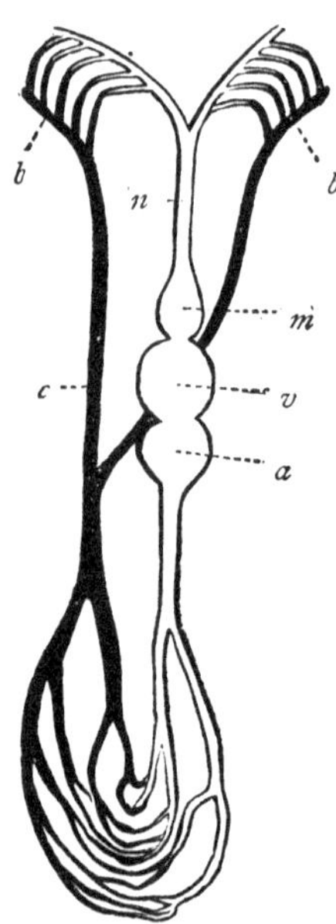

Fig. 104.—Diagram of the Circulation in a Fish. The arterial system is represented black; the venous system is left light. *a* Auricle, receiving the venous blood from the body; *v* Ventricle; *m* Bulbus arteriosus; *n* Branchial artery, carrying the venous blood to the gills (*b*); *c* Great systemic vessel, carrying the pure blood to the tissues.

While the respiration of all fishes is truly aquatic, most

are, nevertheless, furnished with an organ which doubtless corresponds to (or is *homologous* with) the lungs of the higher *Vertebrata.* This is known as the "air" or "swim bladder," and is a sac filled with gas and situated between the alimentary canal and the kidneys. In most cases, the sac contains only a single cavity, but, in many instances, it is variously divided by partitions. In most fresh-water fishes, the gases in the swim-bladder are mainly composed of nitrogen, but, in the sea fishes, it is chiefly filled with oxygen. The sac of the swim-bladder is often closed, but, in other cases, it opens into the gullet by means of a duct which corresponds to the windpipe. In the great majority of fishes, the functions of the air-bladder are mainly hydrostatic, that is to say, it serves to maintain the necessary agreement between the specific gravity of the fish and that of the surrounding water. In the singular Mud-fish (*Lepidosiren*), the air-bladder is composed of two distinct sacs, divided into a number of cellular compartments, and opening into the gullet by a tube. In this fish it acts as a respiratory organ, and is, therefore, not only in structure, but also in function, the representative of the lungs of the other Vertebrates.

The nervous system of fishes is of an inferior type of organization, the brain being of comparatively small size, and consisting mainly of a collection of ganglia. As regards the organs of the senses, two peculiarities deserve notice. In the first place, though fishes possess the essential parts of the organ of hearing, they possess no external ears, and in no case is there any direct communication between the ear and the outer world. In the second place, the organs of smell consist of a double cavity lined by a mucous membrane folded into numerous plaits, into which water is admitted, usually by two distinct apertures or nostrils. Behind, however, the nasal sacs are closed, and they do not communicate by any aperture with the throat, as they do in all the higher Vertebrates. The only exceptions to this rule are the Hag-fishes and their allies (*Myxinoids*) and the Mud-fish (*Lepidosiren*).

As regards their reproductive system, most fishes are truly *oviparous*, and the ovaries are familiarly known as the "roe." Some fishes are ovo-viviparous, retaining their eggs within the body till the young are hatched. The male organs of reproduction are commonly spoken of as the "milt" or "soft roe."

CHAPTER XXV.

ORDERS OF FISHES.

THE number of different kinds of fishes is so enormous that nothing further will be attempted than merely to give an outline of the leading peculiarities which distinguish the different orders. The classification here adopted is the one proposed by Prof. Huxley, who divides the class *Pisces* into the following six orders:

1. *Pharyngobranchii.*
2. *Marsipobranchii.*
3. *Teleostei.*
4. *Ganoidei.*
5. *Elasmobranchii.*
6. *Dipnoi.*

ORDER I. PHARYNGOBRANCHII (Gr. *pharugx*, the upper part of the gullet, and *bragchia*, gills).—This order of fishes includes only a single animal, the anomalous *Amphioxus*, or Lancelet, the organization of which differs in almost all its important points from that of all the other members of the class. In fact, the Lancelet presents us with the lowest type of organization as yet known in the *Vertebrata*. The Lancelet is an extraordinary little fish, from one and a half to two inches long, which burrows in sand-banks in various seas, but is especially abundant in the Mediterranean. The body is lanceolate in shape, and is provided with a narrow membranous border, of the nature of a median fin, which runs along the whole of the dorsal and a portion of the ventral surface, and expands at the tail to form a lancet-shaped caudal fin. There are no true "paired" fins, representing the fore and hind limbs. The mouth is a longitudinal fissure, placed at the front of the head, and completely destitute of jaws, but surrounded by a number of cartilaginous filaments. The throat is provided with several leaf-like filaments, which are richly supplied with blood, and are believed to discharge in part the function of gills. The mouth (Fig. 105, *m*) opens into a dilated chamber, which is believed to represent the pharynx, and is termed the pharyngeal or "branchial" sac. The walls of this chamber (*p*) are strengthened by numerous cartilaginous filaments, between which are a series of transverse slits or clefts, and the whole is covered with a richly-ciliated mucous membrane. The function of this sac is clearly respiratory, the water from without being admitted through the mouth, passing through the branchial clefts into the abdominal cavity, and finally escaping by means of an aperture placed on the ventral surface a little in front of the anus. From the hinder end of the branchial sac proceeds the alimentary canal, which has appended to it a sac-like organ, believed to represent the liver, and which terminates behind in a distinct anal aperture. There is no heart, and the circulation is entirely effected by means of several contractile dilatations, developed upon the great blood-vessels (*h*). The blood itself is colorless. No kidneys have hitherto been discovered, and the reproductive elements are emitted into the abdominal cavity, from which they escape by the pore placed upon the lower surface.

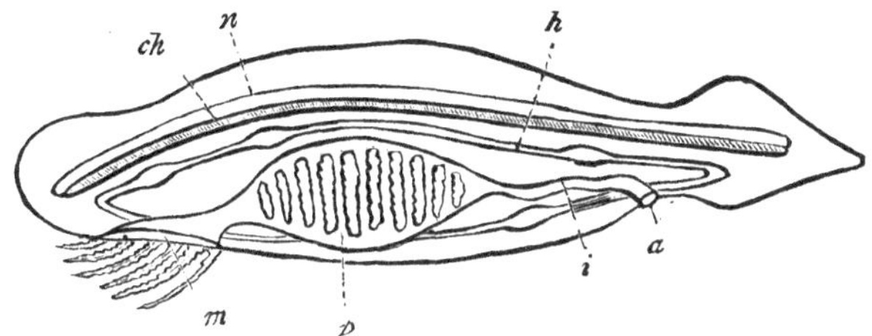

Fig. 105.—Diagram of the Lancelet (*Amphioxus lanceolatus*). *m* Mouth with cartilaginous filaments; *p* Greatly-developed pharynx, or branchial sac, perforated by ciliated apertures; *i* Intestine; *a* Anus; *h* Blood-vessels, with pulsating dilatations in place of a heart; *ch* Notochord; *n* Spinal cord.

There is no skeleton properly so called. The notochord (*ch*) remains throughout life as a semi-gelatinous rod, enclosed in a membranous sheath, and supporting the spinal cord. There is no skull, and the spinal cord (*n*) does not expand in front to form a distinct brain. The brain, however, may be said to be represented, as the front portion of the nervous axis gives off nerves to a pair of eyes, and another branch to a ciliated pit, which is believed to be a rudimentary organ of smell.

ORDER II. MARSIPOBRANCHII (Gr. *marsipos*, a pouch; *bragchia*, gills).—This order includes the Hag-fishes (*Myxinidæ*) and the Lampreys (*Petromyzonidæ*), and it is defined by the following characters: The body is cylindrical and worm-like, and is destitute of limbs. The skull is cartilaginous, there is no lower jaw, and the notochord remains through life, so that there is no vertebral column. The heart is composed of an auricle and a ventricle, but there is no *bulbus arteriosus.* The gills are pouch-like, communicating with the throat on the one hand, and opening externally on the other by means of apertures placed on the sides of the neck.

The Hag-fish (*Myxine*) is an eel-like fish, which agrees with the Lampreys in having neither pectoral nor ventral fins, the representatives of the fore and hind limbs. The mouth is of a very remarkable character, and enables the Hag-fish to lead a very peculiar existence. It is generally found imbedded in the interior of some large fish, into which it has penetrated by means of a single serrated and recurved fang attached to the centre of the palate. The mouth itself is destitute of jaws, and forms a sucking disk or cup. Another remarkable peculiarity of the Hag-fishes is found in the structure of the nose. In all fishes, namely, except these and the Mud-fish (*Lepidosiren*), the nasal chambers are closed behind, and do not communicate with the cavity of the mouth, as they do in the higher Vertebrates. In the Myxinoids, however, such a communication does exist. The nasal sacs are placed in communication with the throat (pharynx) by means of a canal which perforates the palate. A second canal leads from the nasal cavities in front to open by an external aperture (the nostril or "spiracle") on the top of the head behind the mouth.

Another peculiarity, which is best considered in the Lampreys, is to be found in the structure of the respiratory organs, from which the name of the order is derived. When viewed externally, instead of the single great "gill-slit," covered by a "gill-cover," as seen in the ordinary bony-fishes, the side of the neck presents seven round holes placed far back in a line on each side. These holes are the external apertures of the gills (Fig. 106, A), which in these fishes are in the form of sacs or pouches, the lining membrane of which is thrown into numerous folds or plaits, over which the branchial vessels ramify (Fig. 106, B). Internally the sacs communicate with the cavity of the pharynx, by means of a common respiratory tube into which they all open. It follows from this arrangement that the gill-pouches on the two sides of the neck communicate freely with one another through the pharynx. The object of this arrangement is to obviate the necessity for admitting the water to the gills through the mouth,

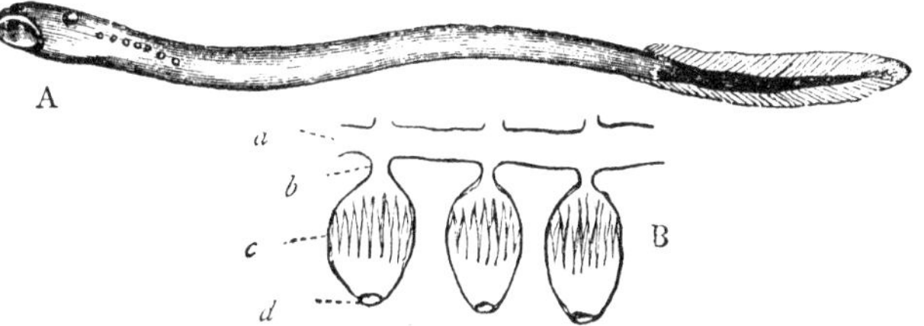

FIG. 106.—A, Lamprey (*Petromyzon*), showing the sucking-mouth and the apertures of the gill-sacs. B, Diagram to illustrate the structure of the gills in the Lampreys. *a* Pharynx; *b* Tube leading from the pharynx into one of the gill-sacs; *c* One of the gill-sacs, showing the lining membrane thrown into folds; *d* External opening of the gill-sac. (In reality the gill-sacs do not open directly into the pharynx, but into a common respiratory tube which communicates with the pharynx; but this is omitted for the sake of clearness.)

as ordinary fishes do. These fishes are in the habit of fixing themselves to foreign objects by means of the suctorial mouth; and, when in this position, it is, of course, impossible that they can obtain the necessary water of respiration through the mouth. As the gill-sacs, however, on the two sides of the neck communicate freely with one another through the pharynx, water can readily pass in and out. The gills are not provided with cilia, but the circulation of water is assisted by a kind of elastic cartilaginous framework upon which the whole respiratory apparatus is supported, and which acts somewhat like the ribs of the higher Vertebrates.

The nasal cavities of the Lampreys, unlike those of the Myxinoids, are closed behind, and do not communicate with the throat. Some of the Lampreys are permanently inhabitants of rivers, but the great sea-lamprey (*Petromyzon marinus*) only quits the salt water and betakes itself to fresh in order to deposit its eggs.

ORDER III. TELEOSTEI (Gr. *teleios*, perfect; and *osteon*, bone).—The fishes comprised in this order, as implied in their name, have a well-ossified or bony skeleton, and they are commonly known as the "bony" fishes. In all the *Teleostei*, the skeleton is bony, the skull is composed of distinct bones, and there is a lower jaw. The vertebral column always consists of more or less completely ossified vertebræ; and the two pairs of limbs, when present, are in the form of fins, supported by rays. The gills are free, comb-like or tufted in shape, and always protected by a bony gill-cover. The *bulbus arteriosus* is not capable of regular contractions, and is separated from the ventricle by only a single valve.

The order *Teleostei* comprises almost all the most familiar fishes, and it will be unnecessary to dilate here upon their structure, as they were taken as the type of the class in describing the fishes generally. It may be as well, however, to recapitulate some of the leading points in the anatomy of the bony fishes. 1. The *skeleton* is always more or less completely ossified, and does not remain cartilaginous throughout life. The notochord is not permanent, and the vertebral column consists of a number of distinct vertebræ. The vertebræ, however, are "amphicœlous," or hollow at both ends, so that there is left between each pair a doubly-conical cavity, which is filled with the cartilaginous or semi-gelatinous remains of the notochord. In this way an extraordinary amount of flexibility is given to the entire vertebral column. In no fish (except the Bony Pike, which belongs to another order) is the conversion of the bodies of the vertebræ into bone carried further than this.

2. The integument usually develops scales, and these in the great majority of cases are of the forms known as "cycloid" and "ctenoid," the former being circular or elliptical horny plates, with plain margins; while the latter have their hinder margins cut into comb-like projections, or fringed with spines (Fig. 100, *a*, *b*).

3. The *anterior and posterior limbs* are usually, but not always, present, and when developed they are always in the form of fins. These fins may be supported by "spinous rays" or "soft rays," or by both. The spinous rays are simple undivided bony spines which taper to a point. The soft rays are doubly divided, splitting up toward their extremities into a number of secondary rays, and being also divided by transverse joints into numerous short pieces.

4. Besides the "paired" fins which represent the limbs, there is also a series of *unpaired* or "median" fins, the rays of which are supported upon a series of dagger-shaped bones, deeply plunged in the flesh in the middle line of the body,

and known as the "interspinous" bones (Fig. 101). The median fins are variable in number, but when fully developed they consist of one or two fins on the back (the dorsal fins), one or two on the ventral surface (the anal fins), and one clothing the posterior extremity of the body (the caudal fin, or tail, Fig. 102). In all the *Teleostei*, the caudal fin has the shape called "homocercal"—that is to say, it consists of two equal lobes—and the vertebral column is not prolonged into the upper lobe (Fig. 103, *a*).

5. The heart consists of two cavities, an auricle and a ventricle, but the *bulbus arteriosus* is not rhythmically contractile, and is separated from the ventricle by only a single pair of valves.

6. The respiratory organs are in the form of free, comb-like, or tufted gills, enclosed in two cavities placed on the sides of the neck. Each of these branchial chambers opens externally by a single aperture, the "gill-slit," which is protected by a chain of bones, forming the "gill-cover," and by a membrane supported by bony rays. Internally the branchial chambers communicate with the throat by a series of clefts or fissures, and the water required in respiration is taken in at the mouth by a process analogous to swallowing.

7. The nasal sacs never communicate behind with the throat (pharynx).

Tabular View of the Main Divisions of the Teleostei.

Sub-order I. Malacopteri.—Usually a complete series of fins, supported by rays, all of which are *soft*, or many-jointed (with the occasional exception of the first rays in the dorsal and pectoral fins). A swim-bladder is always present, and is always connected with the gullet by a duct. The skin is rarely naked, and is mostly furnished with cycloid scales, but sometimes ganoid scales are present.

Among the more important families in this sub-order are the Eels (*Murænidæ*), Herrings (*Clupeidæ*), Pikes (*Esocidæ*), Carps (*Cyprinidæ*), Salmon and Trout (*Salmonidæ*), and Sheat-fishes (*Siluridæ*).

Sub-order II. Anacanthini.—Fins entirely supported by soft rays, and never by spinous rays. Ventral fins either wanting, or placed under the throat, beneath or in advance of the pectorals.

The two leading families in this sub-order are the Cod, Ling, and Haddock family (*Gadidæ*), and the Flat-fishes (*Pleuronectidæ*), comprising the Sole, Turbot, Flounder, and others.

Sub-order III. Acanthopteri.—Fins with one or more of the first rays in the form of undivided, inflexible, spinous rays. Scales mostly ctenoid. Swim-bladder without a duct.

The leading families in this order are the Wrasses (*Cyclo-labridæ*), the Perches (*Percidæ*), the Mackerels (*Scomberidæ*), the Mullets (*Mugilidæ*), and the Gobies (*Gobiidæ*).

SUB-ORDER IV. PLECTOGNATHI.—Certain of the bones of the mouth (the maxillary and præ-maxillary bones) immovably connected on each side of the jaw. Integumentary skeleton in the form of ganoid plates, scales, or spines.

The chief families in this sub-order are the File-fishes (*Balistidæ*), and the Trunk-fishes (*Ostraciontidæ*).

SUB-ORDER V. LOPHOBRANCHII.—Gills arranged in little tufts on the branchial arches. Integumentary skeleton in the form of ganoid scales.

The two families contained in this division are the Sea-horses (*Hippocampidæ*), and the Pipe-fishes (*Syngnathidæ*).

ORDER IV. GANOIDEI (Gr. *ganos*, splendor, or brightness).—The fourth order of fishes is that of the *Ganoidei*, including few living forms, but having a great and varied development in past geological epochs. The Ganoid fishes are distinguished by the imperfect development of the skeleton, which is mostly cartilaginous throughout life, and by having an integumentary skeleton composed of *ganoid* scales, plates, or spines (Fig. 100, *d*). The skull is composed of distinct bones, and there is always a lower jaw. There are usually two pairs of fins (pectoral and ventral), supported by many series of cartilages, and the ventral fins are placed very far back. The first rays in the fins are usually in the form of strong spines. The caudal fin or tail is mostly heterocercal or unsymmetrical (Fig. 103, *b*). The swim-bladder is always present, is often cellular, and is provided with an air-duct. The gills and gill-covers are essentially the same as in the bony fishes. The heart has one auricle and a ventricle; and the *bulbus arteriosus* is rhythmically contractile, is furnished with a distinct coat of muscular fibres, and is furnished with several transverse rows of valves.

The best known of the living Ganoids are the Bony Pike (*Lepidosteus*), the Sturgeon (*Sturio*), and the *Polypterus*. Of these, the Bony Pike is found in the rivers and lakes of North America. It is a large fish, attaining a length of several feet, and it has the body entirely covered with an armor of ganoid scales arranged in obliquely transverse rows. The jaws form a long, narrow snout, armed with a double series of teeth, and the tail is heterocercal. The vertebral column is more perfectly ossified than in any other fish, the bodies of the vertebræ being convex in front and concave behind ("*opisthocœlous*"). The *Polypterus* (Fig. 107, A) inhabits the rivers Nile and Senegal, and is remarkable for the peculiar structure of the dorsal fin, which is broken up into a series of small, detached portions, each composed of a single spine in front, with a soft fin attached to it behind. Some of the species of *Polypterus*

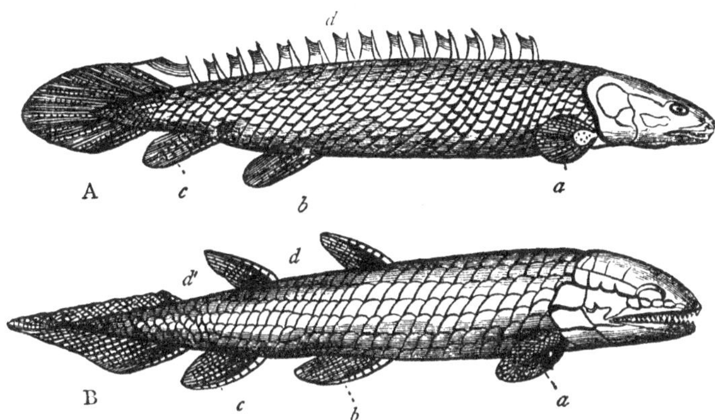

FIG. 107.—Ganoid Fishes. A, *Polypterus*, a living Ganoid. B, *Osteolepis*, a fossil Ganoid (restored): *a* Pectoral fin; *b* Ventral fin; *c* Anal fin; *d d'* Dorsal fins.

are stated to possess *external* gills when young, which they lose when grown up, thus making an approach to the *Amphibia.* Many of the fossil Ganoids are more or less closely allied to the living *Lepidosteus* and *Polypterus.*

Another great group of the Ganoid fishes is represented by the Sturgeons (*Sturionidæ*), in which the skeleton is always very imperfectly ossified, and the head, with more or less of the body, is protected by large ganoid plates, which are often united together at their edges by sutures. The true Sturgeons are chiefly found in the North Sea, the Caspian, and the Black Sea, and they are captured when ascending the great rivers for the purpose of spawning. The swim-bladder of the Sturgeons is one of the chief sources from which *isinglass* is prepared, and the roe is sold as a delicacy under the name of *caviare.* The place of the Sturgeons in North America is taken by the Paddle-fishes (*Spatularia*).

The group of Ganoids represented at the present day by the Sturgeons and Paddle-fishes was formerly represented by numerous remarkable fishes, which are most abundant in the system of rocks known to geologists as the "Old Red Sandstone." The graphic descriptions of Hugh Miller have placed many of these fishes before us as living pictures, but space will not allow of any further notice of them here. One, however, of the more striking forms is figured hereafter (Fig. 108).

ORDER V. ELASMOBRANCHII (Gr. *elasma*, a thin plate; and *bragchia*, gills).—This order includes the Sharks and Rays, and is distinguished by the following characters: The skull and lower jaw are well developed, but the skull is not com-

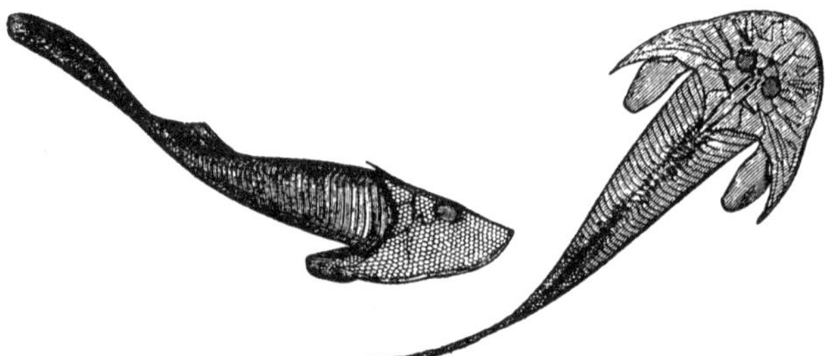

FIG. 108.—*Cephalaspis Lyellii*, from the Old Red Sandstone of Scotland.

posed of distinct bones, and simply forms a kind of cartilaginous box. The vertebral column is sometimes cartilaginous, sometimes composed of distinct vertebræ. The integumentary skeleton is in the form of *placoid* scales (Fig. 100, *c*)—

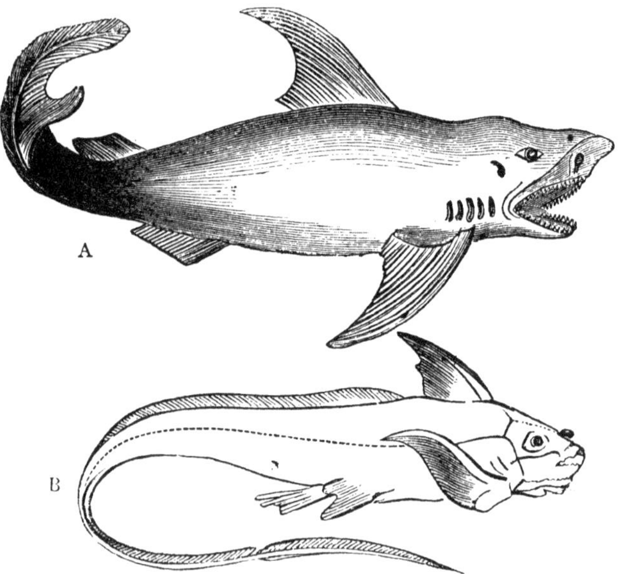

FIG. 109.—Elasmobranchii. A. White Shark (*Carcharias*). B. King of the Herrings (*Chimæra*).

that is to say, of detached grains, tubercles, or plates. There are two pairs of fins, corresponding to the fore and hind limbs, and the ventral fins are placed far back, close to the anus. The heart consists of an auricle and ventricle; and the *bulbus arteriosus* is rhythmically contractile, is provided with a distinct coat of muscular fibres, and is furnished with several transverse rows of valves. The gills are fixed, and form a number of pouches, which open internally into the pharynx, and communicate with the outer world by a series of apertures placed on the side of the neck (Fig. 109). The intestine is very short, but, to compensate for this, the mucous membrane is thrown into a fold, which winds round the intestine in close turns from the pyloric orifice of the stomach to the anus, and which thus greatly increases its absorbing surface.

The best-known members of this order are the Sharks and Rays, but numerous extinct forms testify to its great abundance in past geological epochs.

Tabular View of the Divisions of the Elasmobranchii.

Sub-order I. Holocephali.—The mouth placed at the end of the head, and the external opening of the gills in the form of a single gill-slit.

The best-known member of this sub-order is the *Chimæra monstrosa*, sometimes called the "King of the Herrings."

Sub-order II. Plagiostomi.—Mouth transverse, placed on the under surface of the head; external opening of the gills in the form of several slits on each side of the neck, not protected by a gill-cover.

Fam. a. Cestraphori.—Ex. Port-Jackson Shark.
Fam. b. Selachii.—Ex. Sharks and Dog-fishes.
Fam. c. Batides.—Ex. Rays.

Order VI. Dipnoi (Gr. *di*, double; *pnoē*, breath).—This order is a very small one, and includes only the very singular Mud-fishes (*Lepidosiren*),* which are of great interest from the many points of affinity which they exhibit to the *Amphibia*. The body of the Mud fish (Fig. 110) is completely fish-like, and is protected by a covering of small, horny, overlapping scales, which have the cycloid characters. There are two pairs of limbs, but these are in the form of awl-shaped organs, each supported by a single jointed cartilaginous rod. The pectoral limbs have a membranous fringe inferiorly, and the ventrals are placed very far back. There is also a median

* Recently a singular fish has been discovered in the rivers of Queensland (Australia), which will probably have to be referred to the order *Dipnoi*; but our knowledge about it is still imperfect.

fin behind, forming a continuous fringe round the compressed tail, and supported by cartilaginous rays.

FIG. 110.—*Lepidosiren annectens*, the Mud-fish. *p* Pectoral limbs; *v* Ventral limbs.

The skull is composed of distinct bones, and there is a lower jaw, but the notochord is persistent, and there are no bodies of vertebræ developed. The respiratory organs are *twofold*, consisting, *firstly*, of free filamentous branchiæ or gills, contained in a branchial chamber, which opens externally by a single vertical gill-slit; and, *secondly*, of true lungs, in the form of a double cellular air-bladder communicating with the gullet by means of an air-duct or windpipe. Sometimes, if not always, there are rudimentary *external* gills as well, placed on the side of the neck. The heart consists of a ventricle, and of *two* auricles, divided from one another by an incomplete partition. Lastly, the nasal sacs open behind into the throat, and do not form closed chambers opening only by the nostrils, as they do in all other fishes, except the Myxinoids. The two best-known species are the *Lepidosiren paradoxa* from the Amazons, and the *L. annectens* from the Gambia. They both inhabit marshy tracts, and both appear to be able in the dry season to bury themselves in the mud, and to form a kind of chamber, in which they remain dormant till the rains of the wet season set them free.

ICHTHYOPSIDA.

CHAPTER XXVI.

CLASS II. AMPHIBIA.

THIS class of *Vertebrata* comprises the Frogs and Toads, the Newts and Land-salamanders, the *Cæciliæ*, and some extinct forms, and it may be briefly defined as follows: In all cases gills or branchiæ adapted for aquatic respiration are present during a part or the whole of life; but, in all cases, true lungs adapted for breathing air are ultimately developed, even when the gills are retained through life. All pass through some sort of a metamorphosis after being set free from the egg. The limbs may be absent or there may be only one pair, but in no case are they ever converted into fins. When median fins are present, as is sometimes the case, these are never furnished with fin-rays or interspinous bones, as in the fishes. The skull always articulates with, or is jointed to, the spinal column by *two* articular surfaces or *condyles.* The heart consists of two auricles and a single ventricle. The nasal sacs always open behind into the mouth; and there is a common cavity or "cloaca" which receives not only the termination of the intestine (*rectum*), but also the ducts of the kidneys and of the reproductive organs.

The great and distinguishing character of the *Amphibia* (Gr. *amphi*, both; *bios*, life) is, that they invariably undergo some kind of metamorphosis after birth, though, in some rare cases, the eggs are retained so long within the body of the parent that there is little or no obvious change. In the great majority of cases, however, the Amphibians commence life as water-breathing larvæ, provided with gills; but, in their adult state, they possess true air-breathing lungs, the gills sometimes disappearing when the lungs are developed, but being sometimes retained throughout life. Most Amphibians, therefore, are to a greater or less extent *amphibious*, that is to say, more or less capable of living indifferently either on land or in the water. In the majority of cases, the gills are *external*, placed on the sides of the neck, and not contained in a special cavity, thus differing altogether from the gills of fishes. In the Frogs and Toads, and in some others, there are two sets of gills, one external and the other internal, of which the former is soonest lost. The lungs of the Amphibians never attain a very high state of development, and, in those forms in which the gills are retained throughout life, the chief business of respiration appears to be carried on by the gills. In accordance with the changes in the respiratory process, corresponding alterations take place in the blood-vessels. With the development of the lungs, the vessels which carry blood to them (the pulmonary arteries) increase in size, while the branchial vessels which carry the blood to the gills undergo a proportionate diminution. At first, the condition of the circulation is very much the same as it is in fishes, but ultimately it becomes nearly the same as in the true reptiles.

The *Amphibia* are divided into three living and one extinct order, as follow:

1. *Ophiomorpha.*
2. *Urodela.*
3. *Anoura.*
4. *Labyrinthodontia.*

ORDER I. OPHIOMORPHA (Gr. *ophis*, a serpent; and *morphe*, form).—This order is an extremely small one, and, as its name implies, it comprises certain snake-like Amphibians. The order includes only the curious animals known as *Cæciliæ*, which are found in Java, Ceylon, South America, and Guinea. The body is entirely destitute of limbs, and is enclosed in an integument which is thrown into numerous transverse wrinkles, and sometimes has numerous horny scales imbedded in it. The eyes are concealed by the skin, and are rudimentary. There is no tail, and the anal aperture is placed almost at the extreme end of the body. When adult, respiration is carried on by means of lungs, but gills are present in the young, and there can, therefore, be no doubt as to their being genuine Amphibians.

The *Cæciliæ* are found burrowing in marshy ground, and they are not unlike large earth-worms in appearance, but they sometimes attain a length of several feet.

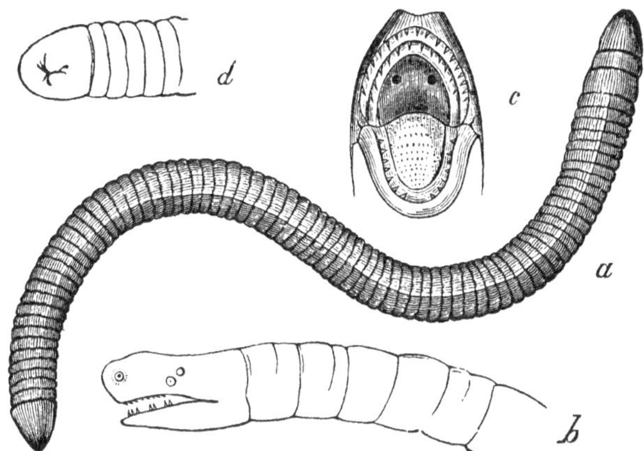

Fig. 111.—*a Siphonops annulatus*, one of the Cæcilians, much reduced; *b* Head of the same; *c* Mouth, showing the tongue, teeth, and internal openings of the nostrils; *d* Tail of the same. (After Dumeril and Bibron.)

Order II. Urodela or Ichthyomorpha (Gr. *ichthus*, a fish, and *morphe*, shape).—In this order are a number of fish-like Amphibians, of which the Newts and Land-salamanders are the most familiar examples. In all the members of this section, the skin is naked, and never develops any hard structures, and in all there is a well-developed, fish-like tail, which is retained throughout life. The vertebræ are sometimes hollow at both ends (*amphicœlous*), sometimes hollow behind and convex or rounded in front (*opisthocœlous*). The ribs are rudimentary and the bones of the forearm (*radius* and *ulna*), and of the shank (*tibia* and *fibula*), are separate, and are not combined so as to form single bones.

The *Ichthyomorpha* are not unfrequently spoken of as the "Tailed" Amphibians (*Urodela*), and they fall into two natural sections, according as the gills are permanently retained throughout life, or are cast off before maturity is attained. The animals belonging to the first section are often called "perennibranchiate," while those belonging to the second are said to be "caducibranchiate."

Among the Perennibranchiate forms, in which the gills are permanently retained after the lungs make their appearance, the best-known examples are Axolotl (Fig. 112), the curious *Proteus anguinus*, and the Mud-eel (*Siren*). The Axolotls inhabit various of the lakes of the American Continent, the best-known species being the *Siredon pisciforme* of the Mexican lakes (Fig. 112). It attains a length of a foot or more, and

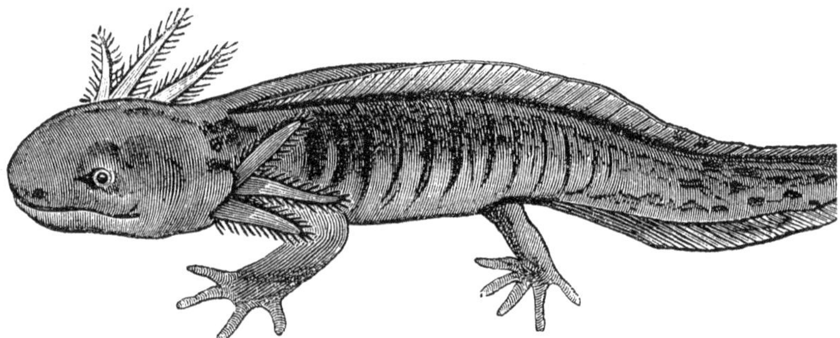

FIG. 112.—The Axolotl (*Siredon pisciforme*). (After Tegetmeier.)

possesses both pairs of limbs, the fore-feet having four toes, the hind-feet five toes. The branchiæ are in the form of three long ramified processes on each side of the head; and the tail is compressed, and fringed by a fin which is prolonged on the back between the shoulders. In a state of nature, the Axolotl is certainly perennibranchiate, and it breeds freely in this condition. It has been shown, however, by Prof. Marsh, of New Haven, that some species, when kept in confinement, lose their gills, and undergo certain other changes, becoming ultimately converted into a Salamandroid, apparently belonging to the genus *Amblystoma*. The *Proteus* is an extraordinary Amphibian which is found inhabiting the waters of caves in Illyria and Dalmatia. It attains a length of about a foot, and is of a pale flesh-color or nearly white. The gills, which are retained throughout life, are of a bright scarlet. Both pairs of limbs are developed, but they are only short and weak, the fore-limbs having three toes each, and the hind-limbs only two. The eyes are extremely small, the animal spending its existence in darkness; and swimming is effected mainly by means of the tail. The *Siren*, or Mud-eel, is a large lizard-like Amphibian, which is found abundantly in the swamps of South Carolina, and attains the great length of three feet. The external branchiæ are retained throughout life, and they are the main organs of respiration. The fore-limbs are present, but the hinder pair of limbs is never developed.

The "caducibranchiate" section of this order is characterized by the fact that both pairs of limbs are always developed, and the branchiæ are never retained throughout life. The most familiar examples are the Water-salamanders or Newts (*Triton*), and the Land-salamanders. The Newts (Fig. 113) are well known as inhabiting pools in many countries,

FIG. 113.—The great Water-Newt (*Triton cristatus*), male. (After Bell.)

and the young lead a strictly aquatic life. When the lungs are developed the external gills wholly disappear, and the respiration becomes strictly aërial, though the animals still spend a great part of their time in the water. The larva or young form is at first destitute of limbs, and the fore-limbs are the first to be developed, the reverse of this taking place in the Frogs. In accordance with their mode of life, the tail is compressed and flattened, so as to form an efficient swimming apparatus. The Water-salamanders are all oviparous, and the young are like the tadpoles of the common frog.

The Land-salamanders, in both their adult and young state, live upon land, and the tail is rounded and cylindrical. The young are not developed in water, but are retained within the body of the parent for a longer or shorter period, so that the reproduction becomes ovo-viviparous, or even viviparous. The best-known Salamanders occur upon the Continent of Europe, and one species is singular in the fact that it inhabits high mountains.

It is important to remember in connection with all these "tailed" Amphibians, that they are wholly distinct from the true *Lizards*, with which they are often confounded. Many of them are completely lizard-like in form, having a long tail and two pairs of legs; all, however, at some time or other in their life, respire by means of gills, and this is never the case with any true Lizard. It must be confessed, however, that a near approach to the Lizards is made by the Land-salamanders, the young of which have sometimes lost their gills before birth.

ORDER III. ANOURA OR THERIOMORPHA (Gr. *ther*, a beast; and *morphe*, shape).—This order is the highest of the *Amphibia*, and comprises the Frogs and Toads. It is sometimes known by the name of *Batrachia* (Gr. *batrachos*, a frog), or *Anoura* (Gr. *a*, without; *oura*, a tail), the latter name being derived from the fact that the adults are "tailless."

The tailless *Amphibia* or *Theriomorpha* are characterized by the fact that while the larva possesses a tail, and is furnished with gills, the adult has no tail, and breathes wholly by lungs. Both pairs of limbs are always developed in the full-grown animal, and the hind-limbs are usually considerably longer than the fore-limbs, and generally have the toes webbed, while those of the fore-limbs are free. The skin is

FIG. 114.—Anoura. Tree-frog (*Hyla leucotænia*). (After Günther.)

soft, and there are rarely any traces of any integumentary skeleton. The spinal column (Fig. 114) is short; the dorsal vertebræ are very long; and the ribs are quite rudimentary, their place being taken by greatly-developed transverse processes. The bodies of the vertebræ are hollow in front and convex behind (*procœlous*). The bones of the forearm (*radius* and *ulna*), and those of the shank (*tibia* and *fibula*), are united together to form single bones. The upper jaw is usually furnished with teeth, and the lower jaw sometimes, but there are no teeth in the Toads. The lungs are well developed, comparatively speaking; and, as there are no ribs by which the cavity of the chest can be expanded, the air is taken into the lungs by a process nearly akin to that of swallowing. There can be no doubt, also, that the skin plays a very important part in the aëration of the blood, and that the frogs, especially, can carry on their respiration by means of the skin without the assistance of the lungs for a very lengthened period. This, however, should not lead to any credence being given to the often-repeated stories of frogs and toads being found in closed cavities in solid rock, no authenticated instance of such an occurrence being known to science. The ova of the frogs and toads are deposited, in masses or strings, in water, and the young or larvæ are familiar to every one as tadpoles. Upon its escape from the egg, the young frog (Fig.

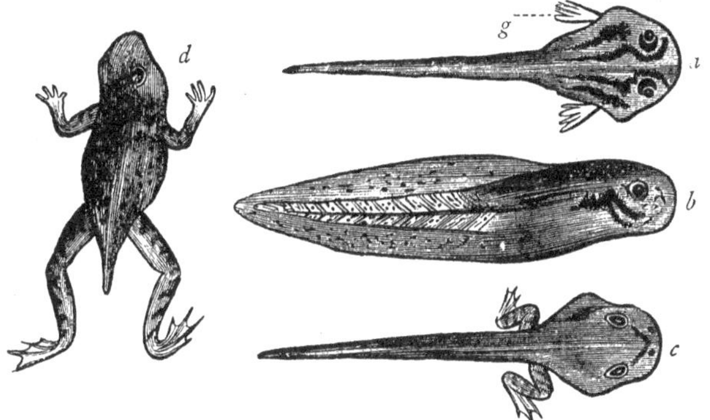

FIG. 115.—Development of the common Frog. *a* Tadpole, viewed from above, showing the external branchiæ (*g*); *b* Side view of a somewhat older specimen, showing the fish-like tail; *c* Older specimen, in which the hind-legs have made their appearance; *d* Specimen in which all the limbs are present, but the tail has not been wholly absorbed. (After Bell.)

115) presents itself as a little fish-like creature with a broad head, a sac-like belly, and a long, compressed tail with which it swims actively. It breathes by means of gills or branchiæ, of which there are two sets, one external, and the other internal; at first there are no limbs; but, as development proceeds, the limbs make their appearance—the hind-legs first, and then the fore-legs. The tail, however (Fig. 115), is still retained as an instrument of progression. Ultimately, when

the limbs are fully developed, and the gills have given place to lungs, the tail disappears, and the animal now takes to the land as a perfect frog.

The development of the Frog is a good illustration of the general zoological law, that the transitory embryonic stages of the higher members of any division of the animal kingdom are often represented by the permanent condition of the lower members of the same division. Thus the transitory condition of the young Frog, in which it breathes by external branchiæ, is to a certain extent permanently represented by the permanent condition of a perennibranchiate Amphibian, such as the *Proteus*. The stage at which the external branchiæ have disappeared, but the tail is still present, and the limbs are developed, is permanently represented in the common tailed Amphibians, such as the Newts.

The order *Anoura* comprises the three families of the Frogs, Toads, and Surinam Toads. The Frogs (*Ranidæ*) are distinguished by having a tongue which is fixed to the front of the mouth, and can be protruded at will, while the upper jaw is always armed with teeth. The typical Frogs have enormously-developed hind legs, the toes of which are united by membrane, or are "webbed." They swim very powerfully, and can take extensive leaps. The Tree-frogs (Fig. 114), on the other hand, are adapted for a wholly different life, inhabiting trees, among which they climb with great ease by the help of suckers developed upon the ends of the toes. They are mostly found in warm countries, especially in America, but one species is European.

In the equally familiar Toads (*Bufonidæ*) the structure of the tongue is the same as in the Frogs, but the jaws are not furnished with teeth. In the Surinam Toads (*Pipidæ*) there is no tongue at all, and usually no teeth.

ORDER IV. LABYRINTHODONTIA.—This, the last order of the *Amphibia*, is not represented by any living forms, and requires to be little more than mentioned. The Labyrinthodonts were Amphibia which were mostly of large size, and of which some must have obtained absolutely gigantic dimensions, the skull of one species being three feet in length and two in breadth. They were first known to science simply by their footprints, which were found in certain Secondary sandstones (*Trias*). These footprints consisted of a series of alternately placed pairs of hand-shaped impressions, the hinder print of each pair being much larger than the fore one. So like were these prints to the shape of the human hand that the unknown animal which had produced them was christened the "*Cheirotherium*" (Gr. *cheir*, hand; *ther*, beast). Further researches, however, showed that these footprints were produced by various species of large Amphibians, to which the name of *Labyrinthodontia* was applied, in consequence of the complicated microscopic structure of the teeth. These extinct Amphibians are known to have existed at the time of the Coal, but they are most characteristic of the period known to geologists as the Trias.

SAUROPSIDA.

CHAPTER XXVII.

CLASS III. REPTILIA.

WE commence now the second great primary division of the *Vertebrata*, namely, that of the *Sauropsida*, comprising the Reptiles and the Birds. These two classes, though very unlike in external appearance, are united by the following characters: There are never at any period of life gills or branchiæ adapted for aquatic respiration; the red corpuscles of the blood are nucleated (Fig. 99, *b*, *c*); the skull articulates with the vertebral column by means of a single articulating surface or condyle; each half of the lower jaw is composed of several pieces, and is jointed to the skull, not directly, but by the intervention of a special bone (the so-called "quadrate bone").

These being the characters by which, among others, Reptiles and Birds are collectively distinguished from other Vertebrates, it remains to see what are the characters by which the Reptiles are distinguished, as a class, from Birds. In all Reptiles the blood is cold—that is to say, very slightly warmer than the temperature of the external medium in which they live. The integument secretes scales, with or without bony plates, but in no case do the integumentary appendages take the form of feathers. The heart consists of two auricles, and a ventricle, which in most is partially divided into two chambers by an incomplete partition, and in a few is completely divided. In any case, however, more or less of the impure venous blood is mixed with the pure arterial blood which circulates over the body. There is no division between the cavities of the thorax and abdomen, and the lungs are not connected with air-sacs placed in various parts of the body. The limbs may be wanting, or rudimentary, but in no case are the fore-limbs constructed upon the type of the "wing" of birds, and in no living Reptile is there the bone which is known in Birds as the "tarso-metatarsus."

The class *Reptilia* includes the Tortoises and Turtles (*Chelonia*), the Snakes (*Ophidia*), the Lizards (*Lacertilia*), and the Crocodiles (*Crocodilia*). With the exception of the Tortoises and Turtles, they are mostly of an elongated cylindrical form, furnished behind with a long tail. The limbs may be altogether absent or quite rudimentary, as in the Snakes, but in almost all the higher members of the class there are two pairs of limbs, which may be either adapted for walking or swimming, and which in some extinct forms support a flying membrane. The internal skeleton is always bony, never cartilaginous or semi-cartilaginous as in many of the fishes. The skull is joined to the spine by a single articulating surface (or *condyle*). The lower jaw is complex, each half being composed of several pieces united by sutures. In Tortoises and Turtles, however, these separate pieces are amalgamated together, and the two halves are also united, so that the whole lower jaw appears to form a single piece. In most Reptiles, on the other hand, the two halves of the lower jaw (Fig. 116) are only loosely united; in the Snakes by ligaments and muscles, in the Lizards by gristle, and in the Crocodiles by suture.

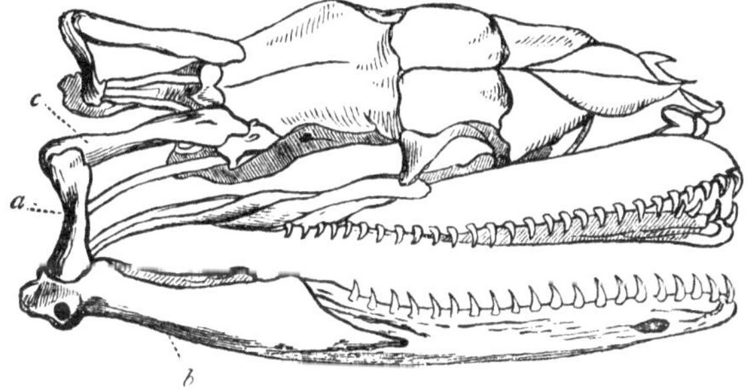

Fig. 116.—Skull of a Serpent (Python). *a* Quadrate bone; *b* Lower jaw, articulating with the movable quadrate bone.

In all, the lower jaw is jointed to the skull by means of a special bone, called the "quadrate bone;" and as this often projects backward, the opening of the mouth is often very extensive, and may even extend backward beyond the base of the skull (Fig. 116, *a*). Teeth are generally present, but these are used chiefly to hold the prey, and not in biting or chewing the food. Except in the Crocodiles, the teeth are not sunk into distinct sockets, and they are usually replaced as fast as shed. They likewise do not differ from one another sufficiently in form or function as to allow of their being divided into different sets, as they can be in the Mammals. Usually the teeth are confined to the jaws proper, but in some cases they are carried by other bones of the mouth. In the Tortoises and Turtles there are no teeth, and the jaws are simply sheathed in horn, so as to constitute a kind of beak, like that of a bird. The integumentary skeleton is in the form of scales, sometimes combined with bony plates. In the Tortoises and Turtles the integumentary skeleton is so united with the true skeleton as to form a kind of bony case or box, in which the body is enclosed.

The digestive system presents little worthy of special notice, except that the termination of the intestine (*rectum*) opens into a cavity called the "cloaca," which receives also the ducts of the urinary and generative organs.

It is, however, in the structure of the circulatory and respiratory organs that the most important characters of the Reptiles are to be looked for. The heart in all Reptiles may be regarded as being, *in function*, three-chambered, being composed of two auricles and a single ventricle, imperfectly divided by an incomplete partition. In the Crocodiles alone the heart is, *structurally*, four-chambered, the ventricle being divided into two by a complete partition. Here, however, the same results are brought about as in the other Reptiles, by means of a communication which subsists between the great vessels which spring from the ventricles thus formed. In the ordinary Reptiles the course of the circulation is as follows (Fig. 117): The impure or venous blood that has circulated through the body is poured by the great veins into the right auricle (*a*). The pure or arterial blood that has been submitted to the action of the lungs is poured by the pulmonary veins into the left auricle (*a'*). Both auricles empty their contents into the ventricle, and, as the partition which divides the ventricle is an incomplete one, it follows that the venous and arterial streams must mix to a greater or less extent in the ventricle. From the ventricle arise the great vessels which carry the blood to the lungs and to all parts of the body, and it follows, as a matter of necessity, that all these parts are supplied with a mixed fluid, consisting partly of impure or venous blood, and partly of pure or arterial blood. In the Crocodiles, in which there are two ventricles completely separated from each other, the same result is brought about by means of a communication which takes place between the great vessels which spring from the ventricles, in the immediate neighborhood of the heart.

From this brief description it will be seen that the peculiarity of the circulation in Reptiles consists in the fact, that the lungs and all parts of the body are supplied with *mixed* blood; whereas in the higher Vertebrates the lungs are supplied with pure venous blood, and the various tissues of the body with pure arterial blood.

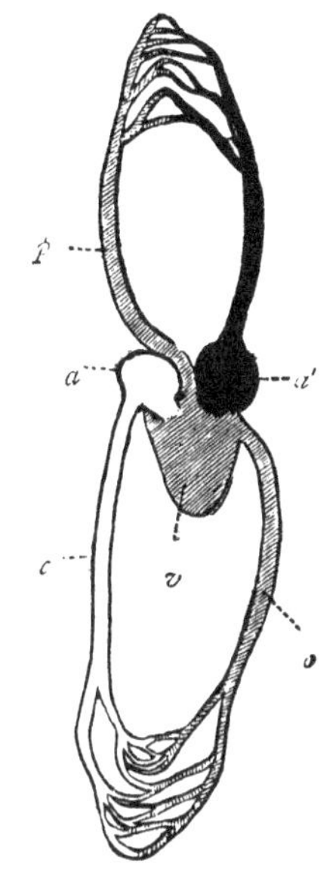

Fig. 117.—Diagram of the circulation of a Reptile. *a* Right auricle, receiving venous blood from the body; *a'* Left auricle, receiving arterial blood from the lungs; *v* Arterio-venous ventricle, containing mixed blood, which is driven by (*p*) the pulmonary artery to the lungs, and by (*o*) the aorta to the body. (The venous system is left light, the arterial system is black, and the vessels containing mixed blood are cross-shaded.)

As regards the structure of the lungs, it is merely to be noted that there is no partition (*diaphragm* or midriff) separating the two cavities of the thorax and abdomen, and that the lungs, therefore, often attain a great proportionate size, sometimes extending through almost the whole length of the cavity of the trunk. There are also no air-sacs communicating with the lungs, as in the Birds.

Lastly, all Reptiles are essentially oviparous, some being ovoviviparous. The egg-shell is usually parchment-like, but in other cases contains more or less calcareous matter.

The class *Reptilia* is divided into four living and five extinct orders, as follows, but the latter require but brief notice:

1. *Chelonia* (Tortoises and Turtles).
2. *Ophidia* (Snakes).
3. *Lacertilia* (Lizards).
4. *Crocodilia* (Crocodiles).
5. *Ichthyopterygia* ⎫
6. *Sauropterygia* ⎪
7. *Pterosauria* ⎬ Extinct.
8. *Anomodontia* ⎪
9. *Deinosauria* ⎭

CHAPTER XXVIII.

DIVISIONS OF REPTILIA.

ORDER I. CHELONIA (Gr. *chelone*, a tortoise).—In this order are included the various Tortoises and Turtles, characterized by having the body enclosed in a bony case or box, and by the fact that the jaws are not provided with teeth, but are encased in horn, so as to form a kind of beak. The case in which the body of a Chelonian is protected is composed partly of integumentary plates, and partly of flattened bones belonging to the true skeleton, and it is composed essentially of two pieces, one placed on the back and the other on the lower surface of the body, firmly united together at their edges. The dorsal shield is more or less convex and rounded, and is called the *carapace* (Fig. 118, *ca*); while the ventral shield is more or less completely flat or concave, and is called the *plastron*. The carapace and plastron, as just said, are united by their edges, but they leave two openings, one in front for the head, and one behind for the tail. The carapace is essentially composed of the flattened and expanded spinous processes of the vertebræ and the greatly-developed ribs, covered by a series of horny plates. These are growths of the integument, and in some cases they constitute the "tortoise-shell" of commerce. The plastron is also composed partly of bony and partly of horny plates, but opinions differ as to whether the bony plates are to be looked upon as formed by an expanded breastbone, or whether they are merely integumentary, the probabilities being in favor of the latter view.

The remaining peculiarities with regard to the skeleton which deserve special mention are: Firstly, that the dorsal vertebræ are immovably connected together, so that this region of the spine is quite inflexible; secondly, that the heads of the ribs are articulated directly to the bodies of the vertebræ; and, thirdly, that the scapular and pelvic arches, supporting respectively the fore and hind limbs, are situated *within* the carapace (Fig. 118, *s* and *p*), so that the shoulder-blade is placed *inside* the ribs instead of outside, as is usually the case.

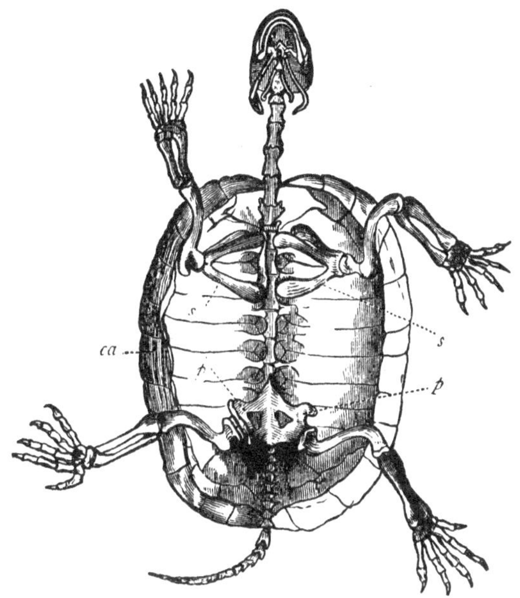

Fig. 118.—Skeleton of a Tortoise (*Emys Europæa*), seen from below, the plastron having been removed. *ca* Carapace, showing the flattened and expanded ribs; *s* Scapular arch, carrying the fore-limbs, and placed in the interior of the carapace; *p* Pelvic arch, carrying the hind-limbs; *r* Ribs.

The *Chelonia* are conveniently divided into groups, according as the limbs are adapted for swimming (natatory), or for progression on land (terrestrial); or, again, enable the animal to lead an amphibious life, sometimes on land and sometimes in the water. Of the strictly aquatic forms the best known are the edible Green Turtle (*Chelonia midas*) and the Hawk's-bill Turtle (*Chelonia imbricata*). The former is found abundantly in many of the seas of warm climates, and is largely imported into Europe as a delicacy. The latter (Fig. 119) is truly a native of warm seas, though an occasional straggler has reached the shores of Britain. It is of comparatively small size—not more than about three feet in length—but is of considerable commercial importance, as it furnishes the "tortoise-shell" of trade, so largely used in various kinds of ornamental work.

FIG. 119.—The Hawk-billed Turtle (*Chelonia imbricata*). (After Bell.)

The Sea-tortoises or Turtles have the carapace much flattened, the legs of unequal length, in the form of solid fins or oars, the toes being conjoined, and hardly distinct from one another.

The Marsh, Pond, and River Tortoises are generally furnished with webbed feet, and lead an amphibious, semi-aquatic existence. The so-called "Soft Tortoises" (*Trionycidæ*) belong here, and are distinguished by the imperfect condition of the carapace, which is simply covered with a leathery skin. A good example is the Soft-shelled Turtle (*T. ferox*) of the Southern States. Here also belong the Snapping-turtles, so well known in the person of the common American species (*Chelydra serpentina*), and the Terrapins (*Emydidæ*), of which many forms are found in all parts of the United States. In the curious little Box-tortoise (*Cistudo Virginea*) the plastron is composed of two movable portions which can be brought into accurate apposition with the carapace, thus completely protecting the animal within.

The Land Tortoises have short legs of nearly equal length, the toes little distinct, and united into a sort of stump, with indistinct, horny claws. Good examples of this group are the common European Tortoise (*Testudo Græca*) and the Indian Tortoise (*T. Indica*), the last attaining a length of over three feet.

ORDER II. OPHIDIA (Gr. *ophis*, a serpent).—This order includes most of the animals which would commonly be called snakes or serpents, and is characterized by the following peculiarities: The body is always more or less elongated, worm-like or cylindrical, and the skin develops horny scales, but never bony plates. There is never any breastbone (*sternum*), nor pectoral arch, nor fore-limbs; nor, as a rule, are there any traces of hind-limbs. In a few cases, however, rudimentary hind-limbs can be detected. The ribs are always very numerous. The two halves of the lower jaw are composed of several pieces each, and they are united to one another in front only by ligaments and muscles (Fig. 120). Hooked, conical teeth

FIG. 120.—The *Naja Haje*, a poisonous Snake of Egypt.

are always present, but they are never lodged in distinct sockets, and are only used to hold the prey, and not in mastication. The lungs and other paired organs are often not symmetrical, one of each pair being usually smaller than the other, or altogether absent.

The most striking of these characters of the snakes is to be found in the nature of the organs of locomotion. The fore-limbs are invariably altogether wanting, and there is no pectoral arch nor breastbone; and the hind-limbs are either totally absent or are at best rudimentary and never exhibit any outward evidence of their existence, beyond the occasional presence of short, horny claws or spurs. In the entire absence, then, or rudimentary condition of the limbs, the snakes progress by means of the ribs, which are always excessively numerous, and, in the absence of a breastbone, are also extremely movable. Their free ends, in fact, are simply attached by muscular fibres to the scales or "scutes," which cover the lower or abdominal surface of the animal. The number of ribs varies from 50 up to 320 pairs, and, by means of this arrangement, the snakes are able to progress rapidly, walking, as it were, upon the ends of the ribs. Their movements are also much assisted by the extreme flexibility of the whole spine, caused by the cup-and-ball articulation of the bodies of the vertebræ, each of which is concave in front and convex behind (*procœlous*).

Of the other characters of the snakes, a few words may be said as to the tongue, the eye, and the teeth—all important structures in this order. The *tongue*, in serpents, is probably more an organ of touch than of taste, and consists of two muscular cylinders, which are united toward their bases. The forked organ thus formed can be protruded and retracted at will, being in constant vibration when protruded, and being in great part concealed by a sheath when retracted. The *eye* of serpents (Fig. 121, A) is not protected by any eyelids, and hence the peculiar stony and unwinking stare for which these reptiles are celebrated. In place of eyelids, the outer layer of the skin is prolonged over the eye as a continuous and transparent film, behind which is a chamber formed by the mucous covering of the eye, into which the tears are discharged. The outer membrane is periodically shed along with the rest of the external or epidermic layer of the integument, and is again renewed. The pupil is round in most serpents, but it forms a vertical slit or fissure in the venomous snakes and in the Boas.

As regards the teeth, it is to be noticed that the snakes are not in the habit of chewing their prey, but of swallowing it whole, and the construction of their dental apparatus is in accordance with this peculiarity. The lower jaw, as before said, articulates with the skull by means of a quadrate bone

(Fig. 117), and this in turn is movably jointed to the cranium. The two halves of the lower jaw are also merely united loosely in front by ligaments and muscles. In consequence of this peculiar arrangement of parts, the serpents have the power of opening the mouth to an extraordinary width, and they can perform the most astonishing feats in the way of swallowing. The teeth are simply fitted for seizing and holding the prey, but not in any way for chewing or dividing it. In the harmless snakes, the teeth are in the form of solid cones, which are arranged in rows round the whole of the upper and lower jaws, a double row existing on the palate as well. In the venomous snakes, on the other hand, the ordinary teeth are usually wanting upon the upper jaws, and these bones are themselves much reduced in size. In place of the ordinary teeth, however, the upper jaws carry the so-called "poison-fangs" (Fig 121, B). These are a pair of long,

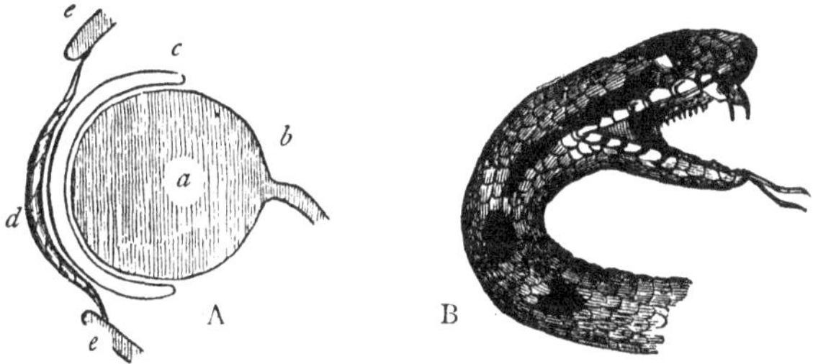

FIG. 121.—A, Diagrammatic Section of the Eye of a Viper (after Cloquet). *a* Eyeball; *b* Optic nerve; *c* Chamber into which the tears are poured; *d* Epidermic layer covering the eye. B, Head of the common Viper (after Bell), showing the poison-fangs.

curved fangs, one on each maxilla or upper jawbone, which, when not in use, are pointed backward, and concealed in a fold of the gum, but can be raised at will by special muscles. Each tooth is perforated by a fine canal or tube, which opens by a distinct aperture at the point of the fang, and is connected with the duct of the "poison-gland." This is a gland, situated under and behind the eye, secreting the poisonous fluid which renders the bites of these snakes dangerous or fatal. When the serpent strikes at any animal, the poison is forced through the poison-fang into the wound, partly by the contractions of the muscular walls of the gland, and partly by the compressive action of the muscles of the jaws. In some other snakes, several of which are not certainly known to be venomous, there are large, grooved fangs placed far back in the mouth upon the upper jaw.

Of the non-venomous, harmless snakes, we have an excellent instance in the common Ringed Snake (*Coluber natrix*), which is of frequent occurrence in most parts of Europe. Like all the snakes, it is strictly carnivorous, having a special liking for frogs, which it swallows whole. It often takes to the water, and can swim rapidly and gracefully, though, in this respect, it is excelled by the true venomous water-snakes (*Hydrophidæ*), which are adapted to an aquatic life by having a compressed swimming-tail. A well-known American example of this group is the common Black Snake (*Bascanion constrictor*). It attains a length of from three to five feet, but is perfectly harmless so far as man is concerned. Other non-venomous snakes, such as the Boas and Pythons, though destitute of poison-fangs, are, nevertheless, highly dangerous and destructive animals. Their bite is harmless, and they seize their prey by coiling themselves round it in numerous folds. By gradually tightening these folds, they reduce their victim to the condition of a shapeless bolus, which they finally proceed to swallow whole. In this way, a large Python or Boa will certainly succeed in disposing of an animal as large as a sheep or calf, and it has been asserted that human beings, and even oxen, can also be swallowed by unusually large specimens of this family.

The Boas and Pythons have a horny spur on each side of the vent, and the tail is prehensile. Their dental apparatus is extremely powerful, giving them a firm hold for the constriction of their prey. They are the largest of all the serpents, attaining a length of thirty to forty feet. The true Boas and Anacondas belong to the New World, but the Pythons are confined to India, Africa, and the Indian Archipelago.

The poisonous snakes are represented by the *Crotalidæ* of the New World and the *Viperidæ* of the Old World. The common Rattlesnake (*Crotalus horridus*) of the United States has the extremity of the tail furnished with a "rattle" or horny appendage composed of several membranous cells of a pyramidal shape articulated one within the other. Before striking its prey, it throws itself into a coil, and shakes its rattle. Another highly-dangerous species is the Copperhead (*Trigonocephalus contortrix*). The common European viper (*Pelias berus*) is hardly fatal to adults, but its bite causes serious inflammation. Highly deadly, however, is the Cobra di Capello or Spectacled Snake (*Naja tripudians*) of India, as also is the nearly-allied *Naja haje* (Fig. 120) of Africa. Other venomous snakes of evil notoriety are the Death-adder (*Acanthophis tortor*) of Australia, the Puff-adder (*Vipera inflata*) of South Africa, the Horned Viper (*Cerastes cornutus*) of Egypt, and the Harlequin-snakes (*Elaps*), but many others are equally dangerous.

ORDER III. LACERTILIA (Lat. *lacerta*, a lizard).—The third order of reptiles is that of the *Lacertilia*, comprising all the animals which are properly known as Lizards, together with some snake-like creatures, such as the Blind-worm. They are distinguished by the following characters: Usually there are two pairs of well-developed limbs, but there may be only one pair, or all the limbs may be rudimentary. In all cases, however, a scapular arch is present. The vertebræ are usually hollow in front (*procœlous*), rarely hollow at both ends (*amphicœlous*). In no living Lacertilian are the teeth lodged in distinct sockets. The eyes are mostly furnished with movable eyelids.

As a general rule, the animals included under this head have four well-developed legs, and would, therefore, be popularly called "Lizards." Some of them, however, such as the common Blind-worm (*Anguis fragilis*) of Europe, exhibit no external indications of limbs, and would, therefore, be generally regarded as Snakes. These snake-like Lizards, however, can be distinguished from the true Ophidians by the consolidation of the bones of the head and jaws, and by the fact that the eyes are generally provided with movable eyelids. Dissection also shows that the shoulder-girdle (or scapular arch) is always present in a rudimentary condition.

Of the snake-like Lizards, a good example is to be found in the common Blind-worm or Slow-worm of Europe. It is completely serpentiform, without any external indications of limbs (Fig. 122), and it is quite harmless. It is remarkable for the fact that, when alarmed, it stiffens its muscles to such an extent that the tail can readily be broken off, as if it were brittle. This same brittleness exists in the Glass-snake (*Ophisaurus ventralis*) of the Southern States, in which also there are no limbs. In other allied genera, there may be fore-feet alone, or hind-feet may be present, or all four limbs exist in a more or less rudimentary condition. In the true Lizards (*Lacerta*), all four limbs are present in a well-developed form; as seen in the common Green Lizard (*L. viridis*) of Europe. The genus *Lacerta* is represented in America by the *Ameivæ*, of which the Striped Lizard (*Ameiva sex-lineata*) of the Southern

States may be taken as a good example. Of all living Lizards, the largest are the Monitors (*Varanidæ*,) which are exclusively confined to the Old World, and attain sometimes a length of from six to eight feet. Very large, too, are some of the

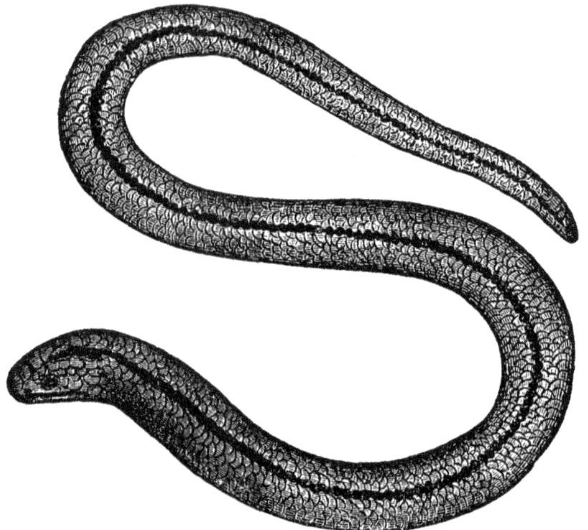

FIG. 122.—Blind-worm (*Anguis fragilis*). (After Bell.)

Iguanas which occur in warm regions in various parts of the world, but especially in South America, where they are often eaten. Related to the Iguanas are the singular Lizards known as the Flying Dragons (*Draco volans*), various species of which inhabit the Indian Archipelago and the East Indies. They are all of small size, living in trees and feeding on insects; and their great peculiarity consists in the fact that certain of the ribs are straightened out, and support a wing-like fold of the skin on each side of the body, by means of which the animal can take very extensive leaps from tree to tree.

The Scincoid Lizards form a very large family, represented by numerous species in all parts of the world. The species figured below is a common form in Egypt and Arabia, and was formerly used as a remedy in various diseases. A nearly-allied species is the Blue-tailed Lizard (*Scincus fasciatus*) of the United States.

The Geckos (*Geckotidæ*) form a large group of night-loving Lizards, which are found in most parts of the world, and chiefly deserve notice from the fact that their eyes are not provided with movable eyelids. The Chameleons, also, cannot be said to possess movable eyelids, for the eye is covered with

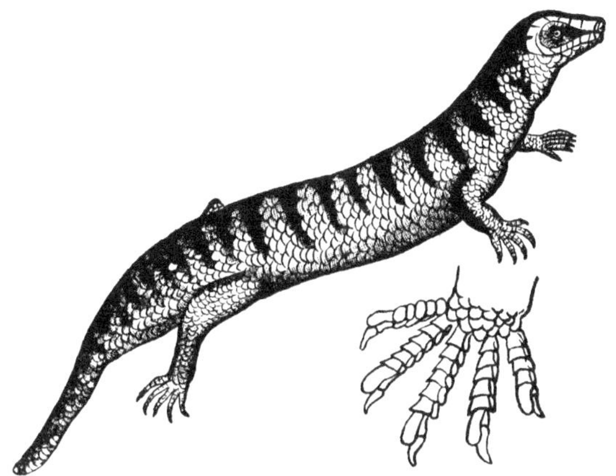

FIG. 123.—The Skink (*Scincus officinalis*)

a single lid, leaving only a central aperture for the pupil. The common species (*Chameleo Africanus*) occurs abundantly in the north of Africa, and has long been known for the changes of color which it has the power of exhibiting. It is a sluggish animal, and catches insects by darting out its long and protrusible tongue with extreme rapidity.

ORDER IV. CROCODILIA.—The last and highest order of the living Reptiles is that of the *Crocodilia*, comprising the Crocodiles, Alligators, and Gavials, and characterized by the following peculiarities: The outer or integumentary skeleton consists partly of horny scales developed by the outer layer of the skin, and partly of large bony plates produced by the inner layer of the skin. The bones of the skull and face are firmly united, and the two halves of the lower jaw are joined by a distinct suture. The teeth form a single row in both jaws, and are implanted in distinct and separate sockets. The front ribs of the trunk are double-headed, and there are no collar-bones. The heart consists of four distinct chambers, two auricles and two ventricles, all completely separated from one another. The mixture of arterial and venous blood, however, which is so characteristic of Reptiles, is provided for by a communication between the great vessels which spring from the two ventricles in the immediate neighborhood of the heart. The eyes are protected by movable eyelids, and the ear by a movable earlid. The tongue is large and fleshy, and is immovably attached to the bottom of the mouth (hence the belief of the ancients that the Crocodile had no tongue). Lastly, the *Crocodilia* agree with the typical Lizards, and differ from the Snakes in having four well-developed limbs.

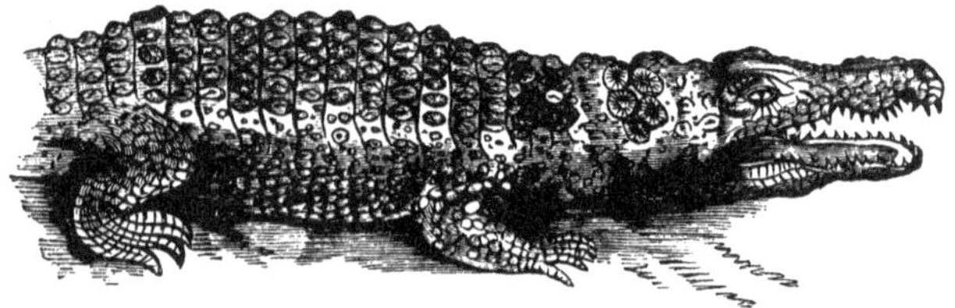

FIG. 124.—Head and fore-part of the body of the common Crocodile (*Crocodilus vulgaris*).

The *Crocodilia* abound in the fresh waters of hot climates, and are the largest of all living Reptiles, not uncommonly attaining a length of sixteen feet or upward. The best known of the *Crocodilia* is the Nilotic Crocodile, which occurs abundantly in Egypt, and was described by both Herodotus and Aristotle.

The true Crocodiles have the feet completely webbed, the hind-legs bordered by a fringe, and the fourth tooth in the lower jaw received in a notch on the side of the upper jaw. They belong mainly to Africa and Asia, but they are also represented in the West Indies and in South America.

The Alligators have the hind-legs simply rounded, and the toes not completely webbed; while the fourth tooth in the lower jaw fits into a cavity in the palate, and is concealed from view when the mouth is shut. Like the Crocodiles they are essentially aquatic in their habits, and lie dormant during the winter in cold climates and the hot season in warm countries. They are extremely voracious, and live upon fish and small Mammals. The best-known species are the common Alligator (*A. Mississippiensis*) of the Southern States, the Caiman (*A. palpebrosus*) of Surinam and Guiana, and the "Jacaré" (*A. sclerops*) of South America.

The Gavial or Gangetic Crocodile occurs in India, and is

distinguished by its narrow, elongated jaws, forming a kind of beak. It attains a length of more than ten feet.

ORDER V. ICHTHYOPTERYGIA (Gr. *ichthus*, fish; *pterux*, wing).—In this order are included a number of gigantic, fish-like Reptiles, which are all extinct, and are characteristic of the Secondary period of geology, and especially of the formation known as the Lias. The chief characters by which they are distinguished have reference to their purely aquatic life, for there can be no doubt that they were inhabitants of the sea. Thus the body was fish-like, without any distinct neck. The vertebræ were hollow at both ends (*amphicœlous*), and the spine thus possessed the flexibility and power of motion so characteristic of the true fishes. The limbs also constituted powerful swimming-paddles (Fig. 125), and it is probable that there was a vertical tail-fin.

Much has been gathered from various sources as to the habits of the *Ichthyosauri*, and their history is one of the most interesting chapters in the geological record. That they chiefly kept to open seas may be inferred from their strong and well-developed swimming apparatus; but the presence of a powerful bony arch supporting the fore-limbs proves that

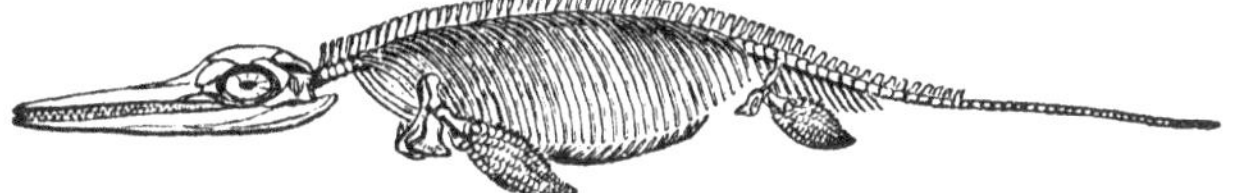

FIG. 125.—*Ichthyosaurus communis*.

they must have occasionally betaken themselves to the land. That they were tenants of stormy waters, or were in the habit of diving in search of prey, has been inferred from the fact that the eyeball is protected from pressure by a ring of bony plates. That they possessed great powers of vision, especially in the dusk, seems to be rendered certain from the size of the pupil and the enormous width of the bony cavities (orbits) which contained the eyes. Lastly, that they were carnivorous and predacious in the highest degree is shown by their wide mouths, long jaws, and numerous powerful and pointed teeth. This is also proved by an examination of their petrified droppings, which are known as "coprolites," and which contain in abundance undigested fragments of fishes and other marine animals.

ORDER VI. SAUROPTERYGIA (Gr. *saura*, lizard; *pterux*, wing).—The Reptiles belonging to this order agree with the last in being all extinct, and in being confined to the Secondary period of geology. The best known are the *Plesiosauri*, which resembled the *Ichthyosauri* in having all the limbs converted into swimming-paddles, but differed in several respects, of which the most obvious is the great elongation of the neck (Fig. 126). The *Plesiosauri* were gigantic marine Reptiles,

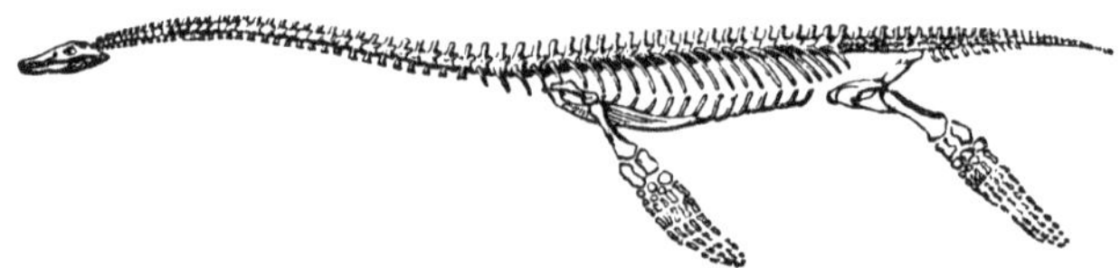

FIG. 126.—*Plesiosaurus dolichodeirus.*

chiefly characteristic of the formations known as the Lias and Oolites. As regards the habits of the *Plesiosaurus*, Dr. Conybeare concludes: "That it was aquatic is evident from the form of its paddles; that it was marine is almost equally so from the remains with which it is universally associated; that it may have occasionally visited the shore, the resemblance of its extremities to those of the Turtle may lead us to conjecture; its movements, however, must have been very awkward on land; and its long neck must have impeded its progress through the water, presenting a striking contrast to the organization which so admirably fits the Ichthyosaurus to cut through the waves." As its breathing-organs are such that it must of necessity have required to obtain air frequently, it may be inferred "that it swam upon or near the surface, arching back its long neck like a swan, and occasionally darting it down at the fish which happened to float within its reach. It may perhaps have lurked in shoal-water along the coast, concealed among the sea-weed, and, raising its nostrils to a level with the surface from a considerable depth, may have found a secure retreat from the assaults of powerful enemies; while the length and flexibility of its neck may have compensated for the want of strength in its jaws and its incapacity for swift motion through the water."

ORDER VII. PTEROSAURIA (Gr. *pteron*, wing; *saura*, lizard).—The Reptiles of this order are all extinct, and, like those of the preceding orders, are exclusively confined to the

Secondary period of geology. The most familiar examples are the so-called *Pterodactyles*, and the distinguishing characters of the order have reference to the fact that they were all adapted for an aërial life. They present, in fact, an extraordinary combination of the characters of birds and reptiles, and they make also some approach to the Mammalian order of the Bats. In the presence of teeth in distinct sockets, and, as we shall see hereafter, in the structure of the fore-limbs, the Pterodactyles differ altogether from all known birds; and there can be little doubt as to their being genuine Reptiles. The only living Reptile which has any power of sustaining itself in the air is the little *Draco volans*, which has been previously mentioned. In this case, however, the animal has no power of true flight, but is simply enabled to take extensive leaps by means of a membranous expansion on each side of the body. In the Bats, again, the power of genuine flight is present; and this is given by means of a leathery membrane which is supported chiefly by certain of the fingers—which are greatly lengthened—and is attached to the sides of the body and hind-limbs.

In the Pterodactyles the power of true flight was present, and this was also conditioned by means of a leathery expanded membrane, attached to the hind-limbs, the sides of the body, and the fore-limbs. In this case, however, the chief support of the flying membrane was derived from the outermost finger of the fore-limb, which was enormously elongated (Fig. 127). That the Pterodactyles passed their existence

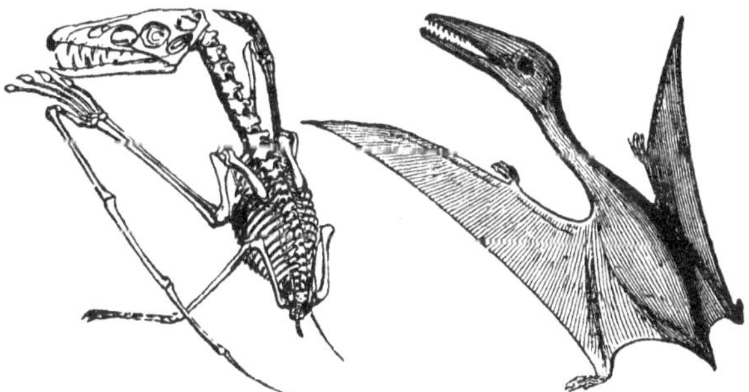

FIG. 127.—*Pterodactylus brevirostris*, the skeleton and the animal restored.

chiefly in the air, and did not simply leap from tree to tree, is shown by two characters in which they agree with the flying birds. Many of the bones, namely, were "pneumatic"—that is to say, were hollow and were filled with air, thus giving the animal the degree of lightness necessary for flight. Secondly, while the shoulder-girdle has many of the characters of birds, the breastbone (*sternum*) is furnished with a prominent ridge or keel, serving for the attachment of the great muscles which work the wings. There can be no doubt, therefore, as to the Pterodactyles having enjoyed the power of genuine flight. Many of them attained no great size, but some of them must have been gigantic, the expanse of wing in one species having been calculated at probably about twenty-seven feet from tip to tip.

Order VIII. Anomodontia (Gr. *anomos*, irregular; *odous*, tooth).—This order comprises a few Reptiles which belong to the Triassic period of geology, and are distinguished by the fact that the jaws were sheathed in horn, so as to form a kind of beak very like that of the Turtles. In some species there appear to have been no teeth at all; but in one genus there were two long tusks, one on each side of the upper jaw. The limbs were fitted for walking and not for swimming, and these singular Reptiles must, therefore, have been terrestrial in their habits.

Order IX. Deinosauria (Gr. *deinos*, terrible; *saura*, lizard).—In this order are included a number of extinct Reptiles, most of which were of gigantic size, and which are confined to the Secondary period of geology. They possessed teeth, sunk in distinct sockets, and the limbs were extremely strong, and adapted for progression on land. In some cases the fore-limbs were very much smaller than the hind-limbs, and there is reason to suppose that some of these extraordinary animals, though of enormous size, walked habitually upon their hind-legs, like Birds. It is also interesting to note that the gigantic footprints of the Sandstones of the Connecticut Valley, formerly regarded as formed by Birds, are now with great probability looked upon as truly the tracks of Deinosaurian Reptiles.

CHAPTER XXIX.

CLASS IV.—AVES.

THE fourth class of the Vertebrates is that of the Birds or *Aves*, which may be shortly defined as being "oviparous Vertebrates, with warm blood, a double circulation, and a covering of feathers" (Owen). The other leading characters which separate the Birds from the other *Vertebrata* are that the red blood-corpuscles are nucleated, the skull articulates with the spine by a single articulating surface (or condyle), the breathing-organs are in the form of lungs, which communicate with a variable number of air-sacs scattered through the body, and the fore-limbs are never terminated (in existing birds) by more than two fingers, ending in claws, and are generally modified so as to form "wings" or organs of flight.

The feathers, which form such a distinctive character of birds, are formed by a modification of the outer layer of the skin (epidermis), and from their non-conducting nature they serve to maintain the high temperature of the body which is so characteristic of the class. A typical feather, such as one of the long feathers of the tail or wing, consists of the following parts: 1. A horny cylindrical tube, which forms the lowest portion of the feather, and is termed the "quill." 2. The "shaft," which forms the central axis of the feather, and which is simply the continuation of the "quill." The under surface of the shaft is always marked by a strong longitudinal groove, and it consists of a horny sheath, filled with a white spongy material, not unlike the pith of a plant. 3. The "webs," which form the lateral expansions of the feather, and are attached to the sides of the shaft. Each web is composed of a number of small branches, called the "barbs," and each barb, in turn, is furnished with a series of smaller fibres called the "barbules." As a rule, the barbs are all kept in connection with one another by means of the barbules, the ends of which are hooked. Toward the base of the shaft, however, the barbs are usually more or less separate and placed at a distance from one another, constituting what is known as the "down." In the Ostriches and the birds allied to them, all the barbs are disunited and placed at a distance, and they are often not at all unlike hairs in appearance. The feathers of birds not only greatly conduce to the high temperature of the body, but also serve to keep out moisture, to which end there is a peculiar oil-gland at the base of the tail, with the secretion of which the bird anoints its plumage.

The skeleton of birds exhibits many points of peculiar interest, mostly in adaptation to an aërial mode of life; but only some of the more important of these can be noticed here. The entire skeleton is at the same time peculiarly compact and singularly light, the compactness being due to the presence of an unusual quantity of phosphate of lime, and the lightness to the fact that many of the bones are filled with air in place of marrow. The cervical region (neck) of the vertebral column is unusually long and flexible, since the fore-limbs are useless as organs of prehension, and all these functions have to be performed by the beak. In all birds the neck is, at any rate, sufficiently long to allow of the application of the beak to the tail, so as to permit of the cleaning and oiling of the whole plumage. The vertebræ which form the back or dorsal region of the spine are generally more or less immovably connected together, so as to give a base of resistance to the wings. In the Ostrich, however, and in other birds in which the power of flight is either very limited or is absent, the dorsal vertebræ are more or less movable one upon the other. The vertebræ which follow the dorsal region of the spine are all amalgamated together to form a single bony mass, which is termed the "sacrum," and this, in turn, is united on both sides with the bones which form the pelvic arch, which carries the hind-limbs. The vertebræ of the tail are more or less movable upon one another; and in almost all living birds, when fully grown, the last joint of the tail (Fig. 129, B, *s*) is a long, slender, ploughshare-shaped bone, which is really composed of several vertebræ united together. It is usually set on at an angle nearly perpendicular to the axis of the body, and it serves to support the great tail-feathers, which act as a rudder during flight. It also serves to support the oil-gland, which supplies the secretion with which the feathers are lubricated. The skull in birds has its several bones generally so amalgamated in the adult, that it forms a bony case of a single piece, the lower jaw alone remaining movable. The head is jointed to the spine by no more than a single articulating surface or condyle. The beak, which forms such a conspicuous feature in birds, is composed of two halves, an upper half or "upper mandible," and a "lower mandible." The lower mandible, like the lower jaw of all the *Sauropsida*, is at first composed of several pieces, but these are all undistinguishably united in the adult, and the two halves of the jaw are also amalgamated together. In no adult bird are *teeth* ever developed in either mandible; but both mandibles are sheathed in horn, constituting the "beak," and the margins of this sheath are sometimes serrated.

The most characteristic points, however, in the skeleton of the birds, are to be found in the structure of the limbs. The cavity of the chest or thorax is bounded behind by the dorsal vertebræ, on the sides by the ribs, and in front by the breastbone or sternum. The ribs vary in number from seven to eleven pairs, and in most birds each rib gives off a peculiar process (Fig. 128, B), which passes over the rib next in succession behind. In front the ribs are jointed to a series of straight bones, which are called the "sternal ribs," and these, in turn, are movably articulated to the breastbone in front. According to Owen, these sternal ribs are "the centres upon which the respiratory movements hinge." In front the cavity of the chest is completed by an enormously-expanded breastbone or sternum (Fig. 128, A), which, in most birds of any powers of flight, extends more or less over the abdominal cavity as well. The sternum of all birds which possess the power of flight is characterized by the presence of a prominent ridge or "keel" (Fig. 128, A, *b*), to which are attached the great muscles (pectoral muscles) which move the wings. As a general rule, the size of this crest or keel gives a tolerably just estimate of the flying powers of the bird to which it belonged. The keel is, of course, most largely developed in those birds which possess the power of flight in its greatest perfection; and in those which do not fly, such as the Ostrich, there is no sternal keel at all. The pectoral arch or shoulder-girdle of birds consists of the shoulder-blades (*scapulæ*), the clavicles or collar-bones, and of two bones, which are distinct in birds, and are called the "coracoid bones." The shoulder-blades (*s s*) are usually long and narrow bones. The coracoid bones (*k k*) correspond with the part of the shoulder-blade which is known in most of the Mammals as the "coracoid

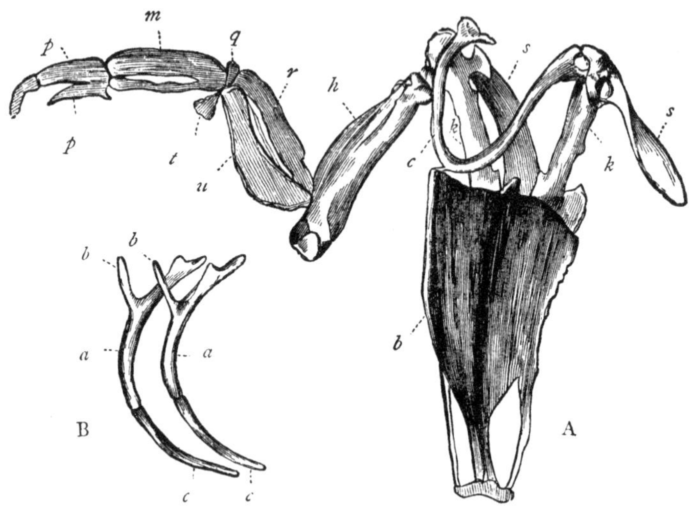

FIG. 128.—A, Breastbone, shoulder-girdle, and fore-limb of Penguin (after Owen); *b* Breastbone (*sternum*), with its prominent ridge or keel; *s s* Shoulder-blades (*scapulæ*); *k k* Coracoid bones; *c* Furculum or Merry-thought, composed of the united collar-bones (*clavicles*); *h* Bone of the upper arm or *humerus*; *r* Radius; *u* Ulna, forming together the forearm; *q* Bones of the wrist or *carpus*; *t* Thumb; *m* Metacarpus; *p* Phalanges of the fingers. B, Ribs of the Golden Eagle; *a a* Ribs giving off processes (*b b*); *c c* Sternal ribs.

process;" and in birds they are not only separate bones, but they are the strongest bones of the pectoral arch. They are more or less nearly vertical, and they form fixed points for the downward stroke of the wing. The collar-bones or clavicles (*c*) in the great majority of birds are united together in front, so as to form a somewhat V-shaped bone, which is technically called the "furculum," but is familiarly called the "merry-thought." The function of this clavicular arch is to keep the wings asunder during their downward stroke, and the strength of the furculum varies, therefore, with the powers of flight enjoyed by each bird. The bones which form the limb proper, or "wing," are considerably modified to suit the special function of flight, but essentially the same parts are present as in the fore-limb of the Mammals. The upper arm is constituted by a single bone, the *humerus* (*h*), which is generally short and stout. The forearm is composed of two bones, the *radius* (*r*) and the *ulna* (*u*), of which the ulna is the bigger. These are followed by the small bones, which form the wrist or *carpus* (*q*), but these are reduced to *two* in number. The carpus is followed by the bones which constitute the root of the hand or *metacarpus* (*m*), but these are also reduced to *two*, instead of being five in number, as they are in most Mammals. The two metacarpal bones are also amalgamated together at both ends, so as to form a single piece, at the base of which, on its outer side, is a rudimentary digit, the "thumb" (*t*), which carries a tuft of feathers, known as the "bastard wing." The metacarpal bones, finally, support each a single finger (*p*), of which one is never composed of more than one bone or *phalanx*, while the other is composed of two or three phalanges. (To understand thoroughly the leading modifications of the limbs of birds, the student will do well to refer to the general description of the limbs of Vertebrates, p. 200, Figs. 96, 97.)

As regards the composition of the hind-limb in birds, the two halves of the pelvic arch (i. e., the *innominate bones*) al-

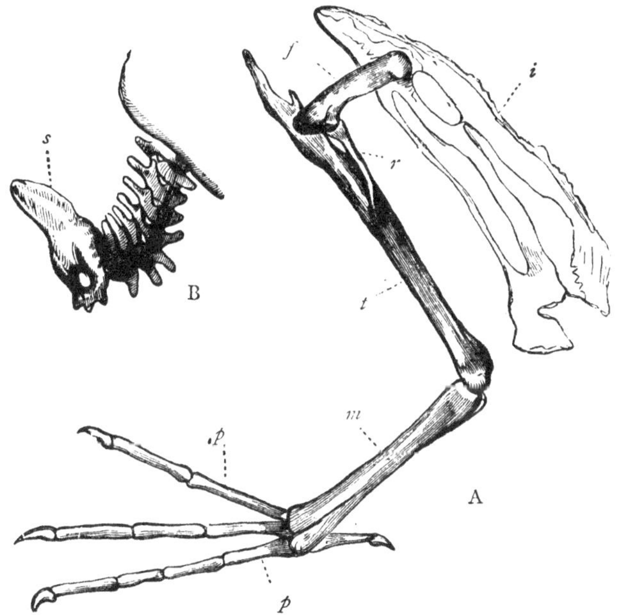

FIG. 129.—A, Pelvis and bones of the Leg of the Loon or Diver (after Owen): *i* Innominate bone; *f* Thigh-bone (*femur*); *t* Tibia, *r* fibula, together forming the shank; *m* Tarso-metatarsus; *p* Phalanges of the toes. B, Tail of the Golden Eagle; *s* Ploughshare-shaped bone, carrying the great tail-feathers.

ways form a single piece each, and they are always firmly united with the sacral region of the spine. With the single exception, however, of the Ostrich, they do not unite below, but remain separate. As in the higher Vertebrates, the lower limb consists of a thigh-bone (*femur*), a shank, composed of two bones (*tibia* and *fibula*), a tarsus, a metatarsus, and phalanges, but some of these parts are obscured by coalescence. The thigh-bone or femur (Fig. 129, *f*) is generally very short, comparatively speaking; and the chief bone of the leg is the *tibia* (*t*), to which a thin and tapering *fibula* (*r*) is attached. In the regular typical limb of a Vertebrate animal the tibia and fibula would be followed by a series of small bones, called the *tarsus*, constituting the ankle-joint (Fig. 97); and the tarsus would in turn be followed by a series of bones constituting the root of the foot, or *metatarsus*. In Birds, however, the tibia and fibula are followed by a single cylindrical bone, which is called the "tarso-metatarsus" (*m*), and which is formed by the amalgamation of the entire metatarsus with the whole or a portion of the tarsus. The most probable view is that only the *lower* portion of the tarsus is present in the tarso-metatarsus, and that the *upper* portion of the tarsus is fused with the lower end of the tibia. In this case the ankle-joint is placed in the middle of the tarsus. The tarso-metatarsus is followed below by the foot, which consists in most birds of four toes, of which three are directed forward and one backward. In no wild birds are there more than four toes; but some domesticated varieties possess a fifth. In most birds with four toes, the toe which is directed backward consists of two phalanges; the innermost of the three forward toes has three phalanges, the next has four, and the outermost toe is composed of five. In many birds, such as the Parrots, the outermost toe is turned backward, so that there are two toes in front and two behind. In the Swifts, again, all the four toes are turned forward. In many of the swimming-birds (*Natatores*) the hinder toe is wanting or rudimentary; and in the Ostrich both this and the next toe are absent, so that the foot consists of no more than two toes.

The digestive system in Birds consists of the beak, tongue, gullet, stomach, intestine, and cloaca, with certain accessory glands. There are no teeth, and the beak is employed, in different birds, for holding and tearing the prey, for prehension, for climbing, and in some cases as an organ of touch, being in these last instances more or less soft, and supplied with nervous filaments. In many birds, too, the base of the bill is surrounded by a circle of naked skin, constituting what is called the "cere," and this, too, serves as an organ of touch. The tongue of birds can rarely be looked upon as an organ of taste, since it is generally cased in horn, like the mandibles. It is principally employed as an organ of prehension, but it is soft and fleshy in the Parrots, and in them, doubtless, acts as an organ of taste. Salivary glands are always present, but they are rarely of large size, and are often of extremely simple structure. In accordance with the length of the neck, the gullet is usually very long in birds, and is generally very dilatable. In the flesh-eating and grain-eating birds the gullet is dilated (Fig. 130, *c*) into a pouch which is called the

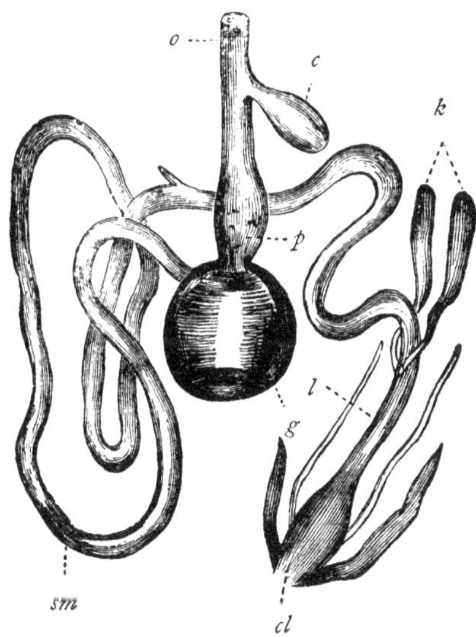

Fig. 130.—Digestive System of the common Fowl (after Owen). *o* Gullet; *c* Crop; *p* Proventriculus; *g* Gizzard; *sm* Small intestine; *k* Intestinal cæca; *l* Large intestine; *cl* Cloaca.

"crop," and is situated in the lower part of the neck, just in front of the merry-thought. This may be simply a dilatation of the tube of the gullet, or it may be a single or double pouch. The function of the crop is to detain the food, for a longer or shorter period, according to its nature, before it is submitted to the action of the proper digestive organs. In the Pigeons, the food which has been previously softened in the crop is returned to the mouth, and supplied to the young in a state suitable for digestion. The gullet, after leaving the crop, shortly opens into a second cavity, called the "proventriculus," which is the true digesting stomach, and is richly supplied with glands which secrete the digestive fluid or gastric juice (*p*). This, in turn, opens into a muscular cavity which is called the "gizzard" (*g*), and which leads into the commencement of the small intestine. The characters of the gizzard vary with the nature of the food. In the birds of prey, which live on a easily-digested animal diet, the walls of the gizzard are thin and membranous. In the grain-eating birds, such as the fowls, whose hard food requires to be crushed before it can be properly digested, the walls of the gizzard are extremely thick and muscular, and the inner lining is hard and horny. In these birds the gizzard constitutes a kind of grinding apparatus, like the stones of a mill; while the "crop" may be compared to the "hopper" of the mill, since it supplies to the gizzard "small successive quantities of food as it is wanted (Owen). The grinding action of the gizzard is further assisted by the small pebbles and gravel which, as is well known, so many birds are in the habit of swallowing. These pebbles take the place of teeth, and there can be no doubt that they are in many cases essential to health, the bird being otherwise unable to triturate its food properly. The intestinal canal extends from the gizzard to the cloaca (*cl*), and is comparatively short. The secretions of the liver and pancreas are poured into the commencement of the small intestine. The commencement of the large intestine is furnished in most birds with two blind tubes or cæca (*k*). These vary considerably in length in different birds, and are sometimes wanting; while their exact function is still questionable. The large intestine is seldom more than a tenth part of the length of the body, and is generally conducted straight from the cæca to the cloaca. . The *cloaca* is a common cavity which in birds, as in Reptiles, receives the termination of the intestine and the ducts of the generative and urinary organs (*cl*).

Respiration is effected in Birds more completely, extensively, and actively, than in any other class of the *Vertebrata*, and, as the result of this, their average temperature is higher than in any other Vertebrates. This extensive development of the respiratory process is due to the fact that air is admitted in Birds not only to the lungs, but also to the interior of a greater or less number of the bones, and to a series of air-receptacles which are scattered through various parts of the body. The lungs are two in number, of a bright-red color, and spongy texture, and they are confined to the back part of the chest. They differ from the lungs of Mammals in not being freely suspended in a membranous bag (*pleura*), but in being fixed to the back wall of the chest. The thoracic and abdominal cavities are not separated from one another by a complete partition (midriff or diaphragm) as the Mammals, but the common thoracico-abdominal cavity is subdivided by means of membranous partitions into a series of cavities or sacs, which are termed the "air-receptacles." These air-sacs are filled with air from the lungs, and vary considerably in number and size in different birds. They not only serve greatly to reduce the specific gravity of the body, but also assist largely in the aëration of the blood. Connected with the air-receptacles, and supplementing their action in both of these respects, is a series of cavities occupying the interior of a greater or less number of the bones, and also containing air. In young birds these air-cavities in the bones do not exist, and the bones are simply filled with marrow, as in the Mammals. In the Penguin, which does not fly, none of the bones contain air-cavities or are "pneumatic;" and in the Ostrich only a few of the bones contain air. In the Pelican and Gannet all the bones of the skeleton, except the phalanges of the toes, are permeated by air; and in the Hornbill even these are pneumatic.

The *heart* in all birds consists of four chambers, and the two sides of the heart are completely separated from one another. In all essential details, as regards the structure of the heart and great vessels, and the course of the circulating fluid, Birds agree with Mammals. The impure venous blood which has traversed the body is returned by the great veins to the right auricle. From the right auricle it passes into the right ventricle, from which it is driven by the pulmonary artery to the lungs. Having been submitted to the action of the air contained in the lungs, and having been thereby changed into arterial blood, the blood is sent back to the left auricle by means of the pulmonary veins. Thence it passes into the left ventricle, by which it is again propelled throughout the whole body, to return again as venous blood to the right side of the heart. The heart, therefore, of birds, differs from that of reptiles in consisting of two sides, each composed of an auricle and ventricle, the right side being wholly concerned with sending the venous blood to the lungs, and the left side being entirely occupied with sending the arterial blood to the body. The right side of the heart is therefore venous, the left side arterial. In all Reptiles, on the other hand, the two circulations—namely, that through the lungs and that through the body—communicate with one another, either in the heart itself or in its immediate neighborhood; so that both the lungs and the body are supplied with a mixture of venous with arterial blood. Though the heart of Birds resembles that of Mammals in general structure, its cavities are "relatively stronger, their valvular mechanism is more perfect, and the contractions of this organ are more forcible and frequent in Birds, in accordance with their more extended respiration and their more energetic muscular actions" (Owen). The urinary organs of birds consists of two elongated kidneys, which open by means of their ducts (the ureters) into the cloaca, along with the termination of the intestine and the ducts of the reproductive organs. As a general rule, the female bird is provided with only a single ovary—that of the left side—and all birds, without exception, are oviparous. The egg is always enclosed in a calcareous shell, and is developed after expulsion from the body, by the process of "incubation" or "brooding"—a process for which birds are especially adapted, in consequence of their very high average temperature. The young bird, when ready for an independent existence, perforates the shell, often by means of a temporary calcareous excrescence developed upon the point of the upper mandible of the bill. In some birds, mostly in the case of those which live upon the ground, the young are able to run about and look for food directly after they come out of the egg, as is seen in the common Fowl. In most birds, however, the young are liberated from the egg in a perfectly helpless and naked condition, and require to be fed by their parents for a longer or shorter time, before they are able to take care of themselves. Most of these birds, such as our common song-birds, reside in trees, and build more or less elaborate nests.

As regards their *nervous system*, the brain of Birds is relatively larger than the brain of Reptiles, but it is destitute of those folds or convolutions which form so marked a feature in the brain of most Mammals. The organs of sense, with the exception of touch and taste, are well developed in Birds, vision especially being generally extremely acute. The eyes are always well developed, and in no bird are they ever wanting or rudimentary. The chief peculiarity of the eye of Birds is, that its anterior portion (*cornea*) forms the segment of a much smaller circle than does the eyeball proper; so that the whole eye assumes a conical shape. Another peculiarity is that the form of the eye is maintained by means of a circle of from thirteen to twenty bony plates, which are placed in the front portion of the fibrous coat of the eye (*sclerotic*). Eye-lashes are almost universally absent; but, in addition to the ordinary upper and lower eyelids, Birds possess a third membranous eyelid—the *membrana nictitans*—which is placed on the inner side of the eye. This nictitating membrane is sometimes transparent, sometimes pearly white, and it can be drawn over the front of the eye like a curtain, moderating the too great intensity of the light. As regards the organ of hearing, the chief point to remark is that Birds have mostly no *external* ear, by means of which the undulations of sound can be collected and transmitted to the internal ear. In some birds, however, as the Ostrich, the external opening af the organ of hearing is provided with a circle of feathers, which can be raised and depressed at will. In the nocturnal Birds, also, (such as Owls), the external opening of the ear is protected by a musculo-membranous valve, foreshadowing the gristly external ear of Mammals. The sense of smell is apparently seldom very acute in Birds, and even the Birds of Prey appear to seek their food mainly by the sight. The external nostrils are usually placed on the sides of the upper mandible, near its base, and form simple perforations which sometimes communicate from side to side. In the curious *Apteryx* of New Zealand, the nostrils are placed at the extreme end of the elongated beak. Sometimes the nostrils are defended by bristles, and sometimes by a cartilaginous scale.

Before passing on to a consideration of the divisions of Birds, a few words may be said on the *migrations* of Birds. In temperate and cold climates, few birds remain constantly in the same region in which they were originally hatched. Those which do so are called "permanent birds." Other birds, such as the Woodpeckers, migrate from place to place without following any very definite course. These are called "wandering birds," and their movements are chiefly conditioned by the scarcity or abundance of food in any particular locality. Other birds, however, at certain seasons of the year, undertake long journeys, usually uniting for this purpose into larger or smaller flocks. Such birds—of which the Swallows are a familiar instance—are properly called "migratory birds," and their movements are conditioned by the necessity of having a certain average temperature, without which they cannot live. Thus the migratory birds of cold climates, when the cold season comes on, travel to warmer countries; but when the hot season of these regions approaches, they migrate back again to temperate zones.

CHAPTER XXX.

DIVISIONS OF BIRDS.

BIRDS may be variously divided, but for our present purpose it is most convenient to regard them as divided into the following eight orders:

I. *Natatores* or *Swimming Birds*, characterized by having the feet webbed, and the legs short and placed far back, while the body is closely covered with feathers and with a thick coating of down next the skin. (*Ex.* Ducks, Geese, Pelicans.)

II. *Grallatores* or *Wading Birds*, characterized by having very long legs, which are destitute of feathers from the lower end of the tibia downward. The toes are usually long and straight, and are never connected to one another by membrane. (*Ex.* Curlews, Snipes, Herons, Storks.)

III. *Cursores* or *Running Birds*, characterized by having very short wings, which are not used in flight; the breastbone is without a ridge or keel; the legs are very robust; and the hind-toe is wanting or rudimentary. (*Ex.* Ostriches and Emeus.)

IV. *Rasores* or *Scratching Birds*, characterized by usually having strong feet, with powerful blunt claws, used for scratching. The upper mandible of the bill is strongly curved and vaulted, and the nostrils are pierced in a membranous space at its base, and are covered by a cartilaginous scale. (*Ex.* Fowls, Pheasants, Pigeons.)

V. *Scansores* or *Climbing Birds*, characterized by having a climbing foot, in which two toes are turned backward and two forward. (*Ex.* Woodpeckers, Parrots, Cuckoos.)

VI. *Insessores* or *Perching Birds*, characterized by having short and slender legs, with three toes in front and one behind, the whole foot being adapted for perching. (*Ex.* Larks, Linnets, Swallows, Crows, Humming-birds.)

VII. *Raptores* or *Birds of Prey*, characterized by having a strong, sharp-edged, and sharp-pointed beak, adapted for tearing animal food, and by their robust legs, armed with four toes, three in front and one behind, all of which are furnished with long, strong, crooked claws or talons. (*Ex.* Eagles, Hawks, Owls.)

VIII. *Saururæ* or *Lizard-tailed Birds*, characterized by having a tail longer than the body, composed of numerous distinct and movable vertebræ, each of which carries a single pair of quill-feathers. (This order includes only the remarkable fossil bird, the *Archæopteryx*.)

ORDER I. NATATORES (Lat. *natator*, a swimmer).—The order of the Swimming Birds comprises birds which are as much at home in the water as upon land, or even more so. In accordance with their aquatic mode of life, the *Natatores* have a boat-shaped body, generally elongated, and usually having a long neck. The legs are short, and are placed behind the cen-

FIG. 131.—Natatores. Penguin (*Spheniscus demersus*).

tre of gravity of the body; this position enabling them to act admirably as swimming-paddles, at the same time that it renders the gait upon dry land comparatively awkward and shuffling. The toes in all the *Natatores* are *webbed* to a greater or less extent, or, in other words, are united by a membrane (Fig. 131). In many the web or membrane between the toes is stretched completely from toe to toe, but in others the membrane is divided between the toes, so that the feet are only imperfectly webbed. As their aquatic mode of life exposes them to great reductions of temperature, the body in the Natatorial birds is closely covered with feathers, with a thick covering of down next the skin. They are further protected against becoming wet while in the water by the great development of the oil-gland at the tail, by means of which the dense plumage is kept constantly oiled. As a rule, the Natatorial birds are polygamous, each male having several females; and the young are hatched in a condition not requiring assistance from their parents, being able to swim about and procure food for themselves the instant they are liberated from the egg.

Among the more important families of the *Natatores* may be enumerated the Penguins (*Spheniscidæ*), the Auks (*Alcidæ*), the Gulls and Terns (*Laridæ*), the Petrels (*Procellaridæ*), the Pelicans (*Pelicanus*), the Cormorants (*Phalacrocorax*), the Gannets (*Sula*), the Ducks (*Anatidæ*), the Geese (*Anserinæ*), and the Swans (*Cygnidæ*).

The Penguins and Auks, with their allies the Divers, Guillemots, and Grebes, have rudimentary, or at any rate small, wings, and are all more at home in the water than upon land. The Gulls, Terns, and Petrels, on the contrary, are all birds of powerful flight, and some of them, such as the Albatross, are habitually found hundreds of miles from the nearest land. The Pelicans, with their allies the Cormorants, Frigate-birds, and Darters, are excellent flyers, and also not uncommonly perch on trees, which few Natatorial birds do. They are distinguished by having the hinder toe directed inward, and united to the innermost of the front toes by a continuous membrane. The Gannets, Ducks, Geese, and Swans, have the bill very much flattened and covered by a soft skin. The edges of the bill are also furnished with a series of transverse plates, which form a kind of fringe or "strainer," by means of which these birds sift the mud in which they habitually seek their food.

ORDER II. GRALLATORES (Lat. *grallæ*, stilts).—The Wading Birds for the most part frequent moist situations, such as marshes and shallow ponds, the shore of the sea, or the banks of rivers or lakes, though some of them keep entirely, or almost entirely, to the dry land. In accordance with their semi-aquatic, amphibious habits, the Waders are distinguished by the great length of their legs—the increase in length being chiefly due to the elongation of the tarso-metatarsus. The legs (Fig. 132) are also unfeathered or naked, as far as the

FIG. 132.—Grallatores. Common Heron (*Ardea cinerea*).

lower end of the tibia, at any rate. There are three anterior toes, and usually a short hind-toe; but the toes are never completely webbed, though they are sometimes partially palmate. The wings are long, and the power of flight is usually considerable; but the tail is very short, and its function as a rudder is chiefly transferred to the long legs, which are stretched out behind in flight. The beak is almost always of great length, generally longer than the head (Fig. 132), and usually more or less pointed, though it is sometimes flattened. In the Avocet the bill is curved upward, instead of being straight, or bent downward, as is generally the case. The typical Waders, as before said, spend most of their time wading about in shallow water, feeding upon small fishes, shell-fish, worms, and insects. Others, such as the Storks, live mostly upon the land, and are more or less exclusively vegetable-feeders.

Among the more important Grallatorial birds are the Rails (*Rallidæ*), Water-hens (*Gallinulæ*), Cranes (*Gruidæ*), Herons (*Ardeidæ*), Storks (*Ciconinæ*), Snipes (*Scolopacidæ*), Sandpipers (*Tringidæ*), Curlews (*Numenius*), Plovers (*Charadriidæ*), and Bustards (*Otidæ*).

The Rails are more or less terrestrial in their habits, but inhabit marshes and fens. Good examples are the Marsh Hen (*Rallus elegans*) and the Virginia Rail (*R. Virginianus*) of North America, and the Corn Crake (*Crex pratensis*) of Europe. The Water-hens (*Gallinula*) and Coots (*Fulica*) are aquatic or semi-aquatic, swimming and diving with the greatest ease. The Cranes are in the main vegetable-feeders, and inhabit dry plains. The Herons, Egrets, Bitterns, and Night Herons, form a beautiful family of wading birds, represented in almost every portion of the known world. Nearly allied to these are the brilliantly-colored Ibises (*Tantalinæ*), which inhabit all warm countries. The *Ciconinæ* are all large birds, and comprise the Storks and Adjutant, while the Spoonbills are mainly separated from them by their flattened, spoon-shaped bill. The *Scolopacidæ*, comprising the Snipes and Woodcocks, the *Tringidæ* (or Sandpipers), the Curlews (*Numenius*), and various other allied birds, are distinguished from the preceding by the possession of a long, soft, slender bill, which is used in probing the ground for food. In the *Charadriidæ* are comprised the Oyster-catchers, Turnstones, Lapwings, Plovers, Thick-knee, and many other familiar birds. Lastly, the *Otidæ* comprise only the Bustards, which are exclusively confined to the Old World, and make a decided approach to the Cursorial Birds.

ORDER III. CURSORES (Lat. *curro*, I run).—The Running or Cursorial Birds, comprising the Ostrich, Cassowary, Emeu, Rhea, and Apteryx, are characterized by the rudimentary condition of the wings, which are useless as organs of flight, and by the compensating length and strength of the legs. In accordance with this condition of the limbs, the bones have few air-cells, and the breast-bone is destitute of the prominent ridge or keel to which the great muscles of the wings are attached. The two sides of the pelvis are united together below in the Ostrich, and in all the pelvic arch has great strength and stability. The legs are extremely powerful, and the hinder toe is wanting in all except the *Apteryx*, in which it is present in a rudimentary condition. The front toes (Fig. 133) are either two or three in number, and are furnished with strong blunt claws or nails. The feathers present the remarkable peculiarity, that the barbs, instead of being connected by means of the barbules, are disconnected and separate from one another, thus coming to resemble hairs in appearance.

FIG. 133.—Cursores. The *Apteryx Australis* (Gould).

The African Ostrich (*Struthio camelus*), which is one of the best-known members of this order, inhabits the desert plains of Africa and Arabia, and is the largest of living birds, attaining a height of from six to eight feet. The head and neck are nearly naked, and the quill-feathers of the wings and tail have their barbs wholly separate, constituting the ostrich-plumes of commerce. The legs are extremely strong, and the feet have only two toes each. The Ostriches run with extraordinary speed, and can outstrip the fastest horse. They are polygamous, each male having several females, and they keep together in larger or smaller flocks. The American Ostriches or Rheas are much smaller than the African Ostrich, and have the head feathered, while the feet are furnished with three toes each. They inhabit the great plains of South America, and are polygamous. The Emeu (*Dromaius*) is exclusively confined to New Holland. In size it nearly equals the African

Ostrich, standing from five to seven feet in height, and it is not uncommonly kept as a domestic pet. The Cassowary (*Casuarius galeatus*) inhabits the Moluccan Islands and New Guinea, and was first brought alive to Europe by the Dutch. It stands about five feet in height, and possesses a singular horny crest upon the head. Another species of Cassowary inhabits Australia, and other species are known to exist in the Indian Archipelago. The last of the living Cursorial birds is the curious bird, the *Apteryx* (Fig. 133) of New Zealand. In this remarkable bird the beak is extremely long and slender, and the nostrils are placed at the extremity of the upper mandible. The legs are comparatively short, and there is a rudimentary hind-toe, provided with a claw. The feathers of the general plumage are long and hair-like, and the wings are altogether rudimentary.

ORDER IV. RASORES (Lat. *rado*, I scratch).—The Scratching Birds—or, as they are often called, the Gallinaceous Birds—are characterized by the fact that the upper mandible of the bill is convex and vaulted (Fig. 134), and has a membranous space at its base, in which the nostrils are pierced. The nostrils are also covered by a cartilaginous scale. The legs are strong and muscular, and are often covered with feathers as far as the ankle-joint. There are four toes (Fig. 134), three in front, and a short hind-toe placed on a higher level than the others. All the toes, in the typical members of the order, are provided with strong, blunt claws, suitable for scratching. The food of the *Rasores* consists chiefly of hard grains and seeds, and, in accordance with this, they have a large crop, and an extremely strong and muscular gizzard. They generally lay their eggs upon the ground, and they are mostly polygamous, each male having several mates. The Doves, however, pair for life. The males take no part in building the nest or in hatching the eggs; and the young are generally precocious, being able to run about and provide themselves with food from the moment they quit the egg. The wings are usually weak, and the flight feeble, and accompanied with a whirring sound; but many of the Pigeons are powerful flyers.

The order *Rasores* is divided into two very well marked sections or sub-orders, called respectively the *Gallinacei* and *Columbacei*. In the *Gallinacei* are all the typical forms of the order, and the characters of this section are therefore the same as those of the order itself. They are distinguished from the *Columbacei* mainly by being less fully adapted for flight, their bodies being much heavier, comparatively speaking, their legs and feet stronger, and their wings shorter. They are also generally polygamous, and the males usually possess "spurs," and are more brilliantly colored than the females.

FIG. 134.—Rasores. Rock-pigeon (*Columba livia*).

The leading families of the Gallinaceous birds are: 1. The *Tetraonidæ* or Grouse family, comprising the true Grouse and Black Game (*Tetrao*), the Ptarmigans (*Lagopus*), the Ruffed Grouse (*Bonasa*), etc. 2. The *Perdicidæ* or Partridge family, comprising the Partridges (*Perdix*), Quails (*Coturnix*), Virginian and Mountain Quails (*Ortyx*), Crested Quails (*Lophortyx*), etc. 3. The *Phasianidæ*, or Pheasant family, comprising the various Pheasants (*Phasianus*), the Domestic and Jungle fowls (*Gallus*), the Turkeys (*Meleagris*), the Guinea-fowls (*Numida*), and the Pea-fowl (*Pavo*). 4. The *Megapodidæ*, or Mound-builders, comprising only some singular Australian and Indian birds, which build enormous mounds, in which they deposit their eggs. 5. The *Cracidæ*, or Curassow family, comprising the large South and Central American birds known as Curassows and Guans.

The *Columbacei* comprise the Pigeons and Doves (Fig. 134), and they are separated from the typical *Rasores* by being much more fully adapted for flight. They are furnished with strong wings and are good flyers; and, in place of being ground-birds, their habits are to a great extent arboreal, in accordance with which the feet are slender and are adapted for perching. They are also not polygamous, and their voice is of a much more gentle, soft, and melancholy character. (Hence the name of *Gemitores* applied to this section, while the *Gallinacei* are called the *Clamatores*.) Besides the true Pigeons and Doves, this sub-order includes also the remarkable extinct bird, the Dodo, which was of gigantic size, comparatively speaking, and inhabited the island of Mauritius up to the commencement of the seventeenth century.

ORDER V. SCANSORES (Lat. *scando*, I climb).—The order of *Scansores* or Climbing Birds is very shortly and easily defined, having no other distinctive and exclusive peculiarity except the fact that the feet have four toes, of which two are turned backward and two forward (Fig. 135). Of the two toes which are turned backward, one is the proper hind-toe, and the other is the outermost toe. This arrangement of the toes enables the Scansorial birds to climb with great ease and readiness. Their powers of flight are usually very moderate, and below the general average, and their food consists of insects and fruits of various kinds. Their nests are usually made in the hollows of old trees, but some (Cuckoos) have the remarkable habit of depositing their eggs in the nests of other birds. They are never polygamous, and the young are born in a naked and helpless condition.

The following families have been established in the *Scansores*: 1. The *Cuculidæ* or Cuckoo family, comprising the true Cuckoos and some allied birds. They are remarkable for the fact that many of them are "parasitic," that is to say, they lay their eggs in the nests of other birds. The Yellow-billed Cuckoo (*C. Americanus*), however, of the United States, builds a nest for itself and brings up its own young. 2. The *Picidæ* or Woodpecker family, comprising many familiar birds, all of which climb and run up trees with the greatest facility. They live mostly on insects, which they catch by darting out their long, worm-like, barbed tongue. 3. The *Psittacidæ* or Parrot family, comprising the true Parrots, the.

Cockatoos, the Lories, the Parrakeets, and the Macaws. They are all natives of hot climates, and are most remarkable for their brilliant plumage, and loud, harsh, and grating voices.

FIG. 135.—Scansores. Purple-capped Lory (*Lorius domicella*).

The beak (Fig. 135) is hooked, and is used as a kind of third foot in climbing, but some move about actively on the ground. 4. The *Rhamphastidæ* or Toucans, distinguished by their enormously large and cellular bills, the sides of which are serrated. They live in deep forests, in small flocks, and are confined to tropical America. 5. The *Trogonidæ* or Trogons, which inhabit the most retired recesses of the forests of the intertropical regions of both hemispheres, and are distinguished by their resplendent plumage.

ORDER VI. INSESSORES (Lat. *insedeo*, I sit upon, or perch). —The sixth order of Birds is that of the *Insessores* or *Perchers*, often spoken of as the Passerine Birds (Lat. *passer*, a sparrow).

They are defined by Owen as follows: "Legs slender, short, with three toes before and one behind, the two external toes united by a very short membrane" (Fig. 136, A, B).

"The *Perchers* form by far the most numerous order of birds, but are the least easily recognizable by distinctive characters common to the whole group. Their feet, being more especially adapted to the delicate labors of nidification" (building the nest), "have neither the webbed structure of those of the *Swimmers*, nor the robust strength and destructive talons which characterize the feet of the *Birds of Rapine*, nor yet the extended toes which enable the *Wader* to walk safely over marshy soils and tread lightly on the floating leaves of aquatic plants; but the toes are slender, flexible, and moderately elongated, with long, pointed, and slightly-curved claws.

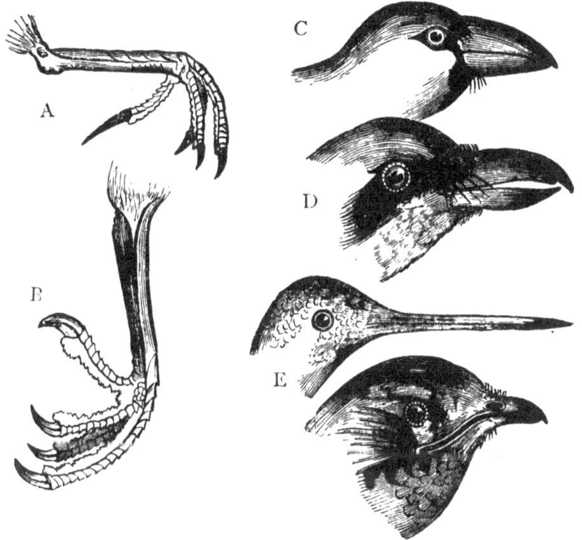

FIG. 136.—Insessores. A, Foot of Yellow Wagtail; B, Foot of Water Ouzel; C, Conirostral beak (Hawfinch); D, Dentirostral beak (Shrike); E, Tenuirostral beak (Humming-bird); F, Fissirostral beak (Swift).

"The *Perchers*, in general, have the females smaller and less brilliant in their plumage than the males; they always live in pairs, build in trees, and display the greatest art in the construction of their nests. The young are excluded in a blind and naked state, and are wholly dependent for subsistence during a certain period on parental care. The brain arrives in this order at its greatest proportionate size; the organ of voice here attains its utmost complexity; and all the characteristics of the bird, as power of flight, melody of voice, and beauty of plumage, are enjoyed in the highest perfection by one or other of the groups of this extensive and varied order."

The structure, then, of the feet gives the definition of the order, but the minor subdivisions are founded on the nature of the beak; this organ varying in form according to the nature of the food, which may be "small or young birds, carrion, insects, fruit, seeds, vegetable juices, or of a mixed kind." In accordance with this character, the *Insessores* have been divided into four great sections, as follows:

1. *Conirostres*—in which the bill is strong and on the whole conical, broad at the base and tapering with considerable rapidity to the point (Fig. 136, C). The upper mandible is not markedly toothed at its lower margin. Good examples of the Conirostral beak are to be found in the common Sparrow, Bullfinch, Crow, or Hawfinch (C). The greater part of the *Conirostres* are omnivorous, eating any thing which may come in their way; but some are granivorous, subsisting upon grains and seeds. To this section belong the Hornbills (*Buceridæ*), the Starlings (*Sturnidæ*), the Crows, Jays, and Magpies (*Corvidæ*), the Cross-bills (*Loxiadæ*), and the numerous Finches, Buntings, Grosbeaks, Tanagers, and Larks (*Fringillidæ*).

2. *Dentirostres.*—The birds of this section are characterized by the fact that the upper mandible of the beak is notched or toothed on its lower margin near the tip (Fig. 136, D). They all feed upon animal food, especially upon insects. In this section are the Shrikes (*Laniidæ*), the Fly-catchers (*Muscicapidæ*), the Thrushes (*Merulidæ*), the Tits (*Paridæ*), and the Warblers (*Sylviadæ*).

3. *Tenuirostres.*—In this section the beak is long and slender, gradually tapering to a point (Fig. 136, E). The toes are generally very long and slender, especially the hinder toe. Many live to a great extent upon vegetable juices, and among these are some of the most fragile and brightly-colored of all the birds. A great many, however, live upon insects, either partially or entirely, and some of these approach nearly to the *Dentirostres* in many of their characters. Among the more important groups included in this section are the Creepers and Wrens (*Certhidæ*), the Honey-eaters (*Meliphagidæ*), the Humming-birds (*Trochilidæ*), and the Hoopoes (*Upupinæ*).

4. *Fissirostres.*—The beak in the Fissirostral birds (Fig. 136, F) is generally short, and remarkably wide in its gape, and the opening of the bill is protected by a number of bristles. This arrangement is in accordance with the habits of the *Fissirostres*, the typical forms of which live upon insects and take their prey upon the wing. The most typical *Fissirostres*, in fact, such as the Swallows and Goat-suckers, fly about with their mouths open, and the insects which they catch in this way are prevented from escaping, partly by the bristles which border the gape, and partly by a sticky secretion within the mouth. The most typical Fissirostral birds are the Swallows and Martens (*Hirundinidæ*), the Goat-suckers (*Caprimulgidæ*), and the Swifts (*Cypselidæ*); but to these the Bee-eaters (*Meropidæ*) and the King-fishers (*Alcedinidæ*) are usually added.

ORDER VII. RAPTORES (Lat. *rapto*, I plunder).—The Birds of Prey are characterized by the form of the beak, which is adapted for tearing animal food (Fig. 137, B). The upper mandible is the longest, hooked at its point, "strong, curved, sharp-edged, and sharp-pointed, often armed with a lateral tooth" (Owen). The body is extremely muscular; the

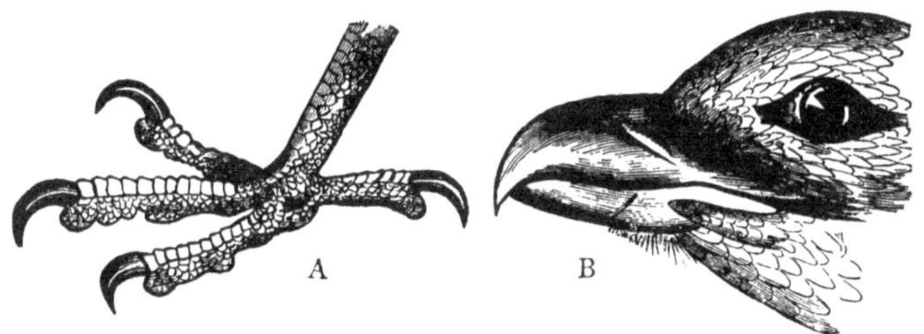

FIG. 137.—Raptores. A, Foot of Peregrine Falcon; B, Head of Buzzard.

legs are robust, short, with three toes in front and one behind; all the toes armed with strong, curved, crooked claws or talons (Fig. 137, A). They all feed upon the flesh of other animals, which they either kill for themselves or find dead, and their flight is generally extremely rapid and powerful. They are not polygamous, and the female is larger than the male. They usually build their nest in lofty and inaccessible situations, and seldom lay more than four eggs. The young are hatched in a naked and helpless condition.

The *Raptores* are divided into two sections—the Nocturnal Birds of Prey, which hunt at night, and the Diurnal Birds of Prey, which hunt by day. In the former section is only the single family of the Owls (*Strigidæ*), in which the eyes are large, and are directed forward; while the plumage is exceedingly soft and loose, so as to render their flight almost noiseless. The Owls (Fig. 138) hunt their prey in the twilight or on moonlight nights, and they live mostly upon field-mice and small birds, but they will also eat insects and frogs. In the

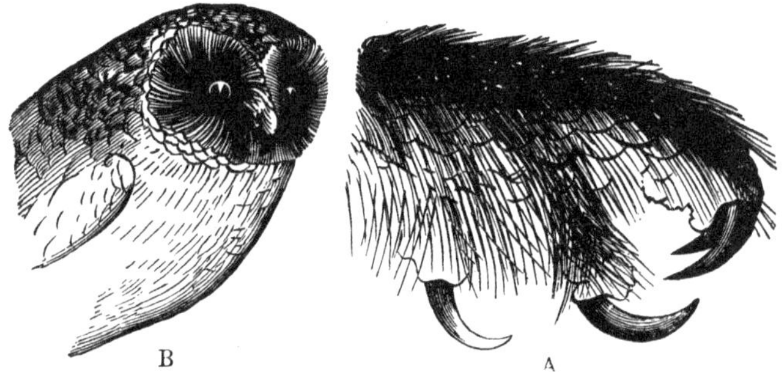

FIG. 138.—A, Foot of Tawny Owl; B, Head of White Owl.

section of the diurnal Raptores are the Falcons and Hawks, the Eagles and the Vultures. In all these the eyes are smaller than in the Owls, and are placed laterally, and the plumage is not soft. They usually possess extraordinary powers of flight. The wings are long and pointed, the sternal keel is greatly developed, the pectoral muscles are of large size, and many of them exhibit powers of locomotion more rapid than those enjoyed by any other members of the animal kingdom.

Of the diurnal Raptores, America has many examples, and some of these are among the most celebrated members of the entire order. Besides many Hawks, Buzzards, and Kites, may be especially mentioned the Bald Eagle, the Californian Vulture, and the Condor. The Bald or White-headed Eagle (*Halietus leucocephalus*) is well known as the national emblem of the United States. It is a fine and courageous bird, and lives to a great extent upon fish, which it either catches for itself, or, more commonly, wrests forcibly from the American Osprey. The Californian Vulture (*Cathartes Californianus*) is the largest of the Birds of Prey, with the single exception of the Condor. It is entirely confined to the Pacific coast. The Condor (*Sarcorhampus gryphus*) has a stretch of wing of from 12 to 14 feet, and is usually seen soaring in majestic circles at great elevations, rising, it is said, to a height of over 20,000 feet. It inhabits the lofty mountain-ranges of the Andes, and lays its eggs at a height of from 10,000 to 15,000 feet.

ORDER VIII. SAURURÆ (Gr. *saura*, lizard; *oura*, tail).—This order includes only the single extinct bird, the *Archæopteryx*, which has been found in the Oolitic rocks of Germany. The *Archæopteryx* was about as big as a common Rook, and shows many singular points of resemblance to the true Reptiles. It differs from all living birds in having two free claws to the wing, and in possessing a long, lizard-like tail. Instead of the ploughshare-shaped bone which terminates the tail in

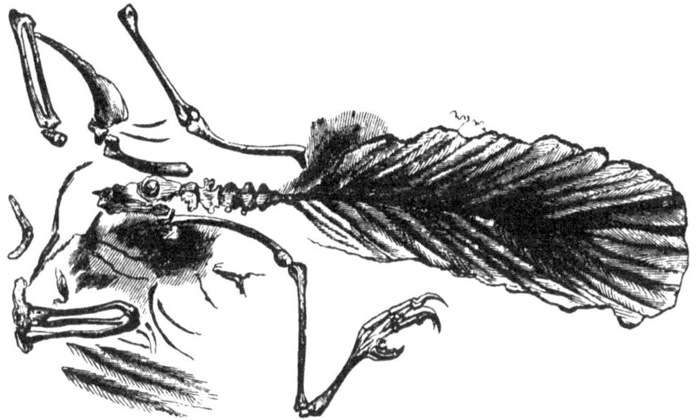

FIG. 130.—Archæopteryx. Tail and detached bones.

living birds (Fig. 129, B), the tail in the *Archæopteryx* is very long, and consists of about twenty distinct and separate vertebræ, each of which supports a pair of quill-feathers. The tail, therefore, except for the presence of feathers, must have been very like that of a *Lizard*.

MAMMALIA.

CHAPTER XXXI.

CLASS V.—MAMMALIA.

THE *Mammalia* include all the ordinary quadrupeds, and may be shortly defined as comprising Vertebrate Animals in which *some part or other of the skin is always provided with hairs, and the young are nourished for a longer or shorter time by means of a special fluid—the milk—secreted by special glands—the mammary glands.* These two peculiarities are of themselves sufficient to separate the Mammals from all other classes of the Vertebrate sub-kingdom. In addition, however, to these two leading characteristics, the following points are of scarcely less importance:

1. The skull is united with the spinal column by means of *two* articulating surfaces or condyles, instead of one, as in the Reptiles and Birds.

2. The lower jaw consists of two halves, each composed of a single piece, and united in front. The lower jaw, also, is always jointed directly with the skull, and there is no quadrate bone.

3. The heart consists—as in Birds—of four distinct chambers, two auricles and two ventricles. The right and left sides of the heart are completely separated from one another, and there is never any direct communication between the blood sent to the lungs and that sent to the body. The red corpuscles of the blood (Fig. 99, *a*) are, in the great majority of cases, in the form of circular disks, and they never contain any internal solid particle or nucleus.

4. The cavities of the chest (thorax) and abdomen are separated from one another by a muscular partition, which is called the midriff or diaphragm, and is the chief agent in respiration.

5. The respiratory organs are in the form of two lungs, placed in the chest, and never communicating with air-receptacles situated in different parts of the body In no case and at no period of life are gills or branchiæ present.

As regards the skeleton of the *Mammalia* it is not necessary to add much to what was said in speaking of the *Vertebrata* generally. With few exceptions, the spinal column is divisible into the same regions as in man—namely, the neck or cervical region, the back or dorsal region, the loins or lumbar region, the sacral region, and the tail or caudal region (*see* Fig. 95). In spite of the great differences observable in the length of the neck in different Mammals, the number of vertebræ which form the cervical region is extraordinarily constant, being almost invariably *seven*. In this respect the Giraffe, which is the longest-necked of Mammals, agrees with the Whale, which can hardly be said to have a neck at all. The vertebræ of the back or dorsal region are mostly thirteen in number, but are often more. In man there are only twelve; and in some cases there are only eleven or ten. The lumbar vertebræ are usually six or seven in number; five in man; rarely less than four. The sacral vertebræ are usually amalgamated to form a single bone—the *sacrum*—but this is wanting in the Whales. The number of vertebræ in the tail or caudal region varies from four to as many as five-and-forty, and they are usually freely movable upon one another. The thoracic cavity or chest in Mammals is always enclosed by a series of ribs; the number of which varies with the number of the dorsal vertebræ. As a rule, the ribs are united to the breastbone or sternum in front, not by bony pieces, as in birds, but by cartilages. Only the front ribs reach the sternum, and these are called the "true" ribs; the hinder ribs fall short of the breastbone, and are called the "false" ribs. The sternum is composed of several pieces, placed one behind the other, but usually amalgamated to form a single bone. It is usually long and narrow in shape, and is only rarely furnished with any ridge or keel, as it is in birds. The regular number of limbs in the Mammals is four, two anterior and two posterior; and for this reason the Mammals are often spoken of as Quadrupeds. Some Mammals, however, such as the Whales and Dolphins, have only the anterior limbs, and many of the Amphibia and Reptiles walk upon four legs. As regards the structure of the fore-limbs (Fig. 96), the general plan of conformation is the same as described in treating of the *Verte-*

brata generally (p. 202). The shoulder-blade or *scapula* is never absent; and the coracoid bone, which is so marked a feature in the Birds, is with hardly an exception amalgamated with the scapula. The clavicles or collar-bones are often wanting or rudimentary, but in no Mammal are they ever united together in front so as to form a merry-thought or "furculum." The regular number of fingers is five, but they vary from one to five, the middle finger being the longest and most persistent of all, and being the only finger left in the Horse. Properly each finger consists of three short bones or *phalanges*, except the thumb, which has two; but this rule is occasionally departed from. While the fore-limbs are never wanting, the hind-limbs are sometimes absent, as in the Whales. Generally speaking, however, the posterior limbs are present, and the pelvic arch has much the same structure as in man. The foot —like the hand—consists regularly of five digits, but it is subject to the same abortion of parts, as we shall see hereafter.

The great majority of Mammals possess *teeth*, but these are only present in the embryo of the whalebone Whales, and are altogether wanting in the scaly and great Ant-eaters. The teeth are also almost invariably implanted in distinct sockets in the jaw. Some Mammals have only a single set of teeth; but in most cases the young Mammal possesses a set of what are called the *milk*-teeth or *deciduous* teeth, which is ultimately replaced by a second set, constituting the *permanent* teeth. No Mammal has ever *more* than two sets of teeth. In man, and in many other Mammals, the teeth are divisible into four groups, which differ from one another in position, appearance, and function. These are termed respectively the *incisors*, *canines*, *præmolars*, and *molars*. It is impossible to describe fully which teeth come under each of these heads without entering into unnecessary details as to the structure of the jaws. It must be sufficient here to point out the general characters and position of these groups in a good illustrative example, such as one of the higher Apes (Fig. 140). The incisors (*i*) vary greatly in size and number, but they are always placed in the front of the mouth, and are the teeth which are used in simply biting or dividing the food. The *canine* or eye-tooth (*c*) is generally larger or more pointed than the other teeth. The canines are sometimes wanting, or are sometimes present in one jaw and not in the other; but there are never more than four altogether—that is to say, one in each jaw on each side. The præmolars and molars (*pm* and *m*) are the so-called "back-teeth," and they vary a good deal in number and function, being sometimes adapted for cutting the food, but more usually for chewing and grinding it down.

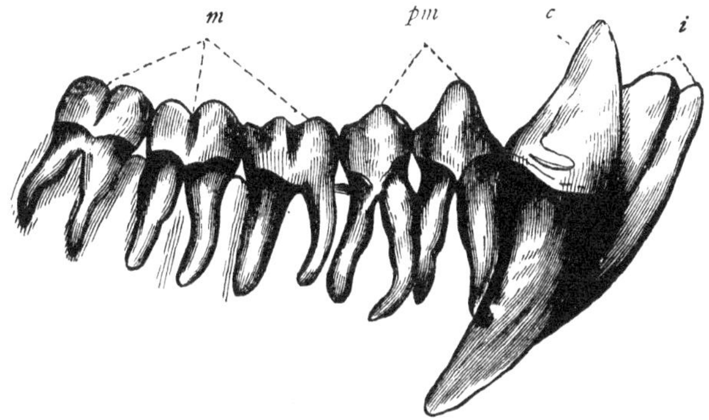

FIG. 140.—Teeth of the right side of the lower jaw of the Chimpanzee (after Owen). *i* Incisors; *c* Canine tooth; *pm* Præmolars; *m* Molars.

All these kinds of teeth are not necessarily present, and the teeth constitute most important characters for separating the various orders of Mammals from one another. For this reason it is usual to express the number of the teeth in any particular animal by an arithmetical formula, called the *dental formula.* For example, the formula for the portion of the jaw of the Chimpanzee figured above (Fig. 140) would be as follows:

$$i\ 2;\ c\ 1;\ pm\ 2;\ m\ 3.$$

But this is only one half of the lower jaw, and the dental formula must include both sides, so that it would be:

$$i\ 2-2;\ c\ 1\quad 1;\ pm\ 2-2;\ m\ 3-3.$$

That this would be the formula is at once evident, when it is remembered that the two sides of the jaw of course contain exactly the same teeth. Still, the formula as given above only includes the lower jaw, and to render it perfect it must take in the teeth of the upper jaw as well. This is effected by placing the figures in two rows separated by short lines, all the figures in the upper row referring to the upper jaw, and those in the lower row to the lower jaw; the short dashes between the figures of each row still indicating the teeth on the two sides of the mouth. The complete formula would therefore run as follows:

$$i\,\frac{2-2}{2-2};\ c\,\frac{1-1}{1-1};\ pm\,\frac{2-2}{2-2};\ m\,\frac{3-3}{3-3}=32.$$

In this way the *dentition*—that is to say, the number and arrangement of the teeth—can be presented in a manner which can be instantly recognized by the eye. It must be remembered, however, that the formula seldom exhibits the regularity of the one of the Chimpanzee given above. The teeth are not necessarily the same in both jaws, and in many cases some may be altogether wanting. To show this there is subjoined the dental formula of a typical Ruminant animal, such as a sheep:

$$i\,\frac{0-0}{3-3};\ c\,\frac{0-0}{1-1};\ pm\,\frac{3-3}{3-3};\ m\,\frac{3-3}{3-3}=32.$$

From this formula it will be seen that the sheep has 32 teeth in both jaws taken together. The upper incisors and canines are wanting, and there are three præmolars and three molars on each side of the upper jaw. In the lower jaw there are six incisors, two canines, and the same number of præmolars and molars as in the upper jaw.

As regards the digestive system of Mammals, the alimentary canal and digestive glands have on the whole the same general structure and arrangement as in man (pp. 203, 204). Some very remarkable modifications, however, in the structure of the stomach and in the termination of the intestine occur in certain Mammals; but these will be noticed in speaking of the orders in which they occur.

The cavity of the abdomen in Mammals is always separated from that of the thorax by a complete muscular partition—the diaphragm. The abdomen contains the greater part of the alimentary canal, the liver, pancreas, kidneys, and other organs. The thorax contains chiefly the heart and lungs. The *heart* is contained in a membranous sac—the pericardium, and consists of two auricles and two ventricles. The heart consists functionally of two sides, each having an auricle and a ventricle, which communicate with one another by apertures, so guarded by valves that the blood can pass from the auricle into the ventricle, but not, under ordinary circumstances, from the ventricle to the auricle. There is in the adult no direct communication between the two sides of the heart. The course of the circulation is indicated in the subjoined diagram, and is shortly as follows: The venous blood, which has become impure by passing through the tissues, is returned by the great veins to the *right* auricle, from which it passes into the right ventricle. From here it is driven through a great vessel, called the pulmonary artery, to the lungs, where it is submitted to the action of the air, and becomes arterial blood. It is then returned to the heart by a series of vessels, called the pulmonary veins, and is poured into the *left* auricle, from which it passes into the left ventricle. From the left ventricle it is propelled to all parts of the body by a great systemic vessel, which is called the aorta (Fig. 141).

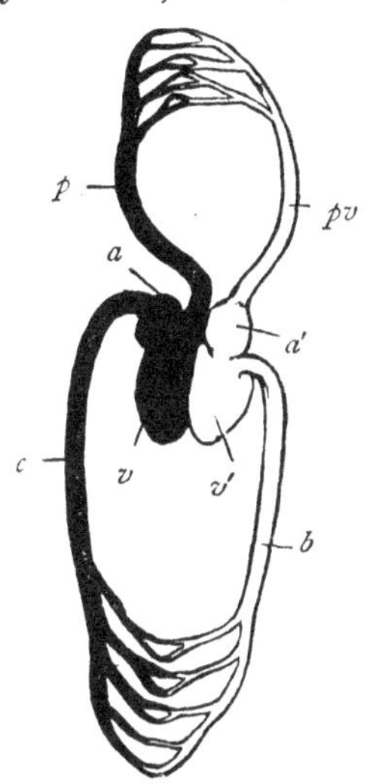

Fig. 141.—Diagram of the circulation in a Mammal. (The cavities containing venous blood are marked black, those containing arterial blood are left white.) *a* Right auricle; *v* Right ventricle; *p* Pulmonary artery carrying venous blood to the lungs; *pv* Pulmonary veins carrying arterial blood from the lungs; *a'* Left auricle; *v'* Left ventricle; *b* Aorta carrying arterial blood to the body; *c* Vena cava carrying venous blood to the heart.

The *lungs* of Mammals are two in number, and differ from those of Birds in being freely suspended in membranous bags. They are spongy and cellular throughout, and they never communicate by apertures on their surface with air-sacs placed in different parts of the body.

The *nervous system* of Mammals is chiefly remarkable for the great proportionate development of the brain, as compared with the spinal cord.

In the higher Mammals, also, the two halves (hemispheres) of the brain proper (*cerebrum*) are connected together by a great band or bridge of nervous tissue, constituting what is known as the *corpus callosum*. This structure is not a conspicuous feature in the two lowest orders of the Mammalia. The senses, as a rule, attain great perfection in the Mammals; and the only sense which can ever be said to be entirely wanting is that of sight. Eyes, however, are always present, though they may be rudimentary; and in those Mammals which are said to be "blind," it is not generally that the eyes are wanting, but that the skin passes unbrokenly over the eyeball, or the optic nerve is degenerated. Even in these cases, however, it is not impossible that there may be some perception of light through the skin. An external ear for collecting sounds is usually present; but it is wanting in the Whales and Dolphins, and in some of the Seals.

The *skin* is invariably furnished over a greater or less part of its surface with the epidermic appendages known as *hairs*, which differ from feathers chiefly in not splitting up as they are produced. In the scaly Ant-eater (*Manis*), the hairs are aggregated together so as to form horny scales; and in the Hedge-hog, Porcupine, and other animals, many of the hairs are developed into long spines or prickles. In other cases, again, as in the Armadillos, the skin is more or less covered by an armor of bony plates. The only apparent exception to the universal presence of hair on some part or other of the integument of all Mammals is constituted by the true Cetaceans (Whales and Dolphins), many of which are without hair when grown up. Some, however, such as the Whales, have a few bristles in the neighborhood of the mouth, even when adult. And the Dolphins, which are totally hairless when grown up, exhibit tufts of hair upon the muzzle before they are born.

The young Mammal is always born in a helpless condition, and is nourished for a longer or shorter time by means of the milk of the mother. The milk is secreted by special organs, called the *mammary glands*, which are present in both sexes, but are normally undeveloped in the male. The number and position of the mammæ vary a good deal in different cases, but they are always placed on the lower surface of the body, and their ducts almost always open upon a special eminence, called the teat or nipple. In one or two cases, however, the mammary glands open by simple slits in the skin of the abdomen, and not by distinct nipples. In ordinary Mammals the milk is obtained by voluntary suction on the part of the young, but in the Marsupials (Kangaroos, Opossums, etc.) the milk is forced into the mouth of the young animal by the action of a special muscle.

So much difference of opinion obtains as to the best foundation upon which to establish a division of the Mammalia into great primary sections, that it has been thought advisable to leave this subject wholly out of consideration. For our present purpose it is enough to adopt the old classification of Mammals into the two great divisions of the *Placental* and *Non-placental* forms. In the Placental Mammals the young is nourished within the body of the mother by means of a structure called the *placenta*, or "after-birth," through which the nutrient materials of the mother's blood reach the young. In consequence of this, the young of the Placental Mammals can be retained within the body for a considerable period, and, when born, they are able to obtain their natural food—the milk—by their own exertions. In the Non-placental Mammals, on the other hand, the young are born at an extremely early period of their development, before there is any necessity that a *placenta* should be formed for the nourishment of the fœtus. In these cases, therefore, the young when born are much more immature and helpless than in the case of the Placental Mammals. So helpless are they, that they are even unable to suck, and have in most cases to be fixed by the mother herself upon the teats, while the milk is forced into their mouths by a muscle which is spread over the mammary gland. Adopting these primary sections as practically sufficient in an elementary work, the whole class of the *Mammalia* may be divided into the following fourteen orders:

Division A.—Aplacental Mammals.

Order 1.—*Monotremata.*
Order 2.—*Marsupialia.*

Division B.—Placental Mammals.

Order 3.—*Edentata.*
Order 4.—*Sirenia.*
Order 5.—*Cetacea.*
Order 6.—*Ungulata.*
Order 7.—*Hyracoidea.*
Order 8.—*Proboscidea.*
Order 9.—*Carnivora.*
Order 10.—*Rodentia.*
Order 11.—*Cheiroptera.*
Order 12.—*Insectivora.*
Order 13.—*Quadrumana.*
Order 14.—*Bimana.*

CHAPTER XXXII.

ORDERS OF MAMMALIA.

Order I. Monotremata (Gr. *monos*, single; *trema*, aperture).—The first and lowest order of the Mammals—that of the *Monotremata*—comprises only two very remarkable animals, both of which are exclusively confined to New Holland. These are the Duck-mole (*Ornithorhynchus*) and the Porcupine Ant-eater (*Echidna*). The *Monotremata* are essentially characterized by the fact that, as in Birds, the termination of the intestine opens into a common chamber or *cloaca*, which receives also the ducts of the urinary and reproductive organs. The jaws are destitute of true teeth; but the *Ornithorhynchus* has a kind of beak, like the bill of a duck, furnished with small horny plates, which act as teeth. The pectoral arch, which supports the fore-limbs, resembles that of Birds in several respects, but especially in the fact that the coracoid bones are distinct, and are not amalgamated with the shoulder-blade. There is no pouch developed on the abdomen of the females, but there are the so-called "marsupial bones." These are two small bones which arise from the front of the pelvis. They are really to be regarded as formed by a conversion into bone of the tendons of one of the muscles of the abdomen. There are no external ears. The mammary glands have no nipples, and the young are said to be devoid of a *placenta*.

The Duck-mole (Fig. 142) is one of the most extraordinary of Mammals, and is found inhabiting the rivers and lakes of Australia and Tasmania. The body resembles that of a small otter, and is covered with a short brown fur. The tail is broad and flattened, and the jaws are sheathed with horn, so as to form a flattened beak, very like the bill of a duck. The legs are short, furnished with five toes each, and webbed, so that the animal swims with great facility. Their food consists chiefly of aquatic insects and mollusks, and they make very extensive burrows in the banks of streams.

FIG. 142.—Monotremata. Duck-mole (*Ornithorhynchus paradoxus*). (After Waterhouse.)

The other member of the *Monotremata* is the Porcupine Ant-eater or *Echidna*, which is not unlike a large hedgehog in appearance. The snout is very long, and is enclosed in a continuous skin till close upon its extremity, where there is a small aperture for the protrusion of a long and flexible tongue. There are no teeth, nor any organs to act as teeth. The feet have five toes each, and are furnished with strong digging-claws, but the toes are not webbed. The skin is covered with strong prickly spines interspersed with bristly hair. The *Echidna* measures from fifteen to eighteen inches in length, and is a nocturnal animal. It lives in burrows, and feeds upon insects, which it captures by protruding its long, sticky tongue.

ORDER II. MARSUPIALIA.—The name of *Marsupials* is derived from the fact that the females of this order are mostly furnished with an abdominal pouch or *marsupium*, within which the nipples are situated. When born, the young are placed by the mother within this pouch, where they adhere to the teats, and can be carried about without injury. Even when further advanced in their development, the young often betake themselves to the shelter of the marsupium. The so-called "marsupial bones" are present, and as they spring from the front of the pelvis they no doubt serve to support the pouch; but this cannot be their sole use, as they exist in the males, and also in the Monotremes, in whom there is no pouch. All Marsupials possess teeth, and the pectoral arch has now the same form as in the higher Mammals, the coracoid bones being now amalgamated with the shoulder-blade. The intestine does not terminate in a cloaca.

Though the *Marsupialia* form an extremely natural order, sharply separated from the other Mammals, they include a large number of varied forms. In fact, this order, from its being the almost exclusive possessor of a continent so large as Australia, has to discharge, in the general economy of nature, functions which are elsewhere performed by several orders. As regards their geographical distribution, with the single

FIG. 143.—Marsupialia. The Koala or Kangaroo-bear (*Phascolarctos cinereus*). (After Gould.)

exception of the family *Didelphidæ* (the true Opossums), the whole order of the Marsupials is exclusively confined to Australia, Van Diemen's Land, New Guinea, and the adjacent islands.

The Marsupials may be primarily divided into the vegetable-eating and rapacious or carnivorous forms—the former characterized by the absence or rudimentary condition of the canine teeth, the molars having broad, grinding crowns; while in the latter there are well-developed canines, and the molars are not adapted for grinding. Of the vegetable-eating forms, the best known are the Kangaroos (*Macropodidæ*), distinguished by the remarkable disproportion between the hind and fore limbs, the former being by far the longest and strongest. By their long hind-legs, assisted by a powerful tail, the Kangaroos can perform astonishing jumps, and, in fact, leaping is their mode of progression when pursued.

The typical Kangaroos live on the great grassy plains of Australia; but the Tree.Kangaroos spend a great part of their time in trees, and the Rock Kangaroos affect mountainous districts. The Kangaroo-bear or Native Sloth (*Phascolarctos cinereus*, Fig. 143) has no tail, and has the body covered with a short, dense fur, while the ears are tufted. The fore-feet can be used as hands, and the toes are all furnished with strong, curved claws. It is a harmless, nocturnal animal, and spends most of its existence in trees. The typical group, however, of the vegetable-eating Marsupials is that of the Phalangers, comprising a large number of small animals which live in trees, and generally possess a prehensile tail. The most familiar example is the Australian "Opossum" (*Phalangista vulpina*), which is largely hunted by the natives. In the so-called "flying" Phalangers, again, the tail is not prehensile, and the animal takes extensive leaps from tree to tree, by means of a fold of skin which stretches between the body and the fore and hind limbs.

Of the carnivorous Marsupials, the Bandicoots (*Perameles*), the Native Devil (*Dasyurus*), the Native Tiger (*Thylacinus*), and the American Opossums (*Didelphidæ*), may be mentioned. The Bandicoots are little, rabbit-like Australian animals, which live upon insects, and seem to fill the place held in the Old World by the Hedgehogs and Shrew-mice. The Native Devil and Thylacine, though both of comparatively small size, are extremely ferocious, and do much mischief to the flocks of the Tasmanian colonists. About twenty species of *Didelphidæ* are known, and they are all exclusively confined to the American Continent. They are all of small size, have prehensile tails, and mostly live among trees. The best-known species is the Virginian Opossum (*Didelphys Virginiana*).

ORDER III. EDENTATA (Lat. *e*, without; *dens*, tooth).—This order of Placental Mammals comprises the Ant-eaters, Armadillos, and Sloths, and is characterized by the fact that the teeth are not covered with enamel, have no complete roots, and are never replaced by a second set. Further, in none of the *Edentates* are there any central incisor teeth, and in all but one there are no incisors at all. In two genera only are there no teeth; so that the name *Edentata* is not a very appropriate one. In all, the toes are furnished with long and powerful claws.

The *Edentata* admit of division into two sections, according as they live upon a vegetable diet and live in trees, or are carnivorous and live upon or below the ground. In the first section are only the Sloths (*Bradypodidæ*), which are exclusively confined to South America, inhabiting the vast primeval forests of this continent. They are in every way adapted for an arboreal life, and are "destined to be produced, to live, and to die on trees." They are excessively awkward when upon the ground; but the feet are furnished with extremely long, curved claws, so that the animal is enabled to move about freely, suspended back downward from the branches of the

Fig. 144.—Edentata. (*Chlamyphorus truncatus.*)

trees. The Armadillos (*Dasypodidæ*) are also exclusively confined to South America; but they are carnivorous, burrowing animals, and are furnished with strong digging-claws. The upper surface of the body is covered with a kind of armor, formed of hard, bony plates or shields, which are united at their edges (Fig. 144). Most of them can roll themselves up into a ball, and they can all bury themselves in the ground when pursued.

The remaining South American Edentates are the hairy Ant-eaters, of which the best known is the great Ant-eater (*Myrmecophaga jubata*). The body in this family is covered with hair, the tail is long, and the teeth are altogether wanting. They feed chiefly upon ants and termites, which they catch by protruding their long and sticky tongues, having previously broken into the nests by means of their strong, curved claws.

The *Edentata* are represented in the Old World by only two genera. One of these is the genus *Manis*, comprising the scaly Ant-eaters or Pangolins, which are exclusively confined to Asia and Africa. In these singular animals the body and tail are covered by a flexible armor, composed of horny plates or scales overlapping like the tiles of a roof. The other genus is *Orycteropus*, comprising only the so-called Ground-hog of South Africa; which also lives upon insects, and burrows by means of its strong digging-claws.

As regards the geographical distribution of the *Edentata*, it is to be remembered that the order has a very limited range at the present day. The true Ant-eaters, the Armadillos, and the Sloths, are exclusively confined to South America, in which country a group of gigantic extinct Edentates existed in the later portion of the Tertiary epoch. The scaly Ant-eater is common to Asia and Africa; and the Ground-hog is confined to South Africa.

ORDER IV. SIRENIA (Gr. *seiren*, a Mermaid).—This order comprises only certain large marine Mammals, known as Dugongs and Manatees, which were long classed with the Whales and Dolphins (*Cetacea*). They agree with the Whales in the adaptation of the body to an aquatic life, especially in the facts that the anterior limbs are converted into swimming-paddles, the hind-limbs are wholly wanting, and the hinder end of the body forms a powerful caudal fin, which is placed so as to strike the water horizontally, and not vertically as in Fishes. They differ from the *Cetacea* in having the nostrils placed at the anterior part of the head, and in having molar teeth with flat crowns, adapted for a vegetable diet. Fleshy lips are present, the upper one usually with a mustache, and the skin is covered with scanty bristles. The head is not disproportionately large as compared with the body, and there is a tolerably distinct neck. They are vegetable-eaters, feeding chiefly upon sea-weeds, and haunting the mouths of rivers and estuaries.

The only existing *Sirenia* are the Manatee (*Manatus*) and the Dugong (*Halicore*), often called "Sea-cows." The Manatees are found on the east coast of America, and on the west coast of Africa. They are large, awkward animals, attaining a length of from eight to ten or fifteen feet, and their flesh is said to be very palatable and wholesome. The Dugongs (Fig. 145) differ little in appearance and habits from the Manatees. They are found on the coasts of the Indian Ocean and the north coast of Australia, and are often killed and eaten. They attain a length of from eighteen to twenty feet. The bones of the skeleton are remarkable for their extreme hardness and density.

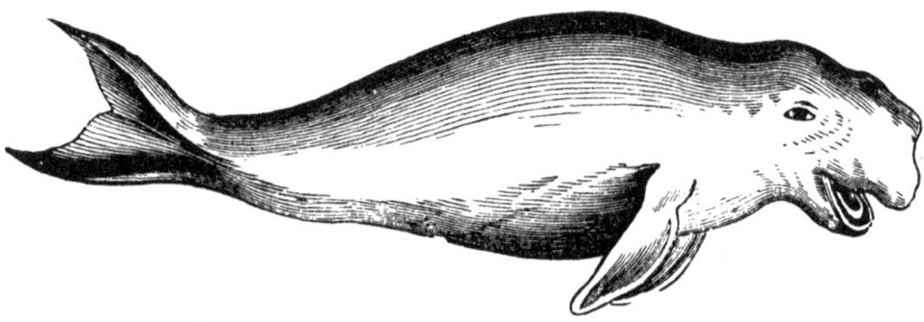

Fig. 145.—Sirenia. The Dugong (*Halicore Indicus*).

Besides these living forms, the *Sirenia* were represented by a gigantic species which formerly inhabited Behring Island on the coast of Kamtchatka. This animal was described by a M. Steller who accompanied Behring on his second expedition, and he named it *Rhytina*. This enormous animal attained a length of twenty-five feet, and a circumference of twenty feet, and it appears to have been completely exterminated, no specimen having been seen for two centuries.

Order V. Cetacea (Gr. *ketos*, a Whale).—This order comprises the Whales, Dolphins, and Porpoises, and it is characterized by the complete adaptation of its members to a watery life. The body is completely fish-like in form, the fore-limbs are converted into swimming-paddles, and the hind-limbs are completely wanting; while the hinder end of the body forms an extremely powerful, horizontal caudal fin. Sometimes there is a dorsal fin as well. The nostrils may be single or double, but always are placed on the top of the head, constituting the "blow-hole." The body is very sparingly furnished with hairs, or is wholly without them in the adult. The head is generally of disproportionately large size as compared with the body, and is rarely separable from the trunk by any distinct constriction or neck. There is no sacrum, and the pelvis is only represented in a rudimentary form. Lastly, the adult is either wholly destitute of teeth, or possesses only a single set, which are always conical in shape, and are never divisible into distinct groups. All the true *Cetacea* are carnivorous, living upon animal food.

Chief among the Cetaceans in importance and zoological interest are the Whalebone Whales (*Balænidæ*), in which the adult is destitute of teeth, though the young whale possesses teeth which never cut the gum. The place of teeth is taken by a series of transverse plates of whalebone or *baleen*, which are used as a kind of screening apparatus or filter to separate from the sea-water the minute Mollusks and Jellyfishes upon which these enormous animals live. The most important member of this family, from a commercial point of view, is the Greenland Whale (*Balæna mysticetus*), which yields most of the whale-oil and whalebone of commerce. The Greenland Whale attains a length of from forty to sixty feet, and of this enormous length about a third is taken up by the head alone. The oil is derived from a thick layer of fat or "blubber," which is situated under the skin, and serves to protect the animal from cold. Though an inhabitant of the sea, the whale is obliged to come to the surface to breathe, and in so doing it ejects from the blow-holes what looks like a column of water, the whole operation being known to the whalers as "blowing." The true nature of this act is still somewhat questionable, but it appears certain that the apparent jet of water is in reality, mainly if not entirely, due to the condensation of the moisture which is contained in the air expelled from the lungs. The old view was that "blowing" consisted in the whale ejecting through the nose the water which had previously been filtered through the baleen-plates of the mouth; but it appears to be quite certain that this view, at any rate, is not the correct one. The Rorquals or Finner Whales resemble the Greenland Whale in most respects, but the skin is furrowed with deep plaits or folds, and there is a dorsal fin, placed on the back. Some of these attain a gigantic size (eighty feet or more), but they are seldom captured, as their commercial value is small.

The Toothed Whales (*Odontoceti*) are best known by the Sperm Whale, an animal as large as, or larger than, the Greenland Whale, but distinguished by having numerous conical teeth, a single blow-hole, and a curiously-truncated head. They yield an excellent oil, and the singular fatty substance which is known as "spermaceti." They also yield the substance called "ambergris," which is used as a perfume; but this is probably a product of disease.

The last family of the *Cetacea* is that of the *Delphinidæ*, comprising the Dolphins (Fig. 146) and Porpoises. They have numerous conical teeth in both jaws, and the nostrils open by a single aperture on the top of the head. The Dolphins are

FIG. 146.—Cetacea. The Common Dolphin (*Delphinus delphis*).

inhabitants of the sea, but two species live in rivers—one in India, and the other in America. The Porpoises are also marine, and occur in all seas. The most remarkable of the *Delphinidæ* is the Narwhal or Sea-Unicorn, which is found in the Arctic seas, and which attains a length of as much as fifteen feet in the body alone. The chief peculiarity of the Narwhal is in the dentition. The females, as a rule, have no teeth, the upper jaw alone having two rudimentary incisors which never cut the gum. In the males, however, while the lower jaw is without teeth, one of the two central incisors of the upper jaw is enormously developed, and grows throughout the life of the animal. It forms a tusk of from eight to ten feet in length, the whole surface of which is spirally twisted. The function of this extraordinary tooth is doubtless offensive.

ORDER VI. UNGULATA (Lat. *ungula*, a hoof).—This order is often spoken of as that of the Hoofed Quadrupeds, and is one of the largest and most important of the orders of Mammalia. The order is characterized by having all the four limbs and by having that portion of the toe which touches the ground encased in a greatly-expanded nail or *hoof*. There are never more than four full-sized toes to each leg, and owing to the presence of hoofs the limbs are useless for grasping, and are only of use in locomotion and in supporting the weight of the body. There are always two sets of teeth, and the molars have broad crowns adapted for grinding vegetable substances.

The *Ungulata* are divided into two great primary sections, according as the toes are even or odd in number:

A. Perissodactyla, or Odd-toed Ungulates, in which the toes are odd in number—either one or three. If horns are present, they are not in pairs.*

B. Artiodactyla, or Even-toed Ungulates, in which the toes are even in number—either two or four; and, if horns are present, they are in pairs.

The living Perissodactyle Ungulates are the Rhinoceroses, the Tapirs, and the Horse and its allies. The *Rhinoceroses* are extremely large and bulky brutes, having a very thick and nearly hairless skin, usually thrown into deep folds. The feet (Fig. 147, D) are furnished with *three* toes each, all en-

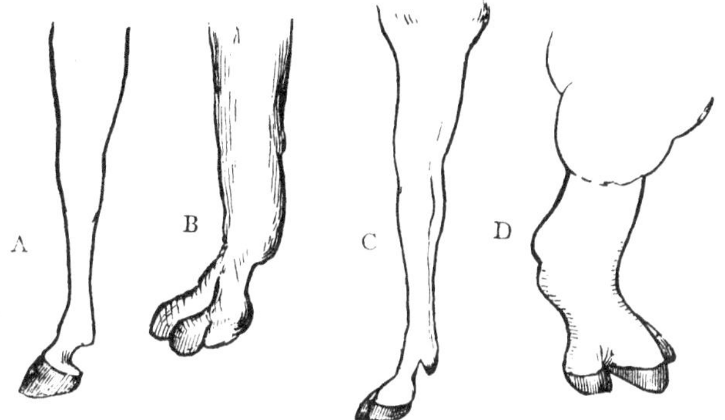

FIG. 147.—Ungulata. A, Perissodactyle foot of Zebra. B, Artiodactyle foot of Llama. C, Artiodactyle foot of Antelope. D, Perissodactyle foot of Rhinoceros.

cased in hoofs. The nose is furnished with one or two horns, composed of longitudinal fibres compacted together, and not having any central core of bone. When there is only one horn, it is, of course, unsymmetrical; and, when there are two, these are not paired, but one is always placed behind the other in the middle line of the head, and the hinder one is much the shorter. The various species of Rhinoceros are found in India, Java, Sumatra, and Africa, inhabiting marshy places and feeding chiefly on the foliage of trees. The *Tapirs* have four toes to each of the fore-legs, but only *three* toes on the hind-legs, so that they are properly odd-toed. The nose forms a short, movable proboscis, used in stripping off the leaves of trees. They are large, clumsy animals, which inhabit South

* The fore-feet of the Tapirs are even-toed, but the hind-feet are perissodactyle.

America, Sumatra, and Malacca. The third and last family of the *Perissodactyla* is that of the *Equidæ*, comprising the Horse, Ass, Zebra, and Quagga. In this family the toes are reduced to *one* to each foot, enclosed in a single broad hoof,

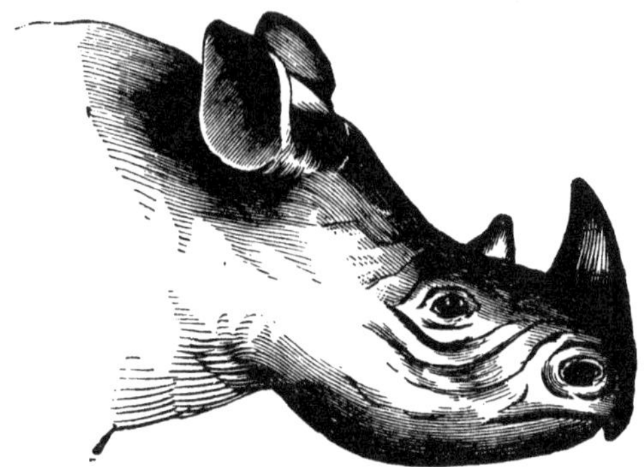

FIG. 148.—Head of Two-horned Rhinoceros (*R. bicornis*).

without any supplementary hoofs. There is a continuous series of incisor teeth in both jaws, and in the males canines are present. The dental formula is:

$$i\,\frac{3-3}{3-3};\quad c\,\frac{1-1}{1-1}\ (\text{or none});\ pm\,\frac{3-3}{3-3};\ m\,\frac{3-3}{3-3} = 40$$

All the varieties of Horses appear to be descended from the single species *Equus caballus*, which seems to have been primitively a native of Central Asia. When the American Continent was discovered it certainly possessed no living horse, but the horse has now become completely naturalized there, and we know that America formerly possessed about twenty species of horses, all of which are now extinct. In the genus *Asinus* are the Asses, Zebras, and Quaggas. The Wild Ass is a native of Asia, and the domestic Ass is probably descended from it. The Zebras and Quaggas are exclusively African, and are distinguished by their beautifully-striped and banded bodies.

The Artiodactyles or Even-toed Ungulates are divided into two groups:

1. *Omnivora*, as the Pig and Hippopotamus.
2. *Ruminantia*, which chew the cud, such as Oxen, Deer, Camels, etc.

Of the *Omnivorous* forms the Hippopotamus or River-horse is characterized by its massive heavy body, short blunt muzzle, and feet with four hoofed toes each. The Hippopotamus is found in the rivers of Abyssinia, and throughout the whole of Africa to the south of this. It reaches a length of from eleven to twelve feet, is nocturnal in its habits, and swims and dives with great facility. It lives upon vegetable food, and is tolerably harmless unless attacked or irritated. The Pigs, Peccaries, and Wart-hogs, constitute the family *Suidæ*, and have usually four toes to each foot, though sometimes the hind-feet have only three toes. All the toes are hoofed, but it is only two which support the weight of the body, the remaining toe or toes being placed at some elevation on the back of the foot. The snout is truncated and cylindrical, and is capable of extensive movement. The tail is very short, or is represented only by a tubercle.

Of the Swine the most important and best known is the Wild Boar (*Sus scrofa*), from which it is probable that all our domestic varieties of swine have sprung. Another form is the Babyroussa or Hog-deer (*Sus babyrussa*), which inhabits the Indian Archipelago, and is remarkable for the great size and backward curvature of the upper canine teeth. The Wart-hogs (*Phacochœrus*) are African, and derive their name from the possession of a fleshy wart under each eye. The Peccaries are exclusively American, the best-known species being the Collared Peccary (*Dicotyles torquatus*). They are not at all unlike small pigs both in appearance and habits, and they are generally found in small flocks.

The *Ruminantia* form a most natural group of the *Ungulata*, characterized by the structure of the foot, the dentition, and the structure of the stomach.

The *foot* is "cloven," consisting of a symmetrical pair of toes, encased in hoofs, and looking as if produced by the cleavage of a single hoof. In most cases there are also two small supplementary hoofed toes placed on the back of the foot.

As regards the *dentition*, the typical state of things is that there should be no incisor nor canine teeth in the upper jaw, but that the lower jaw should have six incisors and two canines, which are all similar in size and form, and constitute a continuous and uninterrupted series of *eight* teeth placed in the front of the lower jaw. There are six molar teeth on each side of each jaw, and these have grinding surfaces. The typical dental formula, therefore, for a Ruminant is:

$$i\,\frac{0-0}{3-3};\quad c\,\frac{0-0}{1-1};\quad pm \text{ and } m\,\frac{6-6}{6-6} = 32$$

In the absence of incisor teeth in the upper jaw, the lower incisors bite against a callous pad of hardened gum. The Camel tribe differs in its dentition from the above typical formula, and certain exceptions likewise occur in the males of some other forms, and in one or two other less important instances.

The *stomach* in the Ruminants is complex, and is divided into several compartments, this being in accordance with their mode of eating. They all, namely, "ruminate" or "chew the cud;" that is to say, they first swallow their food unmasticated, and then bring it up again after a longer or shorter period in order to chew it. This is effected as follows (Fig. 149): The gullet opens at a point between the first two com-

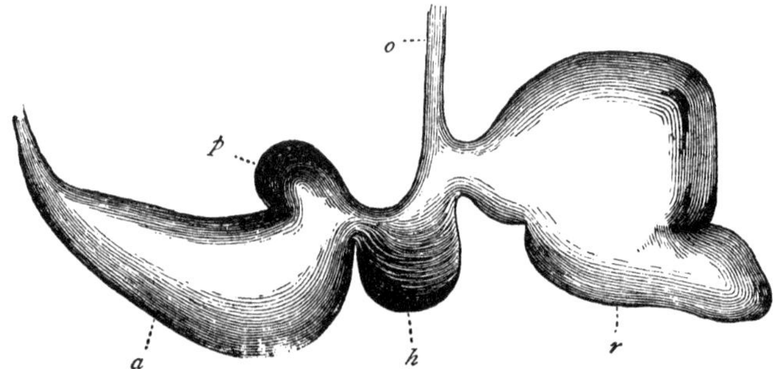

FIG. 149.—Stomach of a Sheep. *o* Gullet; *r* *Rumen* or Paunch; *h* Honeycomb bag or *Reticulum*; *p* Many-plies or *Psalterium*; *a* *Abomasum* or Fourth Stomach.

partments or stomachs, of which the largest lies to the left and is called the "paunch," while the smaller right cavity is called the "honeycomb bag" (*reticulum*). The paunch (*rumen*) is the cavity into which the food is first received, and here it is moistened and allowed to soak for some time. After the food has lain sufficiently long in the paunch, it passes into the "honeycomb bag," where it is made up into little balls or pellets, which are then returned to the mouth by a reversed action of the muscles of the gullet. After having been thoroughly chewed, and prepared for digestion, the food is now swallowed a second time. On this occasion, however, instead of passing into the paunch, the masticated food is conveyed into the third stomach. This is known as the "many-plies" or "*psalterium*," because its lining membrane is thrown into a number of longitudinal folds, like the leaves of a book. The psalterium opens by a wide aperture into the fourth and last stomach, known as the "*abomasum.*" This is a cavity of considerable size, which secretes the true digestive fluid (gastric juice), and it is here that the food is really digested. The abomasum terminates, of course, in the commencement of the small intestine.

The *Ruminantia* include a number of families, of which it is only possible to notice the leading characters of the more important ones—namely, the *Camelidæ*, *Cervidæ*, *Giraffes*, and *Cavicornia.*

The family *Camelidæ* comprises the Camel and Dromedary of the Old World and the Llamas of the New, and is characterized by having no horns, by having two incisors in the upper jaw, and a pair of canines in both jaws; while the foot consists of only two toes, covered with imperfect nail-like hoofs, and destitute of the two supplementary toes. The soles of the feet are covered with a callous horny integument upon which the animal walks. In the Camels the toes are conjoined below by a callous pad, and the back is furnished with one or two fleshy humps. The Arabian Camel or Dromedary has but one hump, and its structure admirably adapts it for a beast of burden in the sandy deserts of Arabia and Africa. One special provision toward this end is the possession of large cells in the paunch, in which a large quantity of water can be stored up, thus enabling the animal to travel for days without drinking. The Bactrian Camel resembles the Dromedary in most respects, but it possesses two humps. The place of the Camels of the Old World is filled in South America by the Llamas and Alpacas (*Auchenia*), which have separate toes, and have no hump. The Llama is extensively used as a beast of burden, but the Alpaca is chiefly of value for its long wool, which is largely manufactured.

The family *Cervidæ* includes the true Deer, and is characterized by the fact that the forehead carries two solid bony antlers, which are not hollow, and are usually much branched. With the single exception of the Reindeer, these appendages are exclusively confined to the males, and they are *deciduous;* that is to say, they are only produced at certain seasons (annually, at the breeding-season), and, when they have fulfilled their purpose, they are shed. They increase in size and in the number of branches every time they are reproduced, till in the old males they may attain an enormous size. Among the more familiar of the Deer may be mentioned the Elk, or Moose (*Alces Americanus*) of Scandinavia and North America, the

Reindeer and Caribou (*Cervus tarandus*) of Northern Europe, Asia, and North America; the Red Deer (*Cervus elaphus*, Fig. 150) of Europe; the Wapiti (*C. Canadensis*) of Canada; and the Roebuck (*Capreolus capræa*) of Northern Europe.

Fig. 150.—Cervidæ.—Head of Stag (*Cervus elaphus*).

Of the Giraffes or *Camelopardalidæ* there is only a single living species, exclusively confined to the African Continent. Both sexes have two pairs of short horns, carried on the forehead; but these are persistent, and are covered with a hairy skin. The neck is extremely long, and the fore-legs much longer than the hind-legs. It is the largest of living Ruminants, and measures as much as from fifteen to eighteen feet in height.

The *Cavicornia* or *Hollow-horned* Ruminants comprise the Oxen, Sheep, Goats, and Antelopes, and are characterized by having horns, which may be present in one or both sexes, and consist of a horny sheath surrounding a central bony axis, or "horn-core." The horns are persistent, and are not periodically shed, and there is usually only a single pair, though sometimes there are two pairs. In their dentition, and in other respects, the *Cavicornia* are to be regarded as being the most typical examples of the *Ruminantia*, and they include a number of animals which are of the highest utility to man. The Antelopes form a very extensive group, closely resembling the true Deer, but distinguished by the possession of *hollow* horns, in place of *solid* antlers. Most of the Antelopes are African, and there are only two European forms (the Chamois being one), while America possesses only the Prongbuck (*Antilope furcifer*). Among the more familiar African species may be mentioned the Gazelle, the Koodoo (Fig. 151),

FIG. 151.—Antelopidæ. Head of the Koodoo (*Strepsiceros Koodoo*).

the Gnu, the Gemsbok, and the Springbok. The Sheep and the Goats (*Ovidæ*) are closely allied to one another, and are well known by their domestic varieties. All the Sheep appear to be natives of the Old World, with the exception of the "Bighorn" (*Ovis montana*) of the Rocky Mountains. Among the true Oxen (*Bovidæ*) the most important species is the domestic Ox (*Bos taurus*) with its innumerable varieties. The true Buffalos (*Bubalus*) are natives of Asia and Africa, and are characterized by their wide horns united at the base. The American Buffalo, or Bison, as it is properly called (*Bison Americanus*), is distinguished by its enormous head, shaggy mane, and conical hump between the shoulders. America also possesses another singular Ox in the person of the Musk Ox (*Ovibos moschatus*), which is found north of the 60th parallel, and is remarkable for its small size and long, woolly coat.

ORDER VII. HYRACOIDEA (Gr. *hurax*, a shrew; *eidos*, form).—This order includes only a single small genus (*Hyrax*), of which only a few species are known. They are all gregarious little animals, living in holes of the rocks, and capable of domestication. One species (*Hyrax Capensis*) occurs commonly in South Africa, and is known to the Dutch colonists as the "Badger." Another species (*Hyrax Syriacus*) occurs in the rocky parts of Arabia and Palestine, and is believed to be the "cony" of Scripture. They present many curious points of resemblance to the gigantic Rhinoceros, and are often placed in the same order, the similarity being especially great as regards the form of the molar teeth. The incisor teeth of the upper jaw are long and curved, with sharp cutting edges, and they grow from a permanent pulp, thus resembling the teeth of the genuine Rodents (such as the Rabbit or Beaver).

Order VIII. Proboscidea (Lat. *proboscis*, the snout).—This order is only represented at the present day by the Elephant, of which there are only two species living. One of these is the African Elephant, which is distinguished by its convex forehead and great flapping ears; the other is the Indian Elephant, which has a concave forehead and small ears. The *Proboscidea* are characterized by having the nose prolonged into a cylindrical trunk or proboscis, at the extremity of which the nostrils are placed (Fig. 152, *n*). The trunk is extremely flexible and highly sensitive, and terminates in a finger-like prehensile lobe. There are no canine teeth; the molars are few in number, large, and transversely ridged, or furnished with tubercles. In the living forms there are no lower incisors, but the upper incisors are two in number, grow from a permanent pulp, and constitute enormous tusks (Fig. 152, *i*). In some of the extinct forms there are two tusk-like lower incisors, and sometimes both the lower and upper incisors are developed into tusks. The feet are furnished with five toes each, but these are only partially indicated externally by the divisions of the hoof. The animal walks upon thick pads of integument, which constitute the soles of the feet. The Indian Elephant inhabits India and the Indian Archipelago and has five hoofs on the fore-feet, but only four on the hind-feet. Like the Ceylon Elephant, which is a mere variety, the males alone possess well-developed tusks. The African Elephant has four hoofs on the fore-feet, and only three on the hind feet, while it is smaller and darker in color than the Indian species. Both sexes also possess tusks, though those of the males are largest. All the Elephants feed upon vegetable matter.

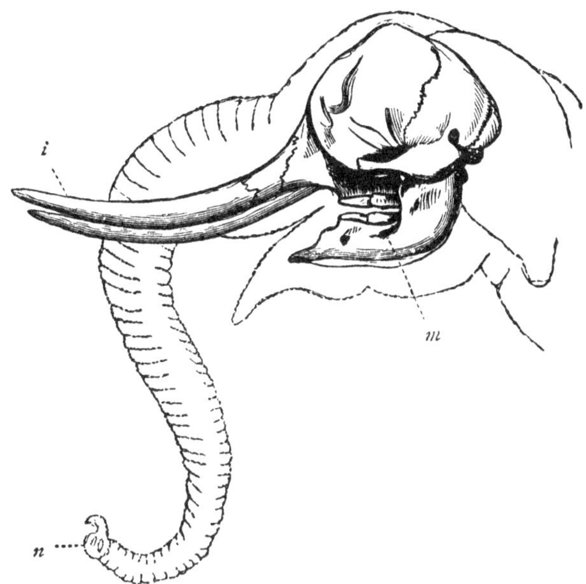

FIG. 152.—Skull of the Indian Elephant (*Elephas Indicus*). *i* Tusk-like upper incisors; *m* Lower jaw, with grinding molars, but without incisors; *n* Nostrils, placed at the extremity of the proboscis.

Though there are now but two living species of Elephant, there is no doubt but that some of the fossil forms have died out since the appearance of man upon the globe. Of these, the best known is the Mammoth, frozen carcases of which have been found in the icy wilds of Siberia.

ORDER IX. CARNIVORA (Lat. *caro*, flesh; *voro*, I devour).—The ninth order of Mammals is that of the *Carnivora* or Beasts of Prey, comprising the Lions, Tigers, Wolves, Dogs, Cats, Hyænas, Seals, Walruses, etc. The *Carnivora* are distinguished by possessing two sets of teeth, which are simply enamelled, and are always of three kinds, incisors, canines, and molars, differing from one another in size and shape. The incisor teeth are generally six in each jaw; the canines are always two in each jaw, and are much longer and larger than the other teeth. The molars are mostly cutting-teeth, furnished with sharp, uneven edges, but one or more of the hinder teeth have tuberculate crowns. The molars, too, graduate from a cutting to a tuberculate form as the diet is strictly carnivorous or becomes more or less miscellaneous.

The dental formula differs considerably in different members of the order, but subjoined is the dental formula of the Cats (*Felidæ*), which are the most typical examples of the Carnivora—

$$i\,\frac{3-3}{3-3};\; c\,\frac{1-1}{1-1};\; pm\,\frac{3-3}{2-2};\; m\,\frac{1-1}{1-1} = 30.$$

Besides the strictly flesh-eating dentition of the *Carnivora*, the order is distinguished by always having the feet provided with strong, curved claws, and the collar-bones (*clavicles*) are either quite rudimentary, or are altogether absent. The *Carnivora* are divided into the following three sections, founded upon the nature of the limbs:

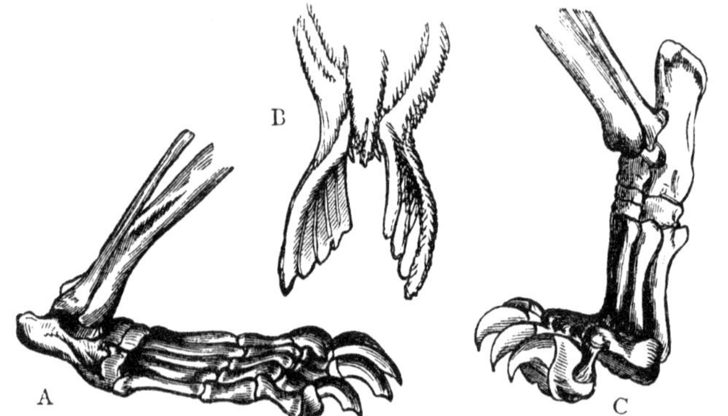

FIG. 153.—Feet of Carnivora (after Owen). A, *Plantigrada*, Foot of Bear; B, *Pinnigrada*, Hind-feet of Seal; C, *Digitigrada*, Foot of Lion.

1. *Pinnigrada* (Fig. 153, B), in which both the fore and hind legs are short, and the feet form broad, webbed, swimming-paddles. The hind-feet are placed very far back, nearly in a line with the axis of the body, and they form with the hinder end of the body a powerful caudal fin. In this section are the Seals and Walruses.

2. *Plantigrada* (Fig. 153, A), comprising the Bears, in which the whole, or nearly the whole, of the foot is applied to the ground, so that the animal walks upon the *soles* of the feet.

3. *Digitigrada* (Fig. 153, C), comprising the Cats, Lions,

Tigers, Dogs, etc., in which the heel is raised from the ground, and the animal walks upon tiptoe.

The Seals and Walruses, forming the family *Pinnigrada*, are distinguished from the other *Carnivora* by their adaptation to an aquatic mode of life. In this respect they agree with the thoroughly aquatic Whales and Dolphins, but they differ from both the *Cetacea* and the *Sirenia*, not only in their dentition, but also in always having well-developed *hind-limbs*. The Seals (Fig. 154) are characterized by having incisor teeth in both jaws, at the same time that the canine teeth are not immoderately developed. They form a very numerous family, of which species are found in most seas out of the limits of

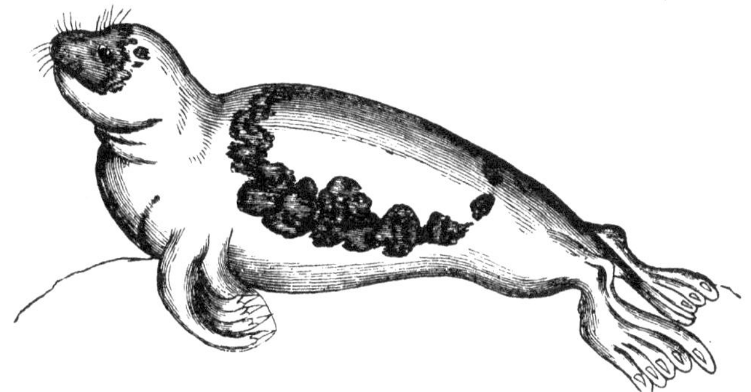

FIG. 154.—Greenland Seal (*Phoca Groenlandica*).

the tropics. They abound, however, especially in the seas of the Arctic and Antarctic regions. They are largely captured both for their oil and for their fur. The Walrus or Morse (*Trichecus*) is distinguished from the true Seals by the fact that in the adult only two of the upper incisors are present; while the upper canines are enormously developed, and form two pointed tusks—fifteen inches or more in length—which are directed downward between the small lower canines, and project considerably below the chin. The Walrus is a large, heavy animal, from ten to fifteen feet in length, which is found in flocks in the Arctic seas, and is hunted both for its blubber and for the ivory of the tusks.

The *Plantigrade Carnivora* apply the whole or the greater part of the sole of the foot to the ground in walking; and this portion of the foot is nearly or altogether destitute of hairs, except in the White Bear. The most typical members of the *Plantigrada* are the Bears (*Ursidæ*), of which the common Brown Bear and the White or Polar Bear are familiar examples. The Bears are much less purely carnivorous than the majority of the order, and, in accordance with their omnivorous habits, the teeth do not exhibit the typical carnivorous characters. The incisors and canines have their usual carnivorous form, but the præmolars and molars are furnished with broad tubercular crowns. The claws are large, curved, and strong, but are not retractile. The tongue is smooth, the ears small and erect, the tail short, the nose mobile, and the pupil circular. Most of the Bears are only carnivorous, in so far that they eat flesh when they can get it; but a great part of their food consists of roots, acorns, honey, and even insects. Nearly related to the true Bears are the familiar Raccoons (*Procyon*) of America, the Coatis (*Nasua*) of South America, and the Wah (*Ailurus*) of India.

The only remaining Plantigrades of importance are the Badgers (*Meles*) of Europe, Asia, and America, the Gluttons or Wolverines (*Gulo*) of the same continents, and the Honey-badgers (*Mellivora*) of Africa.

Forming a kind of transition between the *Plantigrada* and the *Digitigrada* is a group of *Carnivora* which comprises numerous forms, such as the Weasels, Otters, and Civets, which apply part, but not the whole, of the sole of the foot to the ground.

The Weasels (*Mustelidæ*) have short legs and elongated, worm-like bodies, with a stealthy, gliding mode of progression. Good examples are the Pole-cat, the Mink, the Ermine, and the Sable. The two latter furnish the beautiful and valuable furs known by their names. Here also belongs the Skunk (*Mephitis*), celebrated for its intensely disagreeable odor when alarmed or irritated. The Otters are nearly allied to the Weasels, but have webbed feet adapted for swimming. The great Sea-otter yields a very valuable fur. The Civets and Genettes (*Viverridæ*) all belong to the Old World. The true Civet-cat inhabits North Africa, and is furnished with a pouch which secretes the peculiar fatty substance which is used as a perfume under the name of "civet."

The typical group of the *Carnivora* is that of the *Digitigrada*, comprising the three tribes of the Dogs (*Canidæ*), the Hyænas (*Hyænidæ*), and the Cats (*Felidæ*). The family *Canidæ* comprises the true Dogs, the Wolves, the Foxes, and the Jackals, all characterized by their pointed muzzles, smooth tongues, and non-retractile claws, and by the fact that the fore-feet have five toes, while the hind-feet have only four. In the *Hyænidæ*, comprising the Hyænas, there are only four toes to all the feet, the muzzle is rounded, the tongue is rough, and the hind-legs are shorter than the fore-legs. The Hyænas are ill-conditioned, ferocious animals, which occur in Africa, Asia Minor, Arabia, and Persia.

The most highly carnivorous, and therefore the most typical, group of the *Carnivora* is that of the Cats or *Felidæ*, comprising the Lions, Tigers, Leopards, Panthers, Cats, and others. In all these the animal walks lightly upon the tips of the toes, and the soles of the feet are hairy. The jaws are short, and, owing to this and to the great size of the muscles which move the lower jaw, the head assumes a rounded form, with a short muzzle. The molars and præmolars are fewer in number than in any other of the *Carnivora*—hence the shortness of the jaws; and they are all furnished with cutting-edges, except the last molar in the upper jaw, which is tuberculate. The legs are nearly of equal length, and the hind-feet have only four toes, while the fore-feet have five toes each. All the toes are furnished with strong, curved, retractile claws, which, when not in use, are withdrawn within sheaths by the action of elastic ligaments. The tongue is armed with horny eminences, which render it rough and prickly, and adapt it for the office of licking flesh from the bones of the prey. They are all extremely light upon their feet, and excessively muscular; and all have the habit of seizing their prey by suddenly springing upon it. In this section are the Lion (*Felis leo*), the Tiger (*Felis Tigris*), the Jaguar (*Felis onca*), the Puma (*Felis concolor*), the Leopard (*Felis leopardus*), the Lynxes, and the true Cats.

The Lions are entirely confined to the Old World, inhabiting Southern Asia and Africa. The males are maned, and the tail is tufted. The Royal Tiger is exclusively Asiatic, as are most of the Tiger-cats, but some of the latter are American. The Spotted Cats or Leopards are represented, among others, by the Leopard and Cheetah of the Old World, and the well-known Jaguar of the American Continent. The Puma is also American, but its color is uniform. The Lynxes are distinguished by their tufted ears, and are found both in the Eastern and Western hemispheres.

ORDER X. RODENTIA (Lat. *rodo*, I gnaw).—In this order

are a number of small animals, characterized by the absence of canine teeth, and the possession of two long curved incisor teeth in both jaws, which are separated by a wide interval from the molars (Fig. 155). There are seldom more than two

FIG. 155.—A, Skull of the Beaver (after Owen); B, Diagram of one of the incisor teeth of a Rodent, showing the chisel-shaped point. *a* Enamel; *d* Soft tooth-substance (dentine).

incisors in the upper jaw (sometimes four), but there are never more than two in the lower jaw. The molar teeth are few in number (rarely more than four on each side of each jaw). The feet are usually furnished with five toes each.

The most characteristic point about the Rodents is to be found in the structure of the incisor teeth, which are adapted for continuous gnawing. They grow from persistent pulps, and consequently continue growing as long as the animal lives.

FIG. 156.—Hamster (*Cricetus vulgaris*).

They are large, long, and curved, and are covered in front with a layer of hard enamel, so that the softer parts of the tooth are placed behind (Fig. 155, B). The result of this is, that as the tooth is used in gnawing, the softer parts behind wear away more rapidly than the hard enamel in front, and thus the crown of the tooth acquires by use a chisel shape, bevelled away behind, and the enamel forms a persistent cutting-edge. The Rodents are almost all of small size, and are very prolific. They subsist principally, if not entirely, on vegetable matters, especially the harder parts of plants, such as the bark and roots. Many possess the power of building very elaborate nests, and most of them hybernate (*i. e.*, remain torpid throughout the winter). They are very generally distributed over the whole world.

The order *Rodentia* comprises a large number of families, of which little more than the names can be mentioned. The most important families of Rodents are the following: 1. *Leporidæ*, comprising the Hares and Rabbits. The Hares generally occur in temperate regions, but some are African, and one species occurs in the Arctic regions, while the common American Hare (*Lepus Americanus*) extends from Canada to Mexico. 2. *Cavidæ*, comprising the Capybaras, Guinea-pigs, etc. The Capybara is the largest of living Rodents, and is not unlike a small pig. It is a native of South America, and leads an amphibious life. Here also belong the Agoutis (*Dasyprocta*) of South America and the West Indies, and the Pacas of South America. 3. *Hystricidæ*, comprising the Porcupines, and characterized by the fact that the body is covered with longer or shorter spines or quills mixed with bristly hairs. Most of the Porcupines live in burrows, and are much like the Rabbits in their habits, but some are furnished with prehensile tails, and live in trees. 4. *Castoridæ* or Beaver family, comprising the Beaver, Musquash, and Coypu. The Beaver has webbed feet and a scaly tail, and the fur is an article of considerable value. It inhabits both North America and Europe. The Musquash resembles the Beaver in many respects, and is also a native of Northern America; but the Coypu is South American. 5. *Muridæ*, comprising the Mice, Rats, Hamster (Fig. 156), Lemmings, etc. The Rats and Mice are too well known to require more than merely to be mentioned. 6. *Dipodidæ*, comprising the Jerboas of the Old World, and the Jumping Mice of America. 7. *Myoxidæ*, comprising the Dormice, which must not be confounded with the true Mice on the one hand, or with the Shrew-mice on the other hand. 8. *Sciuridæ*, comprising the Squirrels, Flying Squirrels, and Marmots. The Flying Squirrels do not really fly, but, like the "flying" Phalangers, they take long leaps from tree to tree by means of laterally-extended folds of skin. The Marmots, unlike the typical Squirrels, are ground-animals, and live in burrows. An excellent example is afforded by the Prairie-dog (*Arctomys Ludovicianus*) of North America.

ORDER XI. CHEIROPTERA (Gr. *cheir*, hand; *pteron*, wing).—This order is undoubtedly one of the most natural and dis-

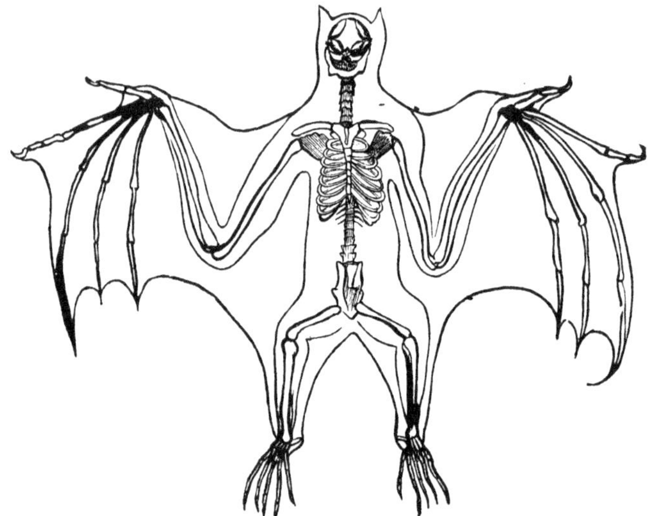

FIG. 157.—Skeleton of a Bat (*Pteropus.*) (After Owen.)

tinctly circumscribed orders in the whole class of the *Mammalia*, comprising only the Bats. In many respects, however, it might be well to regard the order as merely a modified branch of the *Insectivora*, just as the *Pinnigrada* are regarded as a modified offshoot of the *Carnivora*. The *Cheiroptera* or Bats are essentially characterized by the fact that the fore-limbs are much longer than the hind-limbs, and have several of the fingers enormously elongated. These enormously lengthened digits are united by an expanded leathery membrane or "patagium," which not only stretches between the fingers, but is also extended between the fore and hind limbs, and is attached to the sides of the body (Fig. 157). The patagium thus formed often includes the tail, and is nearly or quite naked or destitute of hairs on both sides. It is used as an organ of true flight, and, in accordance with this, there are well-developed collar-bones (clavicles), and the breastbone is furnished with a ridge for the attachment of the pectoral muscles. Of the fingers of the hand at least three are destitute of nails. The mammary glands are placed upon the chest. Teeth of three kinds are always present, and the canines are always well developed.

The Bats are all twilight-loving or nocturnal animals, and they are the only Mammals which possess the power of true flight, though several others can make extended leaps from tree to tree. The eyes are small, but the ears are very large, and their sense of touch is most acute. During the day they retire to caves or crevices in rocks, where they suspend themselves by the short thumbs, which are provided with claws. In their flight, though they can turn with great ease, they are by no means as rapid and active as the true Birds. The tail is sometimes very short, sometimes moderately long, and is usually included in a continuation of the "patagium," which extends between the hind-legs. The body is covered with hair, but the patagium is usually nearly or quite hairless. Most of the Bats hybernate.

The *Cheiroptera* are conveniently divided into the two sections of the Insectivorous and Frugivorous Bats. In the first section are all the bats of temperate climates, most of which are of very small size, and all of which live upon insects. Here also belong the great Vampire-bats (*Phyllostomidæ*) of South America. In the second, or fruit-eating section of the *Cheiroptera*, are only the Fox-bats (*Pteropidæ*), which are especially characteristic of the Pacific Archipelago, inhabiting Australia, Java, Sumatra, Borneo, etc., but occurring also in Asia and Africa. They are among the largest of the Bats, one species—the *Pteropus edulis* or Kalong—attaining a length of from four to five feet from the tip of one wing to the tip of the other.

ORDER XII. INSECTIVORA (Lat. *insectum*, an insect; *voro*, I devour).—The twelfth order of Mammals is that of the *Insectivora*, which comprises a number of small animals, very similar in many respects to the Rodents, but wanting the peculiar incisors of that order, and also being provided with clavicles. All the three kinds of teeth are present, but the dentition is very various, and the only common character is that the crowns of the molar teeth are furnished with small pointed eminences or cusps, adapted for crushing insects. All the toes have claws, there are usually five toes to each foot, and most of the *Insectivora* are *plantigrade*, that is to say, walk upon the soles of the feet. They are all small, and they exist over the whole world, except in Australia and South America, where their place is taken by Marsupials, such as the Opossums.

The *Insectivora* are divided into the three families of the Moles (*Talpidæ*), the Shrews (*Soricidæ*), and the Hedgehogs (*Erinaceidæ*). The Moles (Fig. 158) are distinguished by

FIG. 158.—Insectivora. Mole (*Talpa Europœa*).

having the body covered with hair, the feet short and formed for digging, and the toes furnished with strong, curved claws. There is no external ear, and the eyes are either extremely small, or are completely concealed beneath the fur. They are all nocturnal burrowing animals. The Star-nosed Moles (*Condylura*) are American, but their habits are like those of the European Mole (*Talpa Europœa*, Fig. 158). The Golden Moles (*Chrysochloris*) are African, and are remarkable for the iridescence of their fur. The Shrews are very like the true Mice in external appearance, but they are really widely different. The body is covered with hair, the feet are not adapted for digging, and there are mostly external ears, while the eyes are well developed. No division of the *Insectivora* is more abundant or more widely distributed than the *Soricidæ*, and one of the Shrews is probably the smallest of existing Mammals, not exceeding two and a half inches in length, counting in the tail. Besides the true Shrews (*Sorex*), this family includes also the Elephant Shrews (*Macroscelides*) of Africa, and the common Water-mole (*Scalops aquaticus*) of North

America. The third family includes only the well-known Hedgehogs, which have the power of rolling themselves into a ball at the approach of danger, and which have the upper surface of the body covered with short prickly spines, forming a protective armor. The common European Hedgehog (*Erinaceus Europœus*) is the type of the family, but other species occur in Africa and India. The "Tenrecs" (*Centetes*) of Madagascar are closely allied to the Hedgehogs, but have no power of rolling themselves up. The "Banxrings" (*Tupaia*) of the Indian Archipelago have a long, compressed tail, and live mostly in trees.

Before passing on to the next order, a few words must be said about a curious transitional form, which has been alternately placed in the *Cheiroptera*, the *Insectivora*, or the *Quadrumana*, or has been regarded as the type of a separate order. The animal alluded to is the so-called Flying Lemur (*Galeopithecus volitans*), of which more than one species is known as inhabiting the Indian Archipelago. The leading characteristic in this singular animal is the possession of a flying-membrane, which extends as a broad expansion from the nape of the neck to the arms, from the arms to the hind-legs, and from the hind-legs to the tail. The fingers are not elongated, and do not support a "patagium," so that the animal has no power of true flight, but can simply take extended leaps from tree to tree. The *Galeopithecus* lives chiefly upon small insects and birds, and it should, probably, be regarded as an aberrant form of the *Insectivora*.

ORDER XIII. QUADRUMANA (Lat. *quatuor*, four; *manus*, hand).—The thirteenth order of Mammals is that of the *Quadrumana*, comprising the Apes, Monkeys, Baboons, and Lemurs. The characteristic of this order is that the innermost toe (great-toe) of the hind-limbs can be *opposed* to the other toes, so that the hind-feet become prehensile hands. The term "opposed" simply implies that the toe can be so adjusted, as regards the extremities of the other toes, that any object can be grasped between them, just as the thumb of the human hand can be "opposed" to any of the fingers. The fore-feet may be destitute of a thumb, but, when this is present, it too is generally opposable to the other digits, so that the animal becomes truly four-handed or "quadrumanous."

The *Quadrumana* are divided into three very natural sections, separated from one another both by their anatomical characters and their geographical distribution.

Section A. Strepsirhina.—Characterized by having the nostrils twisted or curved, and placed at the end of the nose, while the second toe of the hind-feet is furnished with a claw. The *Quadrumana* of this section are chiefly referable to Madagascar as their geographical centre, but they spread from Madagascar westward into Africa, and eastward to the Indian Archipelago. In this family are the Aye-Aye (*Cheiromys*), the Loris and Slow Lemurs (*Nycticebidæ*), and the Lemurs (*Lemuridæ*). The Aye-Aye is confined to Madagascar, and is not unlike a large squirrel in appearance, having a long bushy tail. The incisors grow from permanent pulps, like those of Rodents, and there are no canines. The Loris and Slow Lemurs have either no tail or a rudimentary one, and they are confined to Southern Asia, and the great islands of the Indian Archipelago. The true Lemurs are natives of Madagascar, and are often spoken of as "Madagascar cats." They have a soft, woolly fur, and a long tail, which is prehensile. The second toe of the hind-foot has a long and pointed claw.

Section B. Platyrhina.—This section includes those monkeys in which the nostrils are simple, and are placed far apart; the thumbs of the fore-feet are wanting, or, if present, are not opposable; and the tail is generally prehensile. The Platyrhine Monkeys are exclusively confined to South America, occurring especially in Brazil, and they are all adapted for a more or less purely arboreal life. The best-known members of this section are the Marmosets (*Hapalidæ*), and the great family of the *Cebidæ*, comprising the Spider-monkeys, the Howlers, and others. The Howlers (*Mycetes*) are remarkable for having a bony drum at the summit of the windpipe, by which the voice is rendered extraordinarily resonant, and peculiarly weird and terrifying to those who hear it.

Section C. Catarhina.—In this, the highest section of the *Quadrumana*, the nostrils are oblique and placed close together, and the thumbs of all the feet are opposable, so that they are truly "quadrumanous." The dental formula agrees with that of man:

$$i\,\frac{2-2}{2-2};\quad c\,\frac{1-1}{1-1};\quad pm\,\frac{2-2}{2-2};\quad m\,\frac{3-3}{3-3} = 32$$

The incisor teeth, however, are prominent and projecting, and the canines, especially in the males, are large and pointed, while the teeth form an uneven series. The tail is never prehensile, and is sometimes absent. Cheek-pouches are often present. In one single instance (*Colobus*) the thumbs of the fore-limbs are wanting.

FIG. 159.—Quadrumana. Green Monkey (*Cercocebus sabæus*). (After Cuvier.)

With the single exception of a Monkey which occurs on the Rock of Gibraltar, all the Catarhine Monkeys are confined to Africa and Asia. The most typical forms of the section are the *Semnopitheci* and *Macaques* of Asia. Less typical are the Baboons, which inhabit Africa, and are among the most repulsive of all the *Quadrumana.* In these the tail is always short, and often quite rudimentary. The head is large, and the muzzle greatly prolonged, having the nostrils at its extremity. More than any other of the Monkeys they employ the fore-limbs in terrestrial progression, running upon all fours with the greatest ease.

The third family of the Catarhine Monkeys is that of the *Anthropoid* Apes, so called from their making a nearer approach to man in anatomical structure than is the case with any other Mammal. The Anthropoid Apes are distinguished by having no tail, nor cheek-pouches. The hind-limbs are short—shorter than the fore-limbs—and the animal can progress in an erect or semi-erect posture. At the same time the hind-feet are strictly prehensile, since the thumbs are opposable to the other toes. The canine teeth of the males are very long, strong, and pointed, but this is not the case in the females.

In this tribe are the Gibbons, the Chimpanzee, the Orang-outang, and the Gorilla. The Gibbons form the genus *Hylobates*, and they belong to Asia, India, and the Indian Archipelago. The anterior limbs in these monkeys are extremely long, and the hands nearly or quite touch the ground when the animal stands erect. The Orang-outang (*Simia*) has no cheek-pouches, and the hips are covered with hair. The arms are of excessive length, and the hind-legs very short. When young, the head of the Orang-outang is not very different from that of a child, but, as the animal grows, the bones of the face gradually lengthen, while the skull remains much about the same; great bony ridges are developed for the attachment of the muscles which act upon the jaws; the incisors project; the canine teeth of the males become long and pointed, till ultimately the muzzle becomes as pronounced and well marked as in the Carnivorous animals (Fig. 160, A). The best-known

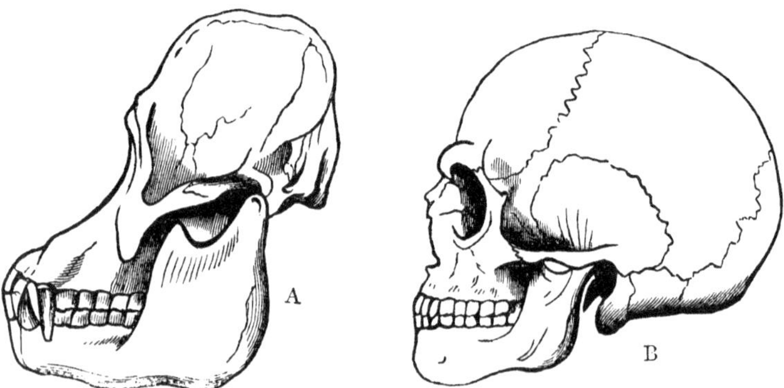

Fig. 160.—A, Skull of the Orang-outang; B, Skull of a European adult.

species of Orang is the *Simia Satyrus*, which inhabits Sumatra, Borneo, and the other larger islands of the Indian Archipelago; but there are probably other species or varieties. The Chimpanzee and Gorilla both belong to Africa, and form the genus *Troglodytes*. The Chimpanzee is a native of Western Africa, and has the arms much shorter proportionately than in the Gibbons and Orangs. Still they are much longer than the hind-limbs, and reach below the knees. The hands are naked to the wrist, and the face is also naked and much wrinkled. The Gorilla is in most respects like the Chimpanzee, but is much larger, attaining a height of fully five feet. It is a native of Lower Guinea and Equatorial Africa, and is enormously strong and very ferocious. It is now generally looked upon as the highest of the Anthropoid Apes.

ORDER XIV. BIMANA (Lat. *bis*, twice; *manus*, hand).—In this order stands Man alone, and little, therefore, needs to be said on this head. Man is distinguished *zoologically* from all other Mammals by his habitually erect posture and progression upon two legs. The lower limbs are exclusively devoted to progression and to supporting the weight of the body. The fore-limbs are shorter than the legs, and have nothing to do with progression. The thumb can be opposed to the other fingers, and the hands are therefore prehensile. The fingers and toes are furnished with nails; but the innermost digit of the foot (the great-toe) is not capable of being opposed to the other toes, so that the foot is useless as a grasping organ. The foot is broad and plantigrade, the whole sole being applied to the ground in walking.

The teeth are thirty-two in number, and they form a nearly even and uninterrupted series, without any gap or interval. The dental formula is:

$$i\,\frac{2-2}{2-2};\quad c\,\frac{1-1}{1-1};\quad pm\,\frac{2-2}{2-2};\quad m\,\frac{3-3}{3-3}\;=\;32$$

The brain is more largely developed, and more richly furnished with large and deep foldings or convolutions, than is the case in any other Mammal. Lastly, Man is the only terrestrial Mammal in which the body is not furnished with a general covering of hair.

The purely *anatomical* distinctions between Man and the other Mammals are thus seen to be not very striking, and *of themselves* they would hardly entitle Man to the position of more than a distinct order in the class Mammalia. When, however, we take into account the vast and unsurmountable *mental* differences, both intellectual and moral, between Man and the highest of the brutes, and when we reflect that this mental difference must have *some* physical correspondence, it becomes a question whether the group *Bimana* should not have the value of a distinct sub-kingdom, while there can be little hesitation in giving Man at least a *class* to himself.

In the words of Dr. Pritchard, " the sentiments, feelings, sympathies, internal consciousness, and mind, and the habitudes of mind and action thence resulting, are the real and essential characteristics of humanity."

GLOSSARY.

Abdomen (Lat. *abdo*, I conceal). The posterior cavity of the body, containing the intestines and others of the viscera. In many Invertebrates there is no separation of the body-cavity into thorax and abdomen, and it is only in the higher *Annulosa* that a distinct abdomen can be said to exist.

Aberrant (Lat. *aberro*, I wander away). Departing from the regular type.

Abnormal (Lat. *ab*, from; *norma*, a rule). Irregular; deviating from the ordinary standard.

Abomasum. The fourth cavity of the complex stomach of the Ruminants.

Abranchiate (Gr. *a*, without; *bragchia*, gills). Destitute of branchiæ or gills.

Acanthocephala (Gr. *acantha*, a thorn; *kephale*, head). A class of parasitic worms in which the head is armed with spines.

Acarina (Gr. *akari*, a mite). A division of the *Arachnida*, comprising the mites and ticks.

Accretion (Lat. *accresco*, I grow larger). The process by which inorganic bodies (such as crystals) grow larger, by the addition of fresh particles from the outside.

Acephalous (Gr. *a*, without; *kephale*, head). Not possessing a distinct head.

Actinosoma (Gr. *aktin*, a ray; *soma*, body). Employed to designate the entire body of any *Actinozoön*, whether this be simple (as in the sea-anemones) or composed of several zoöids (as in most corals).

Actinozoa (Gr. *aktin*, a ray; *zoön*, an animal). That division of the *Cœlenterata* of which the sea-anemones may be taken as the type.

Adductor (Lat. *adduco*, I bring together). The muscles which bring together the valves of the shell of the Bivalve mollusks are known as the "adductors."

Aërial (Gr. *aër*, air). Living in the air; enjoying the power of flight.

Ambulacra (Lat. *ambulacrum*, a place for walking). The perforated spaces or "avenues" in the shell of the *Echinoidea*, through which are protruded the locomotive tube-feet.

Ametabolic (Gr. *a*, without; *metabole*, change). Applied to those insects which do not possess wings when fully grown, and which do not, therefore, pass through any well-marked metamorphosis.

Amœba (Gr. *amoibos*, changing). A species of Rhizopod, so called from the numerous changes of form which it undergoes.

Amorphous (Gr. *a*, without; *morphe*, shape). Not having any definite figure.

Amphibia (Gr. *amphi*, both; *bios*, life). A class of the *Vertebrata* comprising Frogs, Newts, and the like, which have always gills when young, but always develop lungs when fully grown. Most of them, therefore, live indifferently on land or in water.

Amphicœlous (Gr. *amphi*, at both ends; *koilos*, hollow). Applied to vertebræ which are doubly concave, or hollow at both ends.

Amphidiscs (Gr. *amphi*, at both ends; *diskos*, a quoit or round plate). The spicules which surround the reproductive "gemmules" of *Spongilla*, and resemble two toothed wheels united by an axle.

Amphioxus (Gr. *amphi*, at both ends; *oxus*, sharp). The Lancelet, a curious little fish, which alone constitutes the order *Pharyngobranchii*.

Amphipoda (Gr. *amphi*, and *podes*, feet). An order of the *Crustacea*, so called because some of the feet are directed forward and some backward.

Analogous. Applied to parts which perform the same function.

Anarthropoda (Gr. *a*, without; *arthros*, joint; *podes*, feet). The division of *Annulose* animals in which there are no articulated appendages.

Annelida (a Gallicised form of *Annulata*, Lat. *annulus*, a ring). The Ringed worms, which form one of the divisions of the *Anarthropoda*.

Annuloida (Lat. *annulus*, a ring; Gr. *eidos*, form). The sub-kingdom comprising the *Echinodermata* and the *Scolecida*.

ANNULOSA (Lat. *annulus*, a ring). The sub-kingdom comprising the *Anarthropoda* and the *Arthropoda* or *Articulata*, in all of which the body is more or less evidently composed of a succession of rings.
ANOMODONTIA (Gr. *anomos*, irregular; *odous*, tooth). An extinct order of reptiles, called by Huxley *Dicynodontia*.
ANOMURA (Gr. *anomos*, irregular; *oura*, tail). A tribe of Decapod Crustaceans, of which the Hermit-crabs are the type.
ANOPLURA (Gr. *anoplos*, unarmed; *oura*, tail). An order of apterous insects.
ANOURA or ANURA (Gr. *a*, without; *oura*, tail). An order of *Amphibia* comprising the Frogs and Toads, in which the adult is destitute of a tail.
ANTENNÆ (Lat. *antenna*, a yard-arm). The jointed horns or feelers possessed by the majority of the *Articulata*.
ANTENNULES (diminutive of *Antennæ*). Applied to the smaller pair of antennæ in the *Crustacea*.
ANTHROPOID (Gr. *anthropos*, man; *eidos*, form). Applied to those monkeys which make the nearest approach in anatomical structure to man.
APHANIPTERA (Gr. *aphanos*, inconspicuous; *pteron*, a wing). An order of insects, comprising the Fleas.
APLACENTAL (Gr. *a*, without; Lat. *placenta*, a cake). Applied to those Mammals in which the young is destitute of a placenta (*see* Placenta).
APTEROUS (Gr. *a*, without; *pteron*, a wing). Destitute of wings.
AQUATIC (Lat. *aqua*, water). Inhabiting water.
AQUIFEROUS (Lat. *aqua*, water; *fero*, I carry). Applied to the water-carrying canal-system of the sponges.
ARACHNIDA (Gr. *arachne*, a spider). A class of *Articulata*, comprising spiders, scorpions, and allied animals.
ARANEIDA (Lat. *aranea*, a spider). The order of *Arachnida*, comprising the true spiders.
ARCHÆOPTERYX (Gr. *archaios*, ancient; *pterux*, a wing). A fossil bird, which alone constitutes the order *Saururæ*.
ARENACEOUS. Sandy, or composed of grains of sand.
ARTHROPODA (Gr. *arthros*, a joint; *podes*, feet). The division of *Annulosa*, in which the body is furnished with jointed appendages.
ARTICULATA (Lat. *articulus*, diminutive of *artus*, a joint). Sometimes used in the same sense as the term *Arthropoda*.
ARTIODACTYLA (Gr. *artios*, even; *daktulos*, a finger or toe). A division of the hoofed quadrupeds (*Ungulata*) in which each foot has an even number of toes (two or four).
ASCIDIOIDA (Gr. *askos*, a bag or a leather bottle; *eidos*, form). Sometimes employed to designate the *Tunicata*, an order of the lower *Mollusca*, from the resemblance of the body in many cases to a two-necked jar.
ASEXUAL. Applied to modes of reproduction in which the sexes are not concerned (e. g., gemmation and fission).
ASTEROIDEA (Gr. *aster*, a star; and *eidos*, form). An order of *Echinodermata*, comprising the Star-fishes, characterized by their rayed form.
ASTOMATOUS (Gr. *a*, without; *stoma*, mouth). Not possessing a mouth.
ATRIUM (Lat. a hall). Applied to the large chamber or "cloaca" into which the intestine opens in the *Tunicata*.
AURICLE (Lat. *auricula*, a little ear). Applied to the cavity of the heart which drives the blood into the ventricle.
AVES (Lat. *avis*, a bird). The class of the Birds.

BALANCERS. The knobbed filaments which represent the posterior pair of wings in Dipterous Insects; also called "poisers."
BALANIDÆ (Gr. *balanos*, an acorn). A family of *Cirripedia*, commonly spoken of as "Acorn-shells."
BALEEN (Lat. *balæna*, a whale). The horny plates which are placed in the upper surface of the mouth of the true or "whalebone" Whales.
BATRACHIA (Gr. *batrachos*, a frog). Generally applied loosely to any of the *Amphibia;* but sometimes used to designate either the entire class of the Amphibians or the single order of the *Anoura*.
BIFID. Cleft into two parts; forked.
BIMANA (Lat. *bis*, twice; *manus*, hand). The order of *Mammalia* constituted by Man alone.
BIVALVE (Lat. *bis*, twice; *valvæ*, folding-doors). Composed of two plates or valves; applied to the shell of the *Lamellibranchiata* and *Brachiopoda*, and to the carapace of certain *Crustacea*.
BLASTOIDEA (Gr. *blastos*, a bud; *eidos*, form). An extinct order of *Echinodermata*, often called *Pentremites*.
BRACHIOPODA (Gr. *brachion*, the arm; *podes*, feet). A class of the lower *Mollusca*, often called "Lamp-shells," characterized by having two fleshy ciliated "arms" attached to the sides of the mouth.
BRACHYURA (Gr. *brachus*, short; *oura*, tail). A tribe of Decapod Crustaceans with short tails, commonly known as Crabs.

BRACT. Overlapping appendages or plates which protect the polypites in many of the Oceanic *Hydrozoa*.
BRANCHIA (Gr. *bragchia*, the gill of a fish). A respiratory organ adapted for breathing air dissolved in water.
BRANCHIATE. Possessing gills.
BRONCHI (Gr. *brogchos*, the windpipe). The branches of the windpipe (trachea) by which the air is conveyed to the lungs.
BRUTA (Lat. *brutus*, heavy, stupid). Sometimes used to designate the Mammalian order of the *Edentata*.
BUCCAL (Lat. *bucca*, mouth or cheeks). Connected with the mouth or cheeks.
BYSSUS (Gr. *bussos*, flax). The silky threads by which many bivalve Mollusks (such as the Mussels) attach themselves to foreign objects.

CADUCIBRANCHIATE (Lat. *caducus*, falling off; Gr. *bragchia*, gills). Applied to those Amphibians in which the gills fall off before maturity is reached (e. g., Newts and Frogs).
CÆCAL (Lat. *cæcus*, blind). Terminating blindly, or in a closed extremity
CÆCUM (Lat. *cæcus*). A tube which ends in a blind extremity.
CALCAREOUS (Lat. *calx*, lime). Composed of carbonate of lime.
CALICE. The little cup in which the polype of a coral-producing Zoophyte (*Actinozoön*) is contained.
CALYCOPHORIDÆ (Gr. *kalux*, a cup; *phero*, I carry). An order of the Oceanic *Hydrozoa*, so called from their possessing bell-shaped swimming-organs.
CALYX (Lat. a cup). Applied to the cup-shaped body of the Bell-animalcule (*Vorticella*) or of the *Crinoidea*.
CAMPANULARIDA (Lat. *campanula*, a bell). A group of Sertularian Zoophytes.
CANINE (Lat. *canis*, a dog). The eye-tooth in the jaw of Mammals; so called because it is particularly well developed in Dogs and other carnivorous animals.
CARAPACE. A protective shield. Applied to the upper shell of Lobsters, Crabs, and many other Crustaceans; and to the upper half of the bony case in which the Tortoises and Turtles are enclosed.
CARNIVORA (Lat. *caro*, flesh; *voro*, I devour). An order of the *Mammalia*.
CARNIVOROUS. Living upon animal food.
CARPUS (Gr. *karpos*, the wrist). The small bones which intervene between the forearm and hand.
CATARHINA (Gr. *kata*, downward; *rhines*, nostrils). A group of the Monkeys (*Quadrumana*).
CAUDAL (Lat. *cauda*, the tail). Connected with the tail, or hinder end of the body.
CAVICORNIA (Lat. *cavus*, hollow; *cornu*, a horn). The "hollow-horned" Ruminants, in which the horn consists of a central bony "horn-core" surrounded by a sheath of horn.
CENTRUM (Gr. *kentron*, the point round which a circle is described by a pair of compasses). The central portion or "body" of a vertebra.
CEPHALIC (Gr. *kephale*, the head). Connected with the head.
CEPHALOPODA (Gr. *kephale*, head; *podes*, feet). A class of *Mollusca* comprising the Cuttle-fishes and their allies, in which there is a series of arms ranged round the head.
CEPHALO-THORAX (Gr. *kephale*, head; *thorax*, chest). The anterior division of the body in many *Crustacea* and *Arachnida*, composed of the amalgamated segments of the head and thorax.
CEREBRAL. Connected with the brain, or in a restricted sense with the "cerebrum."
CEREBRUM. Sometimes employed in a general way to designate the entire brain; but properly restricted to the "hemispheres" of the brain, which are believed to be concerned mainly with the discharge of the mental functions.
CERVICAL (Lat. *cervix*, the neck). Belonging to the neck.
CETACEA (Gr. *ketos*, a whale). An order of the *Mammalia*.
CHÆTOGNATHA (Gr. *chaite*, a bristle; *gnathos*, the jaw). A class of the *Anarthropoda*.
CHEIROPTERA (Gr. *cheir*, hand; *pteron*, wing). An order of *Mammalia*.
CHELÆ (Gr. *chele*, claw). The prehensile claws with which certain of the limbs are terminated in many of the *Crustacea* (such as Lobsters, Crabs, etc.).
CHELONIA (Gr. *chelone*, a tortoise). An order of Reptiles.
CHITINE (Gr. *chiton*, a coat). A peculiar chemical principle allied to horn, which is found in the outer covering of many of the *Invertebrata*, especially in *Arthropoda* (Insects, Scorpions, Crustaceans, etc.).
CHLOROPHYLL (Gr. *chloros*, green; *phyllon*, leaf). The green coloring-matter of leaves.
CHRYSALIS (Gr. *chrusos*, gold). The motionless pupa of Butterflies and Moths, so called because often exhibiting a golden lustre.
CHYLE (Gr. *chulos*, juice). The milky fluid which is the result of the action of the various digestive fluids upon the food.
CHYLIFIC (Gr. *chulos*, juice; Lat. *facio*, I make). Applied to one of the stomachs, when more than one is present.
CHYME (Gr. *chumos*, juice). The acid pasty liquid produced by the action of the gastric juice upon the food.

CILIA (Lat. *cilium*, an eyelash). Microscopic, hair-like filaments, which have the power of lashing backward and forward, thus creating currents in the surrounding fluid, or subserving locomotion in the animal which possesses them.
CIRRHI (Lat. *cirrus*, a curl). Tendril-like appendages, such as the feet of Barnacles and Acorn-shells (*Cirripedia*), the lateral processes on the arms of *Brachiopods*, etc.
CIRRIPEDIA (Lat. *cirrus*, a curl; *pes*, a foot). A division of the *Crustacea* with curled jointed feet.
CLADOCERA (Gr. *klados*, a branch; *keras*, a horn). An order of *Crustacea* with branched antennæ.
CLAVICLE (Lat. *clavicula*, a little key). The collar-bone of the pectoral or scapular arch.
CLOACA (Lat. a sink). The common cavity in which open the intestinal canal and the ducts of the generative and urinary organs, in many Invertebrates, and some Vertebrates.
CNIDÆ (Gr. *knide*, a nettle). The urticating cells ("thread-cells"), whereby many of the *Cœlenterata* obtain their power of stinging.
COCOON (Fr. *cocon*, the cocoon of the silk-worm). The outer covering of silky hairs with which the pupa or chrysalis of many insects is protected.
CŒLENTERATA (Gr. *koilos*, hollow; *enteron*, the bowel). The sub-kingdom which includes the *Hydrozoa* and *Actinozoa*, the most typical members of Cuvier's division of the *Radiata*.
CŒNOSARC (Gr. *koinos*, common; *sarx*, flesh). The common organized medium or stem by which the separate zoöids of a compound *Hydrozoön* or *Actinozoön* are united together.
COLEOPTERA (Gr. *koleos*, a sheath; *pteron*, a wing). The order of insects (beetles) in which the anterior wings are hardened, and form protective sheaths for the posterior membranous wings.
COLUMBACEI (Lat. *columba*, a dove). The subdivision of the Rasorial birds containing the doves and pigeons.
CONDYLE (Gr. *kondulos*, a knuckle). The surface by which one bone is jointed or articulated to another. Applied especially to the articular surfaces by which the head is jointed to the vertebral column.
CONIROSTRES (Lat. *conus*, a cone; *rostrum*, a beak). The division of perching birds with conical beaks.
COPEPODA (Gr. *kope*, an oar; *podes*, feet). An order of *Crustacea*.
CORACOID (Gr. *korax*, a crow; *eidos*, form). One of the bones which form the scapular arch in Birds, Reptiles, and Monotremata. In the majority of Mammalia it is a mere process of the scapula, and in man its shape is something like that of the beak of a crow; hence its name.
CORALLUM (Lat. for red coral). The hard structures deposited in or by the tissues of any *Actinozoön*, commonly called a "coral."
CORALLITE. The entire coral of a *simple Actinozoön*, or the portion of a *compound* coral which belongs to and is secreted by a single polype.
CORPUS CALLOSUM (Lat. the "firm body.") The great band of nervous matter which unites the two hemispheres of the brain in the *Mammalia*.
CORPUSCULATED (Lat. *corpusculum*, a little body or particle). Applied to fluids which, like the blood, contain floating solid particles or "corpuscles."
CORTICAL LAYER (Lat. *cortex*, bark). The layer of consistent sarcode which encloses the central "abdominal cavity" in the *Infusoria*, and is covered by the "cuticle."
CORYNIDA (Gr. *korune*, a club). An order of Hydroid Zoophytes.
COSTAL (Lat. *costa*, a rib). Connected with the ribs.
CRANIUM (Gr. *kranion*, the skull). The bony or cartilaginous case in which the brain is contained.
CRASPEDA (Gr. *kraspedon*, a margin or fringe). The long convoluted cords, containing thread-cells, which are attached to the free margins of the mesenteries in a Sea-anemone.
CREPUSCULAR (Lat. *crepusculum*, dusk). Applied to animals which are active in the dusk or twilight.
CRINOIDEA (Gr. *krinos*, a lily; *eidos*, form). An order of *Echinodermata*, comprising forms which are usually stalked, and sometimes resemble lilies in shape.
CROCODILIA (Gr. *krokodeilos*, the crocodile). An order of Reptiles.
CROP. A partial dilatation of the gullet, technically called "ingluvies." Sometimes it has the form of a membranous bag appended to the gullet.
CRUSTACEA (Lat. *crusta*, a crust). A class of Articulate animals, comprising Crabs, Lobsters, etc., characterized by having a hard shell or crust, which they cast periodically.
CTENOID (Gr. *kteis*, a comb; *eidos*, form). Applied to those scales of fishes the hinder margins of which are fringed with spines or cut into comb-like projections.
CTENOPHORA (Gr. *kteis*, a comb; *phero*, I carry). An order of *Actinozoa*, comprising oceanic creatures which swim by means of "ctenophores," or bands of cilia arranged in comb-like plates.
CURSORES (Lat. *curro*, I run). An order of Birds, comprising birds in which there is no power of flight, but the body is formed for running vigorously.

CUTICLE (Lat. *cuticula*, diminutive of *cutis*, skin). The transparent pellicle which forms the outer layer of the body in *Infusoria*. The outermost layer of the integument generally.
CYCLOID (Gr. *kuklos*, a circle; *eidos*, form). Applied to those scales of fishes which have a regularly circular or elliptical outline.
CYSTIC (Gr. *kustis*, a bladder). Applied to the embryonic forms of the Tape-worms, which were originally described as a distinct order of Parasitic Worms under the name of *Cystica*, or "Bladder-worms."
CYSTOIDEA (Gr. *kustis*, a bladder; *eidos*, form). An extinct order of *Echinodermata*.

DECAPODA (Gr. *deka*, ten; *podes*, feet). The division of *Crustacea* with ten legs adapted for walking (e. g., Crabs and Lobsters). Also, the family of Cuttle-fishes (*Cephalopoda*) in which the mouth is surrounded by ten processes or "arms."
DECIDUOUS (Lat. *decido*, I fall off). Applied to parts which fall off or are shed during the life of the animal.
DEINOSAURIA (Gr. *deinos*, terrible; *saura*, a lizard). An extinct order of Reptiles.
DENTAL (Lat. *dens*, a tooth). Connected with the teeth.
DENTIROSTRES (Lat. *dens*, a tooth; *rostrum*, beak). The division of the Perching Birds in which the upper mandible of the bill has its lower margin toothed.
DIAPHRAGM (Gr. *diaphragma*, a partition). The "midriff;" or the muscle which in *Mammalia* separates the cavity of the chest from that of the abdomen.
DIBRANCHIATA (Gr. *dis*, twice; *bragchia*, gills). The order of *Cephalopoda* (comprising the Cuttle-fishes, etc.) in which only two gills are present.
DIGIT (Lat. *digitus*, a finger). A finger or toe.
DIGITIGRADA (Lat. *digitus*, a finger; *gradior*, I walk). The division of *Carnivora* in which the animal walks upon the tips of the toes.
DIPNOI (Gr. *dis*, twice; *pnoë*, breath). The order of fishes represented by the *Lepidosiren*.
DIPTERA (Gr. *dis*, twice; *pteron*, a wing). An order of insects characterized by the possession of two wings only.
DISCOPHORA (Gr. *diskos*, a quoit or round plate; *phero*, I carry). The *Medusæ* or Jelly-fishes, so called from their form. Also, the order of the Leeches (*Hirudinea*), from the suctorial disks which they possess.
DISTAL. Applied to the quickly-growing end of the hydrosoma of a *Hydrozoön;* also to the end of a limb, or of any portion of a limb, farthest removed from the trunk.
DIURNAL (Lat. *diurnus*, daily, by day). Applied to animals which are active during the day.
DORSAL (Lat. *dorsum*, the back). Connected with the region of the back.

ECHINODERMATA (Gr. *echinos*, a hedgehog; *derma*, skin). A class of animals comprising the Sea-urchins, Star-fishes, and others, most of which have spiny skins.
ECHINOIDEA (Gr. *echinos*, a hedgehog; *eidos*, form). An order of *Echinodermata*, comprising the Sea-urchins.
ECTOCYST (Gr. *ektos*, outside; *kustis*, a bladder). The external investment of the polypide of a *Polyzoön*.
ECTODERM (Gr. *ektos*, outside; *derma*, skin). The external integumentary layer of the *Cœlenterata*.
EDENTATA (Lat. *e*, without; *dens*, tooth). An order of *Mammalia* in which some or all of the teeth are wanting. Often called *Bruta*.
ELASMOBRANCHII (Gr. *elasma*, a thin plate; *bragchia*, gill). An order of fishes, including the Sharks and Rays.
ELYTRA (Gr. *elutron*, a sheath) The hard chitinous anterior pair of wings in the Beetles, which form protective cases beneath which the posterior membranous wings can be folded.
ENDOCYST (Gr. *endon*, within; *kustis*, a bag). The inner membrane of the polypide of a *Polyzoön*.
ENDODERM (Gr. *endon*, within; *derma*, skin). The inner integumentary layer of the *Cœlenterata*.
ENDOSKELETON (Gr. *endon*, within; *skeletos*, dry). The internal hard structures, such as bones, which serve for the attachment of muscles or the protection of organs, and which are not merely produced by a hardening of the integument.
ENTOMOSTRACA (Gr. *entoma*, insects; *ostrakon*, shell). Literally "Shelled Insects;" applied to a division of the *Crustacea*.
ENTOZOA (Gr. *entos*, within; *zoön*, animal). Animals which live parasitically in the interior of other animals.
EQUILATERAL (Lat. *æquus*, equal; *latus*, side). Having its sides equal. Usually applied to the shells of the *Brachiopoda*.
EQUIVALVE (Lat. *æquus*, equal; *valvæ*, folding-doors). Applied to shells which, like those of most of the *Lamellibranchiata*, are composed of two equal pieces or valves.
ERRANTIA (Lat. *erro*, I wander). An order of *Annelida*, comprising forms which are capable of active locomotion.

EURYPTERIDA (Gr. *eurus*, broad; *pteron*, wing). A group of extinct *Crustacea*.
EXOSKELETON (Gr. *exo*, outside; *skeletos*, dry). Under this term are comprised all structures which are produced by a hardening of the integument.

FEMUR. The thigh-bone, intervening between the pelvis and the bones of the leg proper (*tibia* and *fibula*).
FIBULA (Lat. a brooch). The outermost of the two bones of the leg in the higher *Vertebrata*.
FISSION (Lat. *findo*, I cleave). Multiplication by means of a process of self-division.
FISSIROSTRES (Lat. *fissus*, cleft; *rostrum*, beak). A sub-order of the perching birds, in which the beak can be opened to a great width.
FLAGELLUM (Lat. a whip). The whip-like appendages of certain *Infusoria*, thence called "Flagellate."
FOOT. The muscular organ developed on the under surface of the body in the higher *Mollusca*, and subserving locomotion.
FOOT-JAWS. The limbs of *Crustacea*, which are so modified as to subserve mastication.
FOOT-TUBERCLES. The unjointed lateral appendages of the *Annelida*, which subserve locomotion, but are not articulated to the body.
FORAMINIFERA (Lat. *foramen*, an aperture; *fero*, I carry). An order of *Protozoa*, usually characterized by having a shell perforated by numerous holes or "foramina," through which the pseudopodia are emitted.
FRUGIVOROUS (Lat. *frux*, fruit; *voro*, I devour). Living upon fruits.
FUNNEL. The stomach-sac of the *Ctenophora*; the muscular tube of the Cuttle-fishes by which the water filling the mantle-cavity is ejected.
FURCULUM (Lat. diminutive of *furca*, a fork). The V-shaped bone or "merry-thought" of birds formed by the united clavicles.

GALLINACEI (Lat. *gallina*, a fowl). Often applied to the entire order of the Rasorial birds, but properly restricted to that section of the *Rasores* of which the common fowl is a typical example.
GANGLION (Gr. *gagglion*, a knot). A mass of nervous matter containing nerve-cells and giving origin to nerve-fibres.
GANOID (Gr. *ganos*, splendor; *eidos*, form). Applied to those scales of fishes which are composed of bone, with an outer layer of polished enamel.
GANOIDEI. An order of fishes, now mostly extinct.
GASTEROPODA (Gr. *gaster*, belly; *podes*, feet). The class of the *Mollusca* comprising the ordinary univalves in which locomotion is usually effected by creeping about on a flattened "foot."
GEMMATION (Lat. *gemma*, a bud). The production of fresh structures by a process of budding.
GEPHYREA (Gr. *gephura*, a bridge). A class of the *Anarthropoda*.
GIZZARD. A muscular division of the stomach in insects, birds, etc.
GONOPHORE (Gr. *gonos*, offspring; *phero*, I bear). The generative buds or receptacles of the reproductive elements in the *Hydrozoa*, whether these become detached or not.
GONOSOME (Gr. *gonos*, offspring; *soma*, body). Applied collectively to the assemblage of the reproductive buds of any *Hydrozoön*.
GRALLATORES (Lat. *grallæ*, stilts). The order of the wading birds.
GRANIVOROUS (Lat. *granum*, a grain or seed; *voro*, I devour). Living upon grains or other seeds.
GRAPTOLITIDÆ (Gr. *grapho*, I write; *lithos*, stone). An extinct sub-class of the *Hydrozoa*.
GREGARINIDÆ (Lat. *gregarius*, occurring in numbers together). A class of the *Protozoa*.
GULLET. The tube which leads from the throat to the stomach.

HÆMAL (Gr. *haima*, blood). Connected with the blood-vessels, or with the circulatory system.
HALLUX (Lat. *allex*, the thumb or great-toe). The innermost of the five digits which normally compose the *hind* foot of a Vertebrate animal. The great-toe of man.
HECTOCOTYLUS (Gr. *hekaton*, a hundred; *kotulos*, a cup). One of the "arms" of the male cuttle-fishes, metamorphosed for reproductive purposes.
HEMELYTRA (Gr. *hemi*, half; *elutron*, a sheath). The wing of certain insects (*Hemiptera*), in which the inner portion of the wing is hardened by chitine, and resembles the elytron of a beetle.
HEMIMETABOLIC (Gr. *hemi*, half; *metabole*, change). Applied to those insects which undergo an incomplete metamorphosis.
HEMIPTERA (Gr. *hemi*, half; *pteron*, wing). An order of insects, so called because the anterior wings are sometimes in the form of "hemelytra."
HERMAPHRODITE (Gr. *Hermes*, Mercury; *Aphrodite*, Venus). Possessing the organs of both sexes combined.
HETEROCERCAL (Gr. *heteros*, diverse; *kerkos*, tail). Applied to the tail of fishes when it is unequally lobed.

HETEROGENEOUS (Gr. *heteros*, diverse; *genos*, kind). Unlike or dissimilar in kind.
HETEROMORPHIC (Gr. *heteros*, diverse; *morphe*, shape). Differing in form or shape.
HETEROPODA (Gr. *heteros*, diverse; *podes*, feet). An order of Gasteropodous Mollusks.
HILUM (Lat. *hilum*, a little thing). A small aperture or depression.
HIRUDINEA (Lat. *hirudo*, a horse-leech). The order of *Annelida* comprising the Leeches.
HOLOCEPHALI (Gr. *holos*, whole; *kephale*, head). A sub-order of the *Elasmobranchii*.
HOLOMETABOLIC (Gr. *holos*, whole; *metabole*, change). Applied to those insects which undergo a complete metamorphosis.
HOLOTHUROIDEA (Gr. *holothourion*). The order of *Echinodermata*, comprising the sea-cucumbers.
HOMOCERCAL (Gr. *homos*, same; *kerkos*, tail). Applied to the tail of fishes when the two lobes are equal.
HOMOGENEOUS (Gr. *homos*, same; *genos*, kind). Of like kind or nature.
HOMOLOGOUS (Gr. *homos*, same; *logos*, a discourse). Applied to parts which are constructed upon the same fundamental plan.
HOMOMORPHOUS (Gr. *homos*, same; *morphe*, form). Having a similar external appearance or form.
HUMERUS. The bone of the upper arm in *Vertebrata*.
HYDATIDS (Gr. *hudatis*, a vesicle). The bladder-worm of the tape-worm of the dog.
HYDRA (Gr. *hudra*, a water-serpent). The generic name of the fresh-water polypes.
HYDROIDA (Gr. *hudra;* and *eidos*, form). The sub-class of the *Hydrozoa*, containing the animals most nearly allied to the *Hydra*. Often spoken of as the Hydroid Zoophytes.
HYDRORHIZA (Gr. *hudra;* and *rhiza*, root). The adherent base or proximal extremity of any *Hydroid*.
HYDROSOMA (Gr. *hudra;* and *soma*, body). The entire organism of any *Hydrozoön*.
HYDROTHECA (Gr. *hudra; theke*, a case). The little chitinous cups in which the polypites of the *Sertularida* and *Campanularida* are protected.
HYDROZOA (Gr. *hudra; zoön*, animal). The class of the *Cœlenterata* which comprises animals constructed after the type of the *Hydra*.
HYMENOPTERA (Gr. *humen*, a membrane; *pteron*, a wing). An order of insects (comprising Bees, Ants, etc.), with four membranous wings.
HYOID (Gr. *U; eidos*, form). A bone which supports the tongue, and which derives its name from its resemblance in man to the letter U.
HYRACOIDEA (Gr. *hurax*, a shrew; *eidos*, form). An order of *Mammalia*.

ICHTHYOMORPHA (Gr. *ichthus*, a fish; *morphe*, shape). An order of *Amphibia*, comprising the fish-like Newts, etc.
ICHTHYOPHTHIRA (Gr. *ichthus*, a fish; *phtheir*, a louse). An order of *Crustacea*.
ICHTHYOPSIDA (Gr. *ichthus*, fish; *opsis*, appearance). The primary division of the *Vertebrata*, which includes the two classes of the *Amphibia* and *Pisces*.
ICHTHYOPTERYGIA (Gr. *ichthus*, a fish; *pterux*, a wing). An extinct order of Reptiles.
ICHTHYOSAURIA (Gr. *ichthus*, a fish; *saura*, a lizard). An extinct order of Reptiles.
IMAGO (Lat. an image or apparition). The perfect insect, after it has passed through all its metamorphoses.
INCISOR (Lat. *incido*, I cut). The cutting teeth fixed in the front of the upper jaw, and the corresponding teeth in the lower jaw of the *Mammalia*.
INEQUILATERAL. Having the two sides unequal, as is the case with the shell of the ordinary bivalves (*Lamellibranchiata*).
INEQUIVALVE. Composed of two unequal pieces or valves; as is the case with the shells of the *Brachiopoda*.
INFUSORIA (Lat. *infusum*, an infusion). A class of *Protozoa*, so called from their frequent occurrence in organic infusions.
INSECTA (Lat. *inseco*, I cut into). The class of Articulate animals commonly known as Insects.
INSECTIVORA (Lat. *insectum*, an insect; *voro*, I devour). An order of *Mammalia*.
INSECTIVOROUS. Living upon insects.
INTERAMBULACRA (Lat. *inter*, between; *ambulacrum*, a place to walk in). The rows of plates in an *Echinus*, which are not perforated for the emission of the tube-feet.
INTUSSUSCEPTION (Lat. *intus*, within; *suscipio*, I take up). The act of taking foreign matter into a living being.
INVERTEBRATA (Lat. *in*, without; *vertebra*, a bone of the back). Animals without a spinal column or backbone.
ISOPODA (Gr. *isos*, equal; *podes*, feet). An order of *Crustacea*, in which the feet are equal and like one another.

LABIUM (Lat. for lip). Restricted to the lower lip of Articulate animals.
LABRUM (Lat. for lip). Restricted to the upper lip in Articulate animals.
LABYRINTHODONTIA (Gr. *laburinthos*, a labyrinth; *odous*, a tooth). An extinct order of *Amphibia*.
LACERTILIA (Lat. *lacerta*, a lizard). An order of *Reptilia*, comprising the Lizards and Slow-worms.

Laemodipoda (Gr. *laimos*, throat; *dis*, twice; *podes*, feet). An order of Crustacea, with two legs under the throat.

Lamellibranchiata (Lat. *lamella*, a plate; Gr. *bragchia*, gill). The class of *Mollusca*, comprising the ordinary bivalves with leaf-like gills.

Larva (Lat. a mask). The insect in its first stage after its emergence from the egg, when it is usually very different from the adult.

Lepidoptera (Gr. *lepis*, a scale; *pteron*, a wing). The order of Insects comprising the Butterflies and Moths, characterized by having four wings usually covered by minute scales.

Lingual (Lat. *lingua*, tongue). Connected with the tongue.

Lucernarida (Lat. *lucerna*, a lamp). An order of the *Hydrozoa*.

Lumbar (Lat. *lumbus*, a loin). Connected with the loins.

Macrura (Gr. *makros*, long; *oura*, tail). A tribe of Decapod *Crustaceans* with long tails (e. g., the Lobster, Shrimp, etc.).

Madreporiform. Perforated by small holes like a coral (or *madrepore*); applied to the spongy tubercle by which the water-vascular system of Echinoderms mostly communicates with the exterior.

Malacodermata (Gr. *malakos*, soft; *derma*, skin). Applied to a group of the *Actinozoa*, comprising the soft-skinned Sea-anemones.

Mallophaga (Gr. *mallos*, a fleece; *phago*, I eat). An order of Insects, comprising the Bird-lice.

Mammalia (Lat. *mamma*, the breast). The class of Vertebrate animals which suckle their young.

Mandible (Lat. *mandibulum*, a jaw). The upper pair of jaws in Insects; one of the pairs of jaws in *Crustacea* and Spiders; the beak of *Cephalopoda;* the lower jaw of Vertebrate animals.

Mantle. The external integument of most of the *Mollusca*, which is largely developed and forms a cloak for the internal organs. Technically called the "pallium."

Manubrium (Lat. a handle). The central polypite which is suspended from the roof of the swimming-bell of a *Medusa* or medusiform gonophore among the *Hydrozoa*.

Marsipobranchii (Gr. *marsipos*, a pouch; *bragchia*, gill). An order of Fishes comprising the Hag-fishes and Lampreys, with pouch-like gills.

Marsupialia (Lat. *marsupium*, a pouch). An order of Mammals, in which the females are usually furnished with an abdominal pouch in which the young are carried.

Masticatory (Lat. *mastico*, I chew). Adapted for chewing.

Medusæ. A group of *Hydrozoa*, commonly known as Jelly-fishes, so called because of the resemblance of their tentacles to the snaky hair of the Medusa.

Medusiform. Resembling a *Medusa* in shape.

Medusoid. Like a *Medusa*. Used as a noun to designate the medusiform generative buds (gonophores) of the *Hydrozoa*.

Membrana nictitans (Lat. *nicto*, I wink). The third eyelid present in Birds, etc.

Merostomata (Gr. *mēron*, thigh; *stoma*, mouth). An order of *Crustacea*.

Mesenteries (Gr. *mesos*, intermediate; *enteron*, intestine). The membrane by which the intestine is attached to the walls of the abdomen. In a restricted sense, the vertical plates which divide the somatic cavity of a Sea-anemone into chambers.

Metacarpus (Gr. *meta*, after; *karpos*, the wrist). The bones which form the "root of the hand," and intervene between the wrist and the fingers.

Metamorphosis (Gr. *meta*, denoting change; *morphe*, form). The changes of form which certain animals undergo in passing from their younger to their fully-grown condition.

Metatarsus (Gr. *meta*, after; *tarsos*, the instep). The bones which intervene between the instep (tarsus) and the digits in the hind-foot of the higher Vertebrates.

Molars (Lat. *mola*, a mill). The "grinders" in man; or the teeth in Mammals which are not preceded by milk-teeth.

Mollusca (Lat. *mollis*, soft). The sub-kingdom which includes the true Shell-fish, the *Polyzoa*, the Sea-squirts, and the Lamp-shells.

Molluscoida (*Mollusca*, and Gr. *eidos*, form). The lower division of the *Mollusca*, comprising the *Polyzoa*, Sea-squirts (*Tunicata*), and the Lamp-shells (*Brachiopoda*).

Monothalamous (Gr. *monos*, single; *thalamos*, a chamber). Consisting of only a single chamber. Applied to the shells of *Foraminifera* and *Mollusca*.

Monotremata (Gr. *monos*, single; *trema*, an aperture). The order of Mammals, comprising the Duck-mole and the Spiny Ant-eater, in which the intestinal canal opens into a "cloaca" common to the ducts of the urinary and generative organs.

Morphology (Gr. *morphe*, form; *logos*, discourse). The science of the external *form* and internal *structure* of the various parts and organs of different animals.

Multivalve (Lat. *multus*, many; *valvæ*, folding-doors). Applied to shells which are composed of more than two pieces or valves.

Myriapoda (Gr. *murios*, countless; *pous*, foot). A class of Articulate animals comprising the Centipedes and their allies, characterized by the possession of numerous feet.

Natatores (Lat. *nare*, to swim). The order of the Swimming Birds.

NATATORY. Adapted for swimming.
NAUTILOID. Shaped like the shell of the *Nautilus*.
NECTOCALYCES (Gr. *necho*, I swim; *kalux*, a cup). The swimming-bells of the Oceanic *Hydrozoa*.
NEMATODA or NEMATOIDEA (Gr. *nema*, a thread; *eidos*, form). The division of the *Scolecida*, comprising the Thread-worms and Round-worms.
NERVURES (Lat. *nervus*, a sinew). The ribs which support the membranous wings of insects.
NEURAL (Gr. *neuron*, a nerve). Connected with the nervous system.
NEUROPTERA (Gr. *neuron*, a nerve; *pteron* a wing). An order of Insects, in which the wings are membranous and have numerous interlacing nervures.
NOCTURNAL (Lat. *nox*, night). Applied to animals which are active at night.
NORMAL (Lat. *norma*, a rule). Conforming to the ordinary standard.
NOTOCHORD (Gr. *notos*, the back; *chordé*, a string). A cellular rod which is developed in the embryo of Vertebrates immediately beneath the spinal cord, and which is usually replaced in the adult by the vertebral column.
NUCLEOLUS. The minute solid particle found in the interior of the nucleus of some cells. Also, the minute body attached to the exterior of the "nucleus" of certain of the *Infusoria*.
NUCLEUS (Lat. a kernel). The solid or vesicular body found in the interior of many cells; also, the solid rod or band-shaped body found in the interior of many of the *Protozoa*.
NUDIBRANCHIATA (Lat. *nudus*, naked; Gr. *bragchia*, gill). An order of the *Gasteropoda* in which the gills are naked.

OCEANIC. Applied to animals which inhabit the open ocean.
OCTOPODA (Gr. *okto*, eight; *pous*, foot). The tribe of Cuttle-fishes with eight arms round the head.
ODONTOPHORE (Gr. *odous*, a tooth; *phero*, I carry). The so-called "tongue" or masticatory apparatus of the *Gasteropoda*, *Pteropoda*, and *Cephalopoda*.
ŒSOPHAGUS. The gullet, or the tube by which the food is conveyed from the mouth to the stomach.
OLIGOCHAETA (Gr. *oligos*, few; *chaite*, bristles). An order of the *Annelida* comprising the Earth-worms.
OMASUM (Lat. bullock's tripe). The third stomach of Ruminants, commonly called the "psalterium."
OMNIVOROUS (Lat. *omnia*, every thing; *voro*, I devour). Feeding indiscriminately upon all kinds of food.
OPERCULUM (Lat. a lid). The shelly or horny plate by which the shell of a Univalve Mollusk is closed when the animal has retired within it; also, the chain of flat bones which covers the gills in many fishes.
OPHIDIA (Gr. *ophis*, a serpent). The order of Reptiles comprising the Snakes.
OPHIOMORPHA (Gr. *ophis*, a serpent; *morphe*, shape). The order of *Amphibia*, comprising the *Cæciliæ*.
OPHIUROIDEA (Gr. *ophis*, a serpent; *oura*, tail; *eidos*, form). An order of *Echinodermata*, comprising the Brittle-stars and Sand-stars.
OPISTHOCŒLOUS (Gr. *opisthen*, behind; *koilos*, hollow). Applied to vertebræ, the bodies of which are hollow or concave behind, and convex in front.
ORAL (Lat. *os*, the mouth). Connected with the mouth.
ORTHOPTERA (Gr. *orthos*, straight; *pteron*, a wing). An order of Insects.
OSCULA (Lat. diminutive of *os*, mouth). The large apertures ("exhalant apertures") by which a sponge is perforated.
OSTRACODA (Gr. *ostrakon*, a shell; *eidos*, form) An order of small Crustaceans which are enclosed in bivalve shells.
OVIPAROUS (Lat. *ovum*, an egg; *pario*, I bring forth). Applied to animals which bring forth eggs, in contradistinction to those which bring forth their young alive.
OVIPOSITOR (Lat. *ovum*, an egg; *pono*, I place). The organ possessed by some insects, by means of which the eggs are placed in a position suitable for their development.
OVISAC. The external bag or sac in which certain of the *Invertebrata* carry their eggs after they are extruded from the body.
OVO-VIVIPAROUS (Lat. *ovum*, an egg; *vivus*, alive; *pario*, I bring forth). Applied to animals which retain their eggs within the body until they are hatched.
OVUM (Lat. an egg). The germ produced within the ovary, and capable under certain conditions of being developed into a new individual.

PACHYDERMATA (Gr. *pachus*, thick; *derma*, skin). An old Mammalian order, constituted by Cuvier for the reception of the Elephant, Rhinoceros, Hippopotamus, etc.
PALLIAL. Connected with the mantle or "pallium."
PALLIUM (Lat. a cloak). The "mantle" of the *Mollusca*.
PALPI (Lat. *palpo*, I touch). Processes, supposed to be organs of touch, developed from certain of the organs of the mouth in various Articulate animals, and from the sides of the mouth in the Bivalve Mollusks.

PAPILLA (Lat. for nipple). A minute soft prominence.
PARAPODIA (Gr. *para*, beside; *podes*, feet). The lateral locomotive processes or "foot-tubercles" of many of the *Annelida*.
PARIETAL (Lat. *paries*, a wall). Connected with the walls of a cavity or of the body.
PARIETO-SPLANCHNIC (Lat. *paries*, a wall; Gr. *splagchnon*, an internal organ). Applied to one of the nervous ganglia of the *Mollusca* which supplies the walls of the body and the viscera.
PATAGIUM (Lat. the border of a dress). Applied to the expansion of the integument by which Bats, Flying-Squirrels, and other animals support themselves in the air.
PECTORAL (Lat. *pectus*, the breast). Connected with the chest.
PEDAL (Lat. *pes*, the foot). Connected with the foot; generally used in connection with the *Mollusca*.
PEDICELLARIÆ (Lat. *pedicellus*, a louse). Curious appendages found in many *Echinoderms*, attached to the surface of the body, and resembling a little pair of pincers supported on a stalk.
PEDIPALPI (Lat. *pes*, foot; *palpo*, I feel). An order of *Arachnida*, comprising the Scorpions, etc.
PEDUNCULATED (Lat. *pedunculus*, a stem). Supported upon a stem or stalk.
PELVIS (Lat. basin). The bony arch with which the lower or posterior pair of limbs is connected in many *Vertebrata*.
PERENNIBRANCHIATA (Lat. *perennis*, perpetual; Gr. *bragchia*, gills). Applied to those *Amphibia* which retain their gills throughout life.
PERISSODACTYLA (Gr. *perissos*, uneven; *daktulos*, finger). Applied to those hoofed Mammals (*Ungulata*) which have an uneven number of toes.
PERIVISCERAL (Gr. *peri*, around; Lat. *viscera*, the internal organs). Applied to the space surrounding the viscera in most animals.
PHALANGES (Gr. *phalanx*, a row). The small bones composing the digits of the higher Vertebrates. Normally each digit has three phalanges.
PHARYNGOBRANCHII (Gr. *pharugx*, the pharynx; *bragchia*, gills). The order of Fishes comprising only the Lancelet.
PHARYNX. The upper part of the gullet.
PHYLLOPODA (Gr. *phullon*, leaf; *podes*, feet). An order of *Crustacea*.
PHYSOPHORIDÆ (Gr. *phusa*, bellows or air-bladder; *phero*, I carry). An order of Oceanic *Hydrozoa*.
PINNÆ (Lat. *pinna*, a feather). Lateral processes, applied especially to the processes of the arms of Crinoids, or of the tentacles of *Alcyonaria*.
PINNIGRADA (Lat. *pinna*, and *gradior*, I walk). The family of *Carnivora*, comprising the Seals and Walruses, adapted for an aquatic life.
PISCES (Lat. *piscis*, a fish). The class of the *Vertebrata* comprising the Fishes.
PLACENTA (Lat. a cake). The "after-birth," or the organ by which a vascular connection is established in the higher *Vertebrata* between the mother and the young animal previous to its birth.
PLACENTAL. Possessing a placenta, or connected with the placenta.
PLACOID (Gr. *plax*, a plate; *eidos*, form). Applied to the irregular bony plates, grains, or spines, which are found in the skin of various fishes.
PLAGIOSTOMI (Gr. *plagios*, transverse; *stoma*, mouth). The Sharks and Rays, in which the mouth is transverse, and is placed on the under surface of the head.
PLANARIDA (Gr. *planē*, wandering). A group of the *Scolecida*.
PLANTIGRADA (Lat. *planta*, the sole of the foot; *gradior*, I walk). The group of the *Carnivora* in which the sole of the foot is applied to the ground.
PLANTIGRADE. Walking upon the soles of the feet.
PLASTRON (Gr. *emplastron*, a plaster). The ventral shield of the case of the Tortoises and Turtles (*Chelonia*).
PLATYRHINA (Gr. *platus*, broad; *rhines*, nostrils). A group of the *Quadrumana*.
PLESIOSAURIA (Gr. *plesios*, near; *saura*, a lizard). An extinct order of Reptiles.
PLUTEUS (Lat. a shed). The larval form of the Sea-urchins (*Echinus*).
PNEUMATIC (Gr. *pneuma*, air). Filled with air.
PNEUMATOPHORE (Gr. *pneuma*, air; *phero*, I carry). The air-bladder of the *Physophoridæ*.
PODOSOMATA (Gr. *pous*, foot; *soma*, body). An order of *Arachnida*.
POISERS. See Balancers.
POLLEX (Lat. the thumb). The innermost of the normal five digits of the fore-foot of the higher *Vertebrata*. The thumb of man.
POLYCYSTINA (Gr. *polus*, many; *kustis*, a bladder). An order of *Protozoa*.
POLYGAMOUS (Gr. *polus*, many; *gamos*, marriage). Applied to cases in which one male consorts with several females.
POLYGASTRICA (Gr. *polus*, many; *gaster*, stomach). The name applied by Ehrenberg to the *Infusoria*, under the belief that they possessed many stomachs.
POLYPARY. The hard, chitinous covering secreted by many of the *Hydrozoa*.
POLYPE (Gr. *polus*, many; *pous*, foot). Restricted in modern usage to the single individual of a *simple Actinozoön*, or to the separate zoöids of a *compound Actinozoön*.
POLYPIDE. The separate zoöid of a *Polyzoön*.

POLYPIDOM. Synonymous with polypary, but often applied to the *Polyzoa*, as well as to the *Hydrozoa*.
POLYPITE. The separate zoöid of a *Hydrozoön*.
POLYTHALAMOUS (*polus*, many; *thalamos*, chamber). Many-chambered. Applied to the shells of *Foraminifera* and *Cephalopoda*.
POLYZOA (Gr. *polus*, many; *zoön*, animal). The Sea-mosses and Sea-mats, forming the lowest class of the *Mollusca*.
PRÆMOLARS (Lat. *præ*, before; *molares*, grinders). The molar teeth which succeed the molars of the milk set of teeth.
PRÆ-ŒSOPHAGEAL. Situated in front of the gullet.
PROBOSCIDEA (Lat. *proboscis*, the snout). The order of Mammals comprising the Elephants.
PROCŒLOUS (Gr. *pro*, front; *koilos*, hollow). Applied to vertebræ, the bodies of which are hollow or concave in front.
PROTOPHYTA (Gr. *protos*, first; *phyton*, plant). The lowest division of plants.
PROTOPLASM (Gr. *protos*, first; *plasso*, I mould). The elementary basis of organized tissues. Sometimes used as identical with the "sarcode" of the *Protozoa*.
PROTOZOA (Gr. *protos*, first; *zoön*, animal). The lowest division of the animal kingdom.
PROXIMAL (Lat. *proximus*, next). The slowly-growing, comparatively fixed extremity of a limb or of an organism.
PSALTERIUM (Lat. a stringed instrument). The third stomach of the Ruminants.
PSEUDOHÆMAL (Gr. *pseudos*, falsity; *haima*, blood). The vascular system of the *Annelida*.
PSEUDO-HEARTS. Contractile cavities connected with the reproductive system of the *Brachiopoda*.
PSEUDOPODIA (Gr. *pseudos*, falsity; *podes*, feet). The temporary extensions of the body-substance which are put forth by the *Rhizopoda* at will, and which serve both for locomotion and for prehension.
PTEROPODA (Gr. *pteron*, a wing; *podes*, feet). A class of *Mollusca* swimming by means of fins attached to the sides of the head.
PTEROSAURIA (Gr. *pteron*, a wing; *saura*, a lizard). An extinct order of Reptiles.
PULMONARY (Lat. *pulmo*, a lung). Connected with the lungs.
PULMONATE. Possessing lungs.
PUPA (Lat. a doll). The state of metamorphosis of an insect immediately preceding its appearance in a perfect condition. In this state the insect is very often motionless, and is often called a "chrysalis."
PYLORUS (Gr. *puloros*, a gate-keeper). The valvular aperture between the stomach and the commencement of the intestine.

QUADRUMANA (Lat. *quatuor*, four; *manus*, hand). The order of *Mammalia* comprising the Monkeys, Baboons, Lemurs, etc.
QUADRUMANOUS. Four-handed.

RADIATA (Lat. *radius*, a ray). Formerly applied to a large number of animals which are now placed in separate sub-kingdoms (e. g., the *Cœlenterata*, *Echinodermata*, *Infusoria*, etc.).
RADIOLARIA (Lat. *radius*, a ray). An order of *Rhizopoda*.
RADIUS. The innermost of the two bones of the forearm of the higher Vertebrates. It carries the thumb or pollex, and corresponds with the tibia of the hind-limb.
RAPTORES (Lat. *rapto*, I plunder). The order of the Birds of Prey.
RAPTORIAL. Applied to animals which live by preying upon other animals.
RASORES (Lat. *rado*, I scrape or scratch). The order of the Scratching Birds (Fowls, Pigeons, etc.).
REPTILIA (Lat. *repto*, I crawl). The class of the *Vertebrata* comprising the Tortoises, Serpents, Lizards, Crocodiles, etc.
RETICULUM (Lat. a net). The second stomach of the Ruminants.
RHIZOPODA (Gr. *rhiza*, a root; *podes*, feet). The division of *Protozoa* comprising all those which are capable of emitting pseudopodia.
RODENTIA (Lat. *rodo*, I gnaw). An order of the *Mammalia*. Often called *Glires* (Lat. *glis*, a dormouse).
ROTIFERA (Lat. *rota*, a wheel; *fero*, I carry). A class of the *Scolecida*, comprising the so-called "Wheel-animalcules."
RUGOSA (Lat. *rugosus*, wrinkled). An extinct order of Corals.
RUMEN (Lat. the throat). The first stomach or "paunch" of Ruminants.
RUMINANTIA (Lat. *ruminor*, I chew the cud). A group of the Hoofed *Mammalia*.

SACRUM. The vertebræ which unite with the haunch-bones to form the pelvis.
SARCODE (Gr. *sarx*, flesh; *eidos*, form). The jelly-like substance of which the bodies of the *Protozoa* are composed.
SARCOID. The separate Amœba-like particles which collectively make up the "flesh" of a sponge.

SAUROPSIDA (Gr. *saura*, lizard; *opsis*, appearance). The name given by Huxley to the two classes of the Reptiles and Birds collectively.
SAUROPTERYGIA (Gr. *saura*, a lizard; *pterux*, a wing). An extinct order of Reptiles.
SAURURÆ (Gr. *saura*, a lizard; *oura*, tail). The order of Birds comprising only the extinct *Archæopteryx*.
SCANSORES (Lat. *scando*, I climb). The order of the Climbing Birds (Parrots, Woodpeckers, etc.).
SCAPULA. The shoulder-blade of *Vertebrata*.
SCLEROBASIC (Gr. *skleros*, hard; *basis*, pedestal). The form of coral which constitutes a central axis surrounded by the soft parts of the animal (e. g., Red Coral).
SCLERODERMIC. Applied to those corals which are secreted within the body of the polypes which produce them.
SCOLECIDA (Gr. *skolex*, a worm). A division of the *Annuloida*.
SCUTA (Lat. *scutum*, a shield). Applied to the shield-like integumentary plates developed in many Reptiles.
SEPTA. Partitions.
SERPENTIFORM. Resembling a serpent in shape.
SERTULARIDA (Lat. *sertum*, a wreath). An order of the *Hydrozoa*.
SESSILE (Lat. *sedo*, I sit). Not supported upon a stalk, but attached by a base.
SILICEOUS (Lat. *silex*, flint). Composed of flint.
SIPHON (Gr. a tube). Applied to the respiratory tubes of many of the *Mollusca;* also to other tubes of different functions.
SIPHONOPHORA (Gr. *siphon*, a tube; *phero*, I carry). A sub-class of the *Hydrozoa*.
SIRENIA (Gr. *seiren*, a mermaid). An order of *Mammalia*, comprising the Dugongs and Manatees.
SOLIDUNGULA (Lat. *solidus*, solid; *ungula*, a hoof). The group of Hoofed Mammals comprising the Horse, Ass, and Zebra.
SOMATIC (Gr. *soma*, body). Connected with the body.
SOMITE (Gr. *soma*). A single segment in the body of an Articulate animal.
SPICULA (Lat. *spiculum*, a point). Pointed needle-shaped bodies.
SPINNERETS. The organs by means of which Spiders and Caterpillars spin threads.
SPONGIDA (Gr. *spoggos*, a sponge). The division of the *Protozoa* commonly known as sponges.
STERNUM (Gr. *sternon*). The breastbone.
STOMAPODA (Gr. *stoma*, mouth; *podes*, feet). An order of *Crustacea*.
STOMATODE (Gr. *stoma*, mouth). Possessing a mouth. The *Infusoria* are thus often called the Stomatode *Protozoa*.
STREPSIPTERA (Gr. *strepho*, I twist; *pteron*, a wing). An order of Insects in which the anterior wings are represented by twisted rudiments.
STREPSIRHINA (Gr. *strepho*, I twist; *rhines*, nostrils). A group of the *Quadrumana*.
SUCTORIAL. Adapted for suction or for imbibing fluids.
SUPRA-ŒSOPHAGEAL. Placed above the gullet or œsophagus.

TABULÆ (Lat. *tabula*, a tablet). Horizontal plates or floors which are found in many corals.
TACTILE (Lat. *tango*, I touch). Connected with the sense of touch.
TÆNIADA (Gr. *tainia*, a ribbon). The order of *Scolecida* comprising the Tape-worms.
TARSO-METATARSUS. The single bone produced in Birds by the union and anchylosis of the lower part of the tarsus with the metatarsus.
TARSUS (Gr. *tarsos*, the flat of the foot). The small bones which form the ankle (or "instep" of man); corresponding with the *carpus* of the anterior limb.
TELEOSTEI (Gr. *teleios*, perfect; *osteon*, bone). An order of Fishes, often spoken of as the Bony Fishes.
TELSON (Gr. a limit). The last joint in the abdomen of the *Crustacea*.
TENUIROSTRES (Lat. *tenuis*, slender; *rostrum*, beak). A group of the Perching Birds characterized by their slender beaks.
TERRESTRIAL (Lat. *terra*, earth). Living upon dry land.
TEST (Lat. *testa*, a shell). The shell of *Mollusca*, which are for this reason sometimes called *Testacea*. Also, the calcareous shell of Sea-urchins. Also, the thick, leathery outer tunic of the Sea-squirts (*Tunicata*).
TETRABRANCHIATA (Gr. *tetra*, four; *bragchia*, gill). The order of *Mollusca* characterized by the possession of four gills.
THALASSICOLLIDA (Gr. *thalassa*, sea; *kolla*, glue). A division of *Protozoa*.
THERIOMORPHA (Gr. *ther*, beast; *morphe*, shape). Employed by Owen to designate the "tailless Amphibians," such as Frogs and Toads.
THORAX (Gr. a breast-plate). The chest.
THYSANURA (Gr. *thusanoi*, fringes; *oura*, tail). An order of Insects.
TIBIA. The shin-bone, corresponding to the *radius* of the fore-limb, and being the innermost of the two bones of the leg.
TRACHEA (Gr. *tracheia*, the rough windpipe). The tube which conveys air to the lungs in the air-breathing Vertebrates. In Insects, Myriapods, and Spiders, the air-tubes which ramify through the body.

TREMATODA (Gr. *trema*, a pore or hole). An order of *Scolecida*.
TRILOBITA (Gr. *treis*, three; *lobos*, a lobe). An extinct order of *Crustacea*.
TUBICOLA (Lat. *tuba*, a tube; *colo*, I inhabit). An order of *Annelida*.
TUBULARIDA (Lat. *tuba*, a tube). Often used instead of *Corynida* to designate an order of the *Hydrozoa*.
TUNICATA (Lat. *tunica*, a cloak). The Sea-squirts, a class of the *Molluscoida*.

ULNA (Gr. *olené*, the elbow). The outermost of the two bones of the fore-arm, corresponding with the *fibula* of the hind-limb.
UMBO (Lat. the boss of the shield). The beak of a bivalve shell.
UMBRELLA. The contractile disk of one of the *Lucernarida*.
UNGULATA (Lat. *ungula*, a hoof). The order of *Mammalia* comprising the Hoofed Quadrupeds.
UNIVALVE (Lat. *unus*, one; *valvæ*, folding-doors). Applied to shells composed of a single piece or valve.
URODELA (Gr. *oura*, tail; *dēlos*, visible). The order of the "tailed" Amphibians.

VACUOLES (Lat. *vacuus*, empty). The little cavities formed in the interior of many of the *Protozoa* by the presence of the particles of food surrounded by a little water.
VENTRAL (Lat. *venter*, the stomach). Relating to the inferior surface of the body.
VENTRICLE (Lat. *ventriculus*, diminutive of *venter*, belly). One of the cavities of the heart.
VERMES (Lat. *vermis*, a worm). Sometimes used at the present day in the same, or nearly the same, sense as *Annuloida*, or as *Annuloida* plus the *Anarthropoda*.
VERMIFORM. Worm-like in shape.
VERTEBRA (Lat. *verto*, I turn). One of the bones composing the spinal column or backbone.
VERTEBRATA. The sub-kingdom comprising animals, almost all of which have a more or less well-developed vertebral column.
VESICLE (Lat. *vesica*, a bladder). A little sac, bladder, or cyst.
VISCERA (Lat. *viscus*). The internal organs of the body.
VIVIPAROUS (Lat. *vivus*, alive; *pario*, I bring forth). Applied to anima's which bring forth their young alive.

XIPHOSURA (Gr. *xiphos*, sword; *oura*, tail). An order of *Crustacea*, comprising the King-Crabs.

ZOÖID (Gr. *zoön*, animal; *eidos*, form). The more or less completely independent organisms produced from a primitive being by gemmation or fission, whether these remain attached to one another or are detached and set free.
ZOOPHYTE (Gr. *zoön*, animal; *phuton*, plant). Loosely applied to many plant-like animals, such as Sponges, Corals, Sea-anemones, Sea-firs, Sea-mats, etc.

QUESTIONS.

1. MENTION some of the characters of living beings.
2. What is understood by "organization?"
3. Define Biology and Zoology.
4. What characters separate the higher animals from the higher plants?
5. How does the nutrition of plants differ from that of animals?
6. What is understood by "classification?"
7. What is the basis of a natural and scientific classification?
8. Explain the terms "morphology" and "physiology."
9. What is understood by "sub-kingdoms," and upon what characters are these founded?
10. What are the great Physiological functions? Define these.
11. Explain the terms "homology" and "analogy," and give examples.
12. What leading characters separate the Invertebrate from the Vertebrate animals.
13. What are the chief characters of the *Protozoa?*
14. What is sarcode?
15. What are cilia, flagella, and pseudopodia?
16. Mention the three great classes of *Protozoa.*
17. What is the structure of a *Gregarina*, and where would you expect to find one?
18. What structures characterize the *Rhizopoda?*
19. Describe an *Amœba.*
20. What is the so-called "contractile vesicle?"
21. What is meant by "fission?"
22. How do the pseudopodia of the *Foraminifera* differ from those of an *Amœba?*
23. What structures are absent in the *Foraminifera*, which occur in the *Amœba?*
24. What is the nature of the shell of the *Foraminifera?*
25. What differences subsist between a perforated and an imperforate shell? Between a simple and a compound shell?
26. Where do *Foraminifera* mostly occur?
27. What is understood by "Distribution in Space" and "Distribution in Time?"
28. Mention one or two remarkable fossil *Foraminifera.*
29. What is Chalk to a great extent composed of?
30. What is the nature of the skeleton of the *Radiolaria?*
31. Mention some example of the *Radiolaria.*

32. Of what two essential elements is a Sponge composed?
33. What is the nature of the "sponge-flesh?"
34. Describe the circulation of water in a Sponge.
35. What are the chief variations in the skeleton of Sponges?
36. Whence are the Sponges of commerce obtained?
37. How do the Infusorian Animalcules derive their name?
38. By what leading character are the *Infusoria* distinguished from the other Protozoa?
39. Why were the *Infusoria* formerly called *Polygastrica?*
40. Describe the Bell-animalcule.
41. What peculiarity of the digestive system characterizes the sub-kingdom *Cœlenterata?*
42. Of what is the body of a Cœlenterate animal composed?
43. What is a "thread-cell?"
44. Into what two classes are the *Cœlenterata* divided, and what characters distinguish these? Mention examples of each.
45. What is understood by "gemmation?" How is a compound animal or colony produced?
46. What is meant by the term "zoöid?"
47. What is scientifically understood by the term "individual?"
48. Define the terms "polypite," "cœnosarc," and "polypary."
49. Give examples of the Hydroid Zoophytes.
50. Describe shortly the structure of the *Hydra.*
51. What is the method of reproduction in the *Hydra?*
52. What peculiarities distinguish the polypary of the *Corynida?*
53. Give an example of the *Corynida.*
54. What two sets of zoöids go to form the colony of a Hydroid Zoophyte?
55. Define the terms "trophosome" and "gonosome."
56. What is a "gonophore?"
57. What is a "medusiform gonophore?" Why is it so called, and what is its general structure?
58. Give an example of the *Sertularida,* and mention the differences which distinguish their polypary from that of the *Corynida.*
59. What is a "hydrotheca?"
60. Describe the general structure of the reproductive bud of a Campanularian.
61. How do the Oceanic Hydrozoa differ from the Hydroid Zoophytes?
62. What is a "nectocalyx," and what is its structure?
63. What is a "bract?"
64. Mention examples of the Oceanic Hydrozoa.
65. What is the "float" or "pneumatophore" of the *Physophoridæ?*
66. Of what real nature are many of the so-called *Medusæ* or Jelly-fishes?
67. What is the general structure of a *Medusa,* and with what structure in the Hydroid Zoophytes does it agree?
68. From what circumstance is the name "naked-eyed" *Medusæ* derived?
69. What are the "marginal bodies" of the *Medusæ?*
70. Describe *Lucernaria.*
71. Of what nature are the great Sea-blubbers?
72. Describe "Hydra-tuba" and its development.
73. What is the structure of a "Hidden-eyed Medusa" or Sea-blubber, and from what circumstance is the former name derived?
74. Differences between the naked-eyed and hidden-eyed Medusæ?
75. Describe the generative bud of *Rhizostoma.*

76. Give the leading characters of the *Actinozoa.*
77. How does the transverse section of a *Hydrozoön* differ from that of an *Actinozoön?*
78. What is a "polype?"
79. Describe a Sea-anemone.
80. What are the "mesenteries" of a Sea-anemone, and what organs do they carry?
81. What is a "coral?"
82. What are the "septa" of a coral, and to what part of the living animal do they correspond?
83. What are coral-reefs, where do they occur most abundantly, and what are the chief varieties which occur?
84. How do the *Alcyonaria* differ numerically from the *Zoantharia?*
85. Mention some example of the *Alcyonaria.*
86. In what *Alcyonarian* is there a well-developed *sclerodermic* coral?
87. Of what nature is the *sclerobasic* coral of the *Gorgonidæ?*
88. Mention a well-known example of the *Gorgonidæ.*
89. Give the leading characters of the *Ctenophora.*
90. What is a "ctenophore?"
91. Is a nervous system present in any of the *Actinozoa*, except in the *Ctenophora?*
92. Mention an example of the *Ctenophora.*
93. What animals belong to the *Echinodermata*, and whence is the name of the class derived?
94. What is understood by "bi-lateral symmetry?"
95. At what time of life are the *Echinoderms* bi-laterally symmetrical, and what is their condition in this respect when adult?
96. What is a water-vascular system, what is it called in this class, and what special function does it generally discharge?
97. What is the arrangement of the nervous system in *Echinoderms?*
98. What distinguishes the Sea-urchins from other Echinoderms?
99. Into how many zones may the test of a living Sea-urchin be divided, and how many rows of plates are contained in each zone?
100. What are the "ambulacral" and "inter-ambulacral areas?"
101. What plates are always placed at the summit of the shell?
102. What is the "madreporiform tubercle?"
103. Describe the general nature of the spines and their function.
104. What are "pedicellariæ?"
105. What are the "tube-feet?" Describe the general arrangement of the "ambulacral system."
106. What is the structure of the circulatory and nervous organs in the *Echinus?*
107. Mention a peculiarity in the development of the Echinus.
108. Give the leading characters of the Star-fishes.
109. What peculiarities distinguish the arms of Star-fishes?
110. What is the structure of the stomach in Star-fishes?
111. How do the Brittle-stars resemble the true Star-fishes, and how are they distinguished?
112. What is the structure of the digestive system in Brittle-stars?
113. What is the essential peculiarity of the *Crinoids?*
114. Mention a living Crinoid which is free when adult, and one which is permanently fixed.
115. What is the general shape of the *Holothurians*, and the nature of their integuments?

116. What is the condition of the ambulacral system, and where is the "madreporiform tubercle" situated?

117. What is the mouth surrounded by?

118. What is the "respiratory tree" of the Holothurians?

119. What characters distinguish the *Scolecida?* How are they separated from the *Echinoderms?*

120. What is meant by the term *Entozoa?*

121. In what relation do the Bladder-worms stand to the Tape-worms?

122. What is the structure of an ordinary Tape-worm?

123. Give the structure of the "head" and of a single joint, and state what is the relation of the head to the joints.

124. State shortly the process of development in a Tape-worm.

125. What is the "measles" of the Pig?

126. What are "hydatids" in man?

127. What are the characters of the *Trematode* worms? Mention an example.

128. What is the "rot" of Sheep caused by?

129. What groups of animals are included in the *Turbellaria?*

130. Give an example of the *Acanthocephala*, and state the character from which the name is derived.

131. Where do the *Gordiacea* spend the earlier part of their existence, and what is their common name?

132. Mention examples of the *Nematode* worms.

133. From what do the Wheel-animalcules get their name?

134. What is the general size of the Rotifers, and where are they found?

135. What marked differences are there between the males and females?

136. What are the functions of the ciliated "wheel?"

137. Give the general anatomy of a Rotifer.

138. Give the leading characters of an *Annulose* animal.

139. Into what great divisions is the sub-kingdom *Annulosa* divided, and what are the characters of these?

140. What is the general structure of one of the rings of an Annelide?

141. What is the "pseudohæmal system" of the *Annelida*, and to what is it believed to correspond?

142. What are the characters of the *Hirudinea?*

143. To what does the Medicinal Leech owe its value?

144. How is locomotion effected in the Leeches?

145. How are the *Oligochæta* distinguished?

146. What are the locomotive organs of the Earth-worm?

147. Of what nature are the breathing-organs of the *Tubicola*, and where are they placed?

148. Mention a common Tubicolous Annelide.

149. To what do the *Errantia* owe their name, and what are their locomotive organs?

150. Where are the gills placed in the *Errantia?*

151. What orders of *Annelida* possess gills, and which have not?

152. Give the general characters of *Articulate* animals.

153. Give the characters of the *Crustacea.*

154. How many segments go to the body of a Crustacean, and into what distinct regions may these be distributed?

155. What is understood by the "cephalothorax?"

156. To what section of *Crustacea* does the Lobster belong?

157. What is the "carapace" of the Lobster?

158. What is the part generally called the "tail," and what is the so-called "head?"
159. What are the "antennæ," and how many are there in the Lobster and in *Crustacea* generally?
160. What are "foot-jaws," and why are they so called?
161. What are "chelæ?"
162. Of what nature are the appendages of the abdomen in the Lobster?
163. What is the last segment of the abdomen called?
164. Describe the gills of the Lobster. Where are they placed?
165. Of what nature is the abdomen of the Hermit-crabs?
166. How are the Crabs distinguished from the Lobsters?
167. By what character does the young Crab approach the Lobster?
168. Give an example of the *Isopoda?*
169. What is the character of the appendages of the mouth in the King-Crabs?
170. What is the structure of a Trilobite?
171. What is the nature of the shell of the Ostracode Crustaceans?
172. What change do the Cirripedes undergo in passing from the larval to the adult condition?
173. What are the two types of the Cirripedes? Give examples.
174. Give the general characters of the *Arachnida?*
175. What is the structure of the mandibles of the Spiders? Of the Scorpions?
176. To what do the mandibles of the *Arachnida* correspond?
177. Of what nature are the breathing-organs of the *Arachnida?*
178. What is the structure of *tracheæ?* Of pulmonary sacs?
179. What are the organs of vision in the *Arachnida?*
180. What are the habits of the Mites? Give examples.
181. By what structure do the Scorpions inflict wounds?
182. What is the condition of the abdomen in Scorpions? In Spiders?
183. By means of what organs do the Spiders spin webs?
184. What are the general characters of the *Myriapoda?*
185. What is the general condition of the young Myriapod?
186. What are the distinctions between the Centipedes and Millipedes?
187. What is the number of legs in *Pauropus?*
188. What are the general characters of Insects?
189. What organs are carried by the head in Insects?
190. How many segments form the thorax, and what appendages do they always carry? What appendages do they sometimes carry?
191. What are "nervures?"
192. How many rings generally go to the abdomen of Insects? What appendages (if any) do these support?
193. What are the chief modifications in the organs of the mouth in Insects?
194. Describe the digestive system of an Insect?
195. How is the circulation carried on?
196. Of what nature are the breathing-organs?
197. Of what nature are the eyes in Insects?
198. What is understood by the "metamorphosis" of an Insect?
199. What are the chief differences in the metamorphoses of Insects?
200. What peculiarity distinguishes the adult state of Insects which undergo no metamorphosis?
201. What is understood by the terms "larva," "pupa," and "imago?"
202. What is a "chrysalis?" A "cocoon?"

203. What are the more important Insects which pass through no metamorphosis?
204. The chief characters of the *Hemiptera?* Give examples.
205. What are "hemelytra?"
206. The chief characters of the *Orthoptera?* Give examples.
207. The chief characters of the *Neuroptera?* Give examples.
208. What members compose a colony of White Ants or Termites?
209. The chief characters of the *Aphaniptera?* Give an example.
210. The chief characters of the *Diptera?* Give examples.
211. The chief characters of the *Lepidoptera?*
212. Characters of the larvæ of *Lepidoptera?*
213. What characters distinguish Butterflies and Moths respectively?
214. Chief characters of *Hymenoptera?* Give examples.
215. Give some account of the social communities of Bees and Ants?
216. What is the condition of the wings in *Strepsiptera?*
217. Chief characters of *Coleoptera?* Give examples.
218. What are "elytra?"
219. Mention a useful Beetle.
220. Chief characters of the *Mollusca?*
221. Condition of nervous system in Mollusks? Of circulatory system? Of breathing-organs? Of digestive system?
222. Primary divisions of *Mollusca*, and the characters of these?
223. Chief characters of the *Polyzoa?*
224. Explain the term "polypide?"
225. How is the polypide of a *Polyzoön* distinguished from the polypite of a *Hydrozoön?*
226. Structure of a single "polypide."
227. What are "bird's-head processes," and to what may they be compared?
228. What general distinction is there between the fresh-water and marine *Polyzoa?*
229. Chief characters of *Tunicata?*
230. Nature of the "test?"
231. Structure of the heart in *Tunicata?*
232. Distinctions between simple, social, and compound Tunicates?
233. Chief characters of *Brachiopoda?*
234. Nature of the shell, as compared with that of Bivalves?
235. Structure and nature of the "arms?"
236. Chief characters of the *Lamellibranchiata?* Give examples.
237. Nature and uses of the "foot?"
238. Nature, uses, and number of the "adductor muscles?"
239. What are the "muscular impressions" and the "pallial line?"
240. Structure and mode of opening, and connection between, the valves?
241. Structure of the respiratory organs?
242. Nature and uses of the "respiratory siphons?"
243. Condition of circulatory system? Of digestive system?
244. Condition of young when first hatched?
245. Chief characters of the *Gasteropoda?* Give examples. Why spoken of as "univalves?"
246. What is the "operculum?"
247. Compare the *Gasteropoda* with the *Lamellibranchiata* as regards the head.
248. What is the nature of the "odontophore?"

249. Condition of the heart and breathing-organs?
250. What divisions of the *Gasteropoda* may be founded on the nature of the breathing-organs?
251. Condition of the young water-breathing Gasteropod?
252. Structure and modifications of the shell in Gasteropods?
253. What are the two leading conditions of the mouth of the shell?
254. General characters of the *Nudibranchiata?*
255. Nature of the foot in the Heteropoda?
256. General characters of the air-breathing Gasteropods?
257. General characters of the *Pteropoda?*
258. General characters of the *Cephalopoda?*
259. Nature of the "arms" and their suckers in the Cuttle-fishes?
260. Structure of the "funnel?"
261. Nature of the ink-bag? What living Cephalopod is without an ink-bag?
262. Nature of the breathing-organs? Of the circulatory organs? Of the respiratory process? Of the nervous system?
263. Peculiarities in the reproductive process in the Cuttle-fishes?
264. Nature of the internal shell of the Cuttle-fishes?
265. What two living Cephalopods possess an external shell, and what are the differences between these?
266. Characters of the Dibranchiate Cephalopods? Give examples.
267. Describe the shell of the Argonaut?
268. Characters of the Tetrabranchiate Cephalopods?
269. Describe the shell of the Pearly Nautilus?
270. Mention some fossil forms allied to the Pearly Nautilus?
271. General characters of the *Vertebrata?*
272. What is the "notochord?"
273. General structure of a "vertebra?"
274. Regions generally recognizable in the vertebral column?
275. General structure of the fore-limb?
276. General structure of the hind-limb?
277. General structure of the digestive system?
278. Source of the blood? Nature of the "blood-corpuscles?"
279. What Vertebrate animal has no heart?
280. General nature of the respiratory organs?
281. What is the difference between a gill and a lung?
282. General structure of the nervous system?
283. Define the terms "oviparous," "viviparous," and "ovo-viviparous."
284. Into what primary sections are the Vertebrata divided by Huxley?
285. What are the five classes of Vertebrates?
286. General characters of Fishes?
287. Chief forms of scales?
288. Nature of the "lateral line?"
289. Form of the vertebræ of a Fish?
290. Position and connections of the ribs?
291. Nature of the "interspinous bones?"
292. Nature and position of the limbs of Fishes?
293. Distinction between the "paired" and "median fins?"
294. Number and names of the median fins?
295. Difference between homocercal and heterocercal tail?
296. What are the "rays?" Difference between "soft rays" and "spinous rays?"
297. What are the "pyloric cæca?"

298. General arrangement of the gills in a Bony Fish?
299. Structure of the heart and course of the circulation in a typical Fish?
300. What is the "swim-bladder," and to what does it correspond?
301. Nature of the swim-bladder in the Mud-fish?
302. Condition of the organ of hearing? of the nose?
303. In what Fishes does the nose open behind into the throat?
304. General characters of the Lancelet?
305. General characters of the Marsipobranchii? Give examples.
306. Nature of the respiratory organs in the Lampreys?
307. General characters of *Teleostei?* Give examples.
308. General characters of the *Ganoidei?* Give examples.
309. Condition of the vertebræ in the Bony Pike?
310. General characters of the *Elasmobranchii?* Give examples.
311. General characters of the *Dipnoi?*
312. Distribution of the Mud-fishes?
313. General characters of the class *Amphibia?*
314. Nature of the metamorphosis in Amphibians?
315. General characters of the *Ophiomorpha?* Give examples.
316. General characters of *Urodela?* Give examples.
317. Explain the terms "perennibranchiate" and "caducibranchiate."
318. Distinctions between Tailed Amphibians and Lizards?
319. General characters of the *Anoura?* Give examples.
320. Phenomena of the development of a Frog? What general zoological law is illustrated thereby?
321. General characters of the *Sauropsida?*
322. General characters of Reptiles?
323. Structure of the lower jaw, and its connections with the skull?
324. General nature of the teeth?
325. In what Reptiles are the teeth sunk in sockets?
326. How does the intestine terminate in Reptiles?
327. Structure of the heart and course of the circulation in Reptiles generally?
328. Condition of the heart in the Crocodiles?
329. General characters of the *Chelonia?* Give examples?
330. Leading peculiarities in the skeleton of *Chelonia?*
331. Chief groups of Chelonians? Give examples.
332. General characters of the *Ophidia?*
333. Condition of the limbs in Snakes?
334. Mode of progression in Snakes?
335. Structure of the tongue? of the eye?
336. Structure and connections of the lower jaw?
337. General structure and function of the teeth?
338. Nature of the teeth in the non-venomous and poisonous Snakes respectively?
339. Examples of harmless Snakes?
340. Examples of poisonous Snakes?
341. General characters of the *Lacertilia?* Give examples.
342. Characters which separate the snake-like Lizards from the true Serpents?
343. Peculiarities of the Flying Dragon?
344. General characters of the *Crocodilia?*
345. Differences between Crocodiles and Alligators, and the geographical distribution of each?
346. Peculiarity of the Gavial, and its geographical distribution?

347. Leading characters of the *Ichthyopterygia?*
348. Characters separating *Plesiosaurus* from *Ichthyosaurus?*
349. Leading characters of *Pterosauria?*
350. General characters of Birds?
351. General structure of a quill-feather?
352. Peculiarity of the feathers of the Ostrich?
353. Characters of the backbone in Birds?
354. Nature and position of the "ploughshare" bone?
355. Structure of the beak?
356. Nature of the "sternal ribs?"
357. Form of the sternum in Birds which fly? In those which do not fly?
358. Structure of the shoulder-girdle?
359. Nature and function of the "merry-thought?"
360. General structure of the fore-limb or wing?
361. General structure of the hind-limb?
362. Nature of the "tarso-metatarsus?"
363. Number and position of the toes?
364. What bird possesses only two toes?
365. What is the "cere?"
366. General structure of the digestive system in Birds?
367. Condition of gizzard in flesh-eating and grain-eating Birds respectively?
368. What are the "intestinal cæca?"
369. Structure and peculiarities of the lungs?
370. Structure and functions of the air-sacs?
371. What is meant by a bone being "pneumatic?"
372. In what cases are the bones not pneumatic?
373. General structure of the heart and course of the circulation in Birds?
374. What is "incubation," and why are Birds specially adapted for this process?
375. What differences subsist in the condition of the young bird at birth?
376. What peculiarities distinguish the eye of Birds?
377. What is the "membrana nictitans?"
378. What peculiarities distinguish the ear of Birds?
379. What Birds have a rudimentary external ear?
380. General characters of the *Natatores?* Give examples.
381. General characters of the *Grallatores?* Give examples.
382. General characters of the *Cursores?*
383. Mention some of the more remarkable Cursorial Birds, and state something as to their peculiarities and geographical distribution.
384. General characters of the *Rasores?*
385. Characters and examples of the Gallinaceous Birds?
386. Characters and examples of the Columbaceous Birds?
387. Mention an extinct Columbaceous Bird.
388. General characters of the *Scansores?*
389. Leading families of the *Scansores?*
390. General characters of the *Insessores?*
391. Mention the four sections into which the *Insessores* are divided, and state the peculiarities distinguishing these.
392. Give examples of each of these sections.
393. General characters of the *Raptores?*
394. Distinctions between Nocturnal and Diurnal Birds of Prey?
395. Characters of the *Saururæ?*

396. For what bird has this order been established?
397. General characters of the *Mammalia?*
398. General structure of the backbone?
399. General number of cervical vertebræ?
400. Distinction between "true" and "false" ribs?
401. General structure of limbs?
402. What Mammals have no teeth?
403. What are the "milk-teeth?"
404. Describe the general groups of teeth in a Mammal.
405. What is the "diaphragm?"
406. General structure of the heart and course of the circulation?
407. General structure of the lungs?
408. What is the "corpus callosum?"
409. What Mammals possess no external ear?
410. Mention some modifications of the integumentary appendages in Mammals.
411. What Mammals are without hair when adult?
412. What are the mammary glands?
413. What is the "placenta?"
414. General characters of the *Monotremata?*
415. Mention the animals included in the *Monotremata*, and state their geographical distribution.
416. What are the "marsupial bones?"
417. General characters of the *Marsupialia?*
418. Geographical distribution of the Marsupials?
419. Give examples of the Marsupials.
420. General characters of the *Edentata?*
421. Geographical distribution of the order?
422. Leading groups of the *Edentata*, and their distinguishing characters?
423. General characters of the *Sirenia?*
424. Existing forms of the *Sirenia?*
425. General characters of the *Cetacea?*
426. Give examples of the Whalebone Whales and Toothed Whales.
427. What is the "blowing" of a Whale?
428. What characters distinguish the Dolphins?
429. Nature of the tusk of the Narwhal?
430. General characters of the *Ungulata?*
431. Divisions of *Ungulata* and their characters?
432. Nature and position of the horns of *Rhinoceros?*
433. Characters and distribution of the Tapirs?
434. Characters, chief forms, and geographical distribution of the *Equidæ?*
435. Characters and distribution of the Hippopotamus?
436. Characters of the *Suida?* Leading forms?
437. General characters of the Ruminants?
438. Structure of the stomach in Ruminants?
439. Characters and distribution of *Camelidæ?*
440. Characters and distribution of *Cervidæ?*
441. Nature of the horns of *Cervidæ?*
442. Characters and distribution of the Giraffe?
443. Characters and leading forms of the *Cavicornia?*
444. General characters and distribution of the *Hyracoidea?*
445. General characters and distribution of *Proboscidea?*

446. Distinctions between the Indian and African Elephants?
447. General characters of the *Carnivora?*
448. Sections into which the *Carnivora* are divided, and the characters of these?
449. Characters of the Seals?
450. Dentition of the Walrus?
451. Characters of the Bears?
452. Other Plantigrade *Carnivora?*
453. Characters and examples of the *Mustelidæ?*
454. Examples of the *Melidæ?*
455. Examples of the *Viverridæ?*
456. Characters and examples of the *Canidæ?*
457. Characters and distribution of *Hyænidæ?*
458. Characters and distribution of the *Felidæ?*
459. General characters of the *Rodentia?*
460. Structure of the incisor teeth of Rodents?
461. Leading families of the *Rodentia?*
462. General characters of the *Cheiroptera?*
463. What is the "patagium" of Bats?
464. Sections of the *Cheiroptera*, and their distribution
465. General characters of the *Insectivora?*
466. Characters and examples of the *Talpidæ?*
467. Characters and examples of the *Soricidæ?*
468. Characters and examples of the *Erinaceidæ?*
469. Other *Insectivora?*
470. Characters and distribution of the Flying-Lemurs?
471. General characters of the *Quadrumana?*
472. Characters and distribution of the *Strepsirhina?*
473. Examples of *Strepsirhina?*
474. Characters and distribution of *Platyrhina?*
475. Examples of Platyrhine Monkeys?
476. Characters and distribution of *Catarhina?*
477. General characters of the Baboons?
478. Characters and examples of the Anthropoid Apes?
479. General characters of the order *Bimana?*

INDEX.

THE END.

Cornell's Physical Geography:

ACCOMPANIED WITH NINETEEN PAGES OF MAPS, A GREAT VARIETY OF MAP QUESTIONS, AND ONE HUNDRED AND THIRTY DIAGRAMS AND PICTORIAL ILLUSTRATIONS; AND EMBRACING A DETAILED DESCRIPTION OF THE

Physical Features of the United States.

By S. S. CORNELL.

1 vol. Large 4to. 104 pages.

The attention of teachers is particularly requested to this new volume by the Author of Cornell's popular Series of Geographies, in the belief that it will be found to embrace all that is valuable and interesting in this important branch of study, and to be, beyond competition, *the best text-book on the subject.* It is no mere rehash of time-honored details, but has been drawn from original sources, and is on a level with the present advanced state of the science. Clearness, adaptation to the school-room, inductiveness of arrangement, and the presentation of one thing at a time and every thing in its proper place—features which have contributed so largely to the success of the other Geographies of the CORNELL Series—are among its striking characteristics.

It is interesting to the learner. The dry statistical style usual in similar text-books has been avoided, and the great wonders of Nature, always fascinating to the inquiring mind, are presented in the most striking manner, so as to rivet the attention and impress the memory.

The illustrations are numerous and beautiful, and are used wherever it was thought they would help to elucidate the text. Maps and diagrams have been liberally introduced. The maps are executed in the finest style of the art—carefully drawn, distinctly engraved, and tastefully colored according to the most approved style. Each map is accompanied with questions in great variety.

The physical features of our own country receive particular attention in a closing chapter. The student is aided by a fine Physical Map of the United States, which (in addition to the features usually presented) shows the mean annual temperature of different parts of the country, the vegetable products of different sections, and their mineral resources, the relative values of the precious metals produced in the several States being clearly represented to the eye by an ingenious plan. A Map of Alaska, on a comparatively large scale, is also presented.

It is believed that the above features, besides others which there is no space here to enumerate, cannot fail to recommend this work to all.

Lockyer's Astronomy.

AMERICAN EDITION.

"This is by far the clearest and best manual of Astronomy we have ever seen. A child may understand it—and yet it contains information which will be new to all who have not time to follow the latest discoveries." — *New York Daily Times.*

The opinion of the critic of the *Times*, given above, is that of all who have examined this sterling school-book, which is winning golden opinions everywhere. Out of hosts of letters, we have space for the following only:—

"Lockyer's Elements of Astronomy is a work of rare excellence. As a text-book for the use of schools *it is unsurpassed* by any work on that subject with which I am acquainted."—PROF. JOHN S. HART, *State Normal School, Trenton, N. J.*

"I think it an excellent work—well calculated for class use by pupils of an academic grade. The arrangement and typography are worthy of especial commendation. It is a decided success."—S. B. HOWE, *Supt. of Schools, Schenectady, N. Y.*

"It is *the best* school-book on Astronomy that I ever saw. The spectra of the sun, stars, and nebulæ, are worth the price of the book. The diagrams are excellent. I deem it superior to all other books on that science with which I am acquainted. Of course I shall use it."—W. H. PITT, *A. M., Princ. Friendship Academy.*

"I have examined with much satisfaction the admirable elementary treatise on Astronomy by Lockyer. It furnishes the reader with the means of learning in a short time the great features of the modern progress of Astronomy. No book (except, perhaps, Youmans's New Chemistry) has appeared, which so easily, yet thoroughly, prepares the reader for the subsequent study of that mighty auxiliary to modern science—spectrum-analysis. Moreover it presents, in a clear and succinct style, the relations of the various parts of the stellar universe, besides inducting the learner into that neglected branch of Astronomy known as Geography of the Heavens. Its illustrations are genuine aids to the comprehension of the subject-matter."—DAVID BEATTIE, *Supt. Troy (N. Y.) City Schools.*

"I have examined it with care, and find it admirably adapted for use in schools. It is so plainly written and so fully illustrated as to render it specially suited for beginners in the science, and, at the same time, profitable for advanced students."—PROF. C. STALEY, *Union College.*

"It is a clear and beautiful unfolding of a profound and fascinating subject. It seems to me admirably adapted to the academic grade of students. Not attempting to discuss those problems and theories of the science for which such pupils have neither the time nor the capacity, the author has given an outline of the subject that is clear, sufficiently complete, and thoroughly modern."—PROF. BRADLEY, *Princ. Albany Free Academy.*

CORNELL'S OUTLINE MAPS.

Series of Outline Maps.

By S. S. Cornell, author of "Cornell's Series of School Geographies."
13 Maps, mounted on Muslin.

The Series consists of the following Maps:

THE WORLD. Size, 32 by 52 inches. Comprising the Eastern and Western Hemispheres, Diagrams of Meridians and Parallels, Tropics and Zones, Northern and Southern Hemispheres, and Heights of the Principal Mountains.
NORTH AMERICA. Size, 27 by 32 inches.
THE UNITED STATES AND CANADA. Size, 32 by 52 inches.
EASTERN AND MIDDLE STATES. Size, 27 by 32 inches. With enlarged plans of the Vicinities of Boston and New York.
SOUTHERN STATES. Size, 27 by 32 inches.
WESTERN STATES. Size, 27 by 32 inches.
MEXICO, CENTRAL AMERICA, AND WEST INDIES. Size, 27 by 32 inches. With enlarged plans of the Isthmus of Nicaragua and the Great Antilles.
SOUTH AMERICA. Size, 27 by 32 inches.
EUROPE. Size, 27 by 32 inches.
BRITISH ISLANDS. Size, 27 by 32 inches.
CENTRAL, SOUTHERN, AND WESTERN EUROPE. Size, 27 by 32 inches.
ASIA. Size, 27 by 32 inches. With enlarged plans of Palestine and the Sandwich Islands.
AFRICA. Size, 27 by 32 inches. With enlarged plans of Egypt, Liberia, and Cape Colony.

Each Map is substantially mounted on cloth, and the set is neatly put up in a portfolio, and accompanied with a complete Key for the teacher's use.

In their production the most recent and reliable authority has been carefully consulted; if close attention and an honest endeavor will accomplish any thing, these Maps are accurate. They are believed to be the only Series that includes separate Maps of the different sections of the Union. By an economical arrangement and judicious disposition of matter, the space has been used to the greatest possible advantage. Every thing is clear and pleasing to the eye. The engraving and printing are bold and distinct. The coloring is neat and durable. They are printed on fine white paper, and strongly backed with muslin. The Key contains the pronunciation of *all* names employed in it.

Quackenbos's Standard Text-Books:

AN ENGLISH GRAMMAR: 12mo, 288 pages.

FIRST BOOK IN ENGLISH GRAMMAR: 16mo, 120 pages.

ADVANCED COURSE OF COMPOSITION AND RHETORIC: 12mo, 450 pages.

FIRST LESSONS IN COMPOSITION: 12mo, 182 pages.

ILLUSTRATED SCHOOL HISTORY OF THE UNITED STATES: 12mo, 538 pages.

ELEMENTARY HISTORY OF THE UNITED STATES: Beautifully illustratad with Engravings and Maps. 12mo, 230 pages.

A NATURAL PHILOSOPHY: Just Revised. 12mo, 450 pages.

APPLETONS' ARITHMETICAL SERIES: Consisting of a Primary, Elementary, Practical, Higher, and Mental Arithmetic.

Benj. Wilcox, A. M., Princ. River Falls Acad., Wis.: "I have taught in seminaries in this State and in New York for more than twenty years, and am familiar with most of the works that have been issued by different authors within that period; and I consider Quackenbos's Text-Books *the most unexceptional* in their respective departments."—**C. B. Tillinghast,** Princ. of Academy, Moosop, Conn.: "I think Quackenbos's books *the nearest perfection* of any I have examined on the various subjects of which they treat."

Pres. **Savage,** Female College, Millersburg, Ky.: "Mr. Q. certainly possesses rare qualifications as an author of school-books. His United States History *has no equal,* and his Rhetoric is really *indispensable.*"—**David Y. Shaub,** Pres. Teachers' Inst., Fogelsville, Pa.: "I approve of all the Text-Books written by Mr. Quackenbos."—Rev. Dr. **Winslow,** N. Y., Author of "Intellectual Philosophy:" "All the works of this excellent author are characterized by clearness, accuracy, thoroughness, and completeness; also by a gradual and continuous development of ulterior results from their previously taught elements."

Rev. Dr. **Rivers,** Pres. Wesleyan University: "I cordially approve of all the Text-Books edited by G. P. Quackenbos."—**W. B. McCrate,** Princ. Acad., E. Sullivan, Me.: "Quackenbos's books need only to be known to be used in all the schools in the State. Wherever they are introduced, they are *universally liked.*"—**Jas. B. Rue,** County Supt. of Schools, Council Bluffs, Iowa: "Any thing that has Quackenbos's name is sufficient guarantee with me."—**Methodist Quarterly Review,** Jan. 1860: "Every thing we have noticed from Mr. Quackenbos shows that the making of books of this class is his proper vocation."

Single copies of the above Standard works will be mailed, post-paid, for examination, on receipt of one-half the retail prices. Liberal terms made for introduction. Address

D. APPLETON & CO., Publishers,

549 & 551 *Broadway, New York.*

Marshall's Book of Oratory. Part I. Part II.

Markham's History of England. Revised by ELIZA ROBBINS. 12mo. 387 pages.

Mangnall's Historical Questions. 12mo.

Nicholson's Text-Book of Zoology. For Schools and Colleges. 12mo. 353 pages.

——— **Text-Book of Geology.** For Schools and Colleges. 12mo. 266 pages.

Otis's Easy Lessons in Landscape-Drawing. In 6 Parts.

——— **Drawing-Books of Animals.** In 5 Parts.

Quackenbos's Standard Text-Books. Consisting of:

- FIRST LESSONS IN COMPOSITION. With Rules for Punctuation, and Copious Exercises. 12mo. 182 pages.
- ADVANCED COURSE OF COMPOSITION AND RHETORIC. A Series of Practical Lessons. 12mo. 450 pages.
- FIRST BOOK IN ENGLISH GRAMMAR. For Beginners. 18mo. 108 pages.
- ENGLISH GRAMMAR. 12mo. 288 pages.
- PRIMARY HISTORY OF THE UNITED STATES. Made easy for Beginners. Child's quarto. 200 pages.
- ELEMENTARY HISTORY OF THE UNITED STATES. With numerous Illustrations and Maps. 12mo. 260 pages.
- ILLUSTRATED SCHOOL HISTORY OF THE UNITED STATES. With Maps, Battle-fields, etc. Brought down to the present Administration. 12mo. 530 pages.
- A NATURAL PHILOSOPHY. Exhibiting the Application of Scientific Principles in Every-day Life. 12mo. 450 pages.
- PRIMARY ARITHMETIC. 16mo. 108 pages.
- ELEMENTARY ARITHMETIC. 12mo. 144 pages.
- PRACTICAL ARITHMETIC. 12mo. 312 pages.
- HIGHER ARITHMETIC. 12mo.
- MENTAL ARITHMETIC. 18mo. 168 pages.

Reid's Dictionary of the English Language.

Robbins's Class-Book of Poetry. 16mo. 252 pages.

——— **Guide to Knowledge.** 16mo. 417 pages.

Shakespearian Reader. By JOHN W. S. HOWS. 12mo.

Schmidt's Course of Ancient Geography. Arranged with special reference to convenience of recitation. By Prof. H. I. SCHMIDT, D. D., of Columbia College. 12mo. 328 pages.

www.ingramcontent.com/pod-product-compliance
Lightning Source LLC
LaVergne TN
LVHW010136110826
845151LV00002B/368

* 9 7 8 1 4 2 5 5 3 9 8 3 2 *